Natural Selections
National Parks in Atlantic Canada, 1935–1970

During the Depression the Canadian National Parks Branch was under pressure to make the park system truly national, to bring the advantages of parks to all provinces. In Atlantic Canada, however, it found itself dealing with an environment that was far different from what it was accustomed to in Western Canada. The land areas were smaller and flatter, and, having been settled for generations, could hardly be considered wild. Wildlife was smaller and there was less of it.

Natural Selections traces the history of the first four parks in Atlantic Canada through the selection, expropriation, development, and management stages. Alan MacEachern shows how the Parks Branch's preconceptions about the landscape and people of the region shaped the parks created there. In doing so he details the evolution of the park system, from the conservation movement early in the century to the rise of the ecology movement. MacEachern analyses Parks Canada's efforts to fulfil its twin mandates of preservation and use, arguing that the agency never favoured one over the other but oscillated between being more or less interventionist in ensuring both.

Touching on a wide range of matters – from landscape aesthetics to tourism promotion, from DDT to Martin Luther King – *Natural Selections* expands our understanding of the relation between nature and culture in the twentieth century.

ALAN MACEACHERN is an environmental historian currently studying the consequences of a century of ecological policy at Banff National Park.

Natural Selections

National Parks in Atlantic Canada, 1935–1970

ALAN MacEACHERN

McGill-Queen's University Press
Montreal & Kingston · London · Ithaca

Legal deposit third quarter 2001
Bibliothèque nationale du Québec

Printed in Canada on acid-free paper

This book has been published with the help of a grant
from the Humanities and Social Sciences Federation of
Canada, using funds provided by the Social Sciences
and Humanities Research Council of Canada.

McGill-Queen's University Press acknowledges the
financial support of the Government of Canada through
the Book Publishing Industry Development Program
(BPIDP) for its activities. It also acknowledges the
support of the Canada Council for the Arts for its
publishing program.

Excerpt from "Cape Breton" from *The Complete Poems
1927–1979* by Elizabeth Bishop. Copyright © 1979, 1983
by Alice Helen Methfessel. Reprinted by permission of
Farrar, Straus and Giroux, LLC.

Canadian Cataloguing in Publication Data

MacEachern, Alan Andrew, 1966–
    Natural selections: national parks in Atlantic Canada,
    1935–1970
    Includes bibliographical references and index.
    ISBN 0-7735-2157-7
    1. National parks and reserves – Atlantic Provinces –
    History.   2. Canada. National Parks Branch – History.
    3. Nature – Social aspects – Canada.   I. Title.
    SB484.C3M32 2001   333.78'3'09715   C00-901079-3

This book was typeset by Typo Litho Composition Inc.
in 10/12 Palatino.

Contents

Photos and Maps

Dedicated to my mentors (in order of appearance): Meredith MacEachern, Arnold MacEachern, Jack Myron, Doug Baldwin, Andy Robb, Ed MacDonald, George Rawlyk, Ian McKay, and Colin Duncan

Acknowledgments

Sitting in a house on the road into Ingonish, just outside Cape Breton Highlands National Park, I listened attentively while Maurice Donovan, eighty-nine, told of being expropriated from the park in 1939. His wife, Emma, sat nearby, quiet at first but then jumping in more and more to correct and clarify, speaking first to her husband and soon directly to me. Their farm had become the ninth and tenth holes of the park golf course, and by the time the Donovans left in 1939 golfers were fading their shots around the house. (Strange that Maurice would interject later in our talk, "Golf is a wonderful game.") Maurice was talking about his confrontation with the course contractor, Bill Stewart, the only person from the park who drove across the farm whenever he saw fit. Maurice remembered,

I came up to the clubhouse one morning and they were all sitting on the patio out front and Bill Stewart saw me and I went up to him. I was polite, but I said, "It's still my property. I'd like to tell you before you come tomorrow morning not to cross my line or you'll get buckshot in the arse." Stewart tore off his sweater, and I tore off my coat. "Come on. Come on."

Maurice sprang from his chair to demonstrate, big grin on his face, ripping off his cardigan, his body poised, his fists ready. He subsided just as quickly: "But Cullen got between us and …" Things were smoothed over, and in time he and Stewart even got along, a respect formed. Maurice kept talking, settled back down in his chair, but my heart was still pounding. In beginning this project I had planned to be the all-seeing observer, the calm chronicler of all sides of national park policy-making. I did not want an eighty-nine-year-old man leaping at me, if only to tell fifty-five-year-old stories.

Most conversations were more restrained than this one, but throughout my work I found strongly held beliefs in the quiet insistences of expropriated landowners, past park staff, and dry Parks Canada documents. Often the argument was of a simple park vs. non-park, us vs. them variety. But underlying the words, it seemed, was a desire to explain personal feelings about how individuals interact, how society works, and how humans treat nature. People's relations to the parks raised all three issues. When it came time for me to write, their words made it impossible to treat the subject in total abstraction. The four parks examined here are still in existence, the communities around them still in love-hate relationships with Parks Canada. I want to thank those who talked to me, and made my study more than just an academic exercise.

In Nova Scotia, thanks to Wilf and Claudette Aucoin, Eva and Ernest Deveau, Angus Leblanc, Wilfred Boudreau, Wilfred Aucoin, Charlie Dan Roach, Neil MacKinnon, Fred Williams, Emma and Maurice Donovan, Gordon Doucette, and Ronald Dauphinee, as well as Superintendent Tim Reynolds of Cape Breton Highlands National Park. In Prince Edward Island, thanks to Ernest and Bernice Smith, Ralph Burdett, Alfred Morrison, and George and Joyce Robinson, as well as Barb MacDonald at Prince Edward Island National Park. In New Brunswick, thanks to Audley Haslam, Bob Keirstead, Winnie Smith, Winnie and Leo Burns, Ann and Dan Keith, Roy Groves, and Larry Hughes, as well as Superintendent Mart Johanson, John Brownlee, and Alain Chevette at Fundy National Park. In Newfoundland, thanks to Art Hefferin, George Squires, Mark Lane, Don Spracklin, Ralph Ford, Dennis Chaulk, Hector Chaulk, Wayne Chaulk, Mildred and Clayton King, and Judy Day, as well as Attilla (Ted) Potter, Greg Stroud, and Superintendent Chip Bird at Terra Nova National Park. Thanks also to those who agreed to be interviewed but asked that their names not be used.

I was given a great deal of assistance from archival and library staff in Atlantic Canada and Ontario. Thanks in particular to Harry Holman and Marilyn Bell at the Provincial Archives of Prince Edward Island, Chris Dennis at the Centre for Newfoundland Studies in St John's, Barry Cahill at the Provincial Archives of Nova Scotia, and Patrick Burden at the National Archives of Canada. Staff in the Queen's University history department, the National Library and National Archives of Canada, and the libraries at Queen's University and the University of Prince Edward Island deserve special thanks for accommodating years of my pesky presence.

This project began as a Ph.D. dissertation at Queen's University. That thesis would not have been written without the financial support of the Social Sciences and Humanities Research Council of Canada, and that of Queen's University's School of Graduate Studies, Department of History, and Grad Club. Thank you to those who aided in the writing of that tome: Jeannie Prinsen, Ed MacDonald, Lynda Jessup, Jeff Brison, Rick Stuart, Robert Malcolmson, Karen Dubinsky, Robin Winks, Brian Osborne, Richard West Sellars, Madeline DeWolf, Pearl Anne Reichwein, Helen Harrison, Peter Campbell, Mike Dawson, Jamie Allum, Bill Waiser, and – above all – my supervisors George Rawlyk, Colin Duncan, and Ian McKay.

My sincere thanks also to those who helped turn this into a book: Jeannie Prinsen, Colin Duncan, and Ian McKay once more, Don Akenson, Joan McGilvray, Roger Martin, and Judy Williams for McGill-Queen's, two anonymous readers, Boyde Beck, Nathan Stairs, Winnie and David Wake, Frances Girling, Del Muise, Ian Joyce, and the Social Sciences and Humanities Council of Canada for continuing support. Cormac McCarthy, Bob Dylan, and Norman Maclean served as inspirations while writing; if in style the book at times resembles the picaresque ramblings of a violent fly fisherman, blame them.

My family – Meredith, Arnold, Scott, Genny, Meredith, Lincoln, Jeannie, Richard, Allison, Errol, Alycia, Elaine, Ruth, Sam, Rebekah, and Beth – deserves far more thanks than just copies of this book as Christmas presents, but isn't that just the way.

Having never acknowledged anything before, I am only now realizing that authors save their most loved of loved ones for last not just to give them special prominence, but as a form of procastination, it being difficult to put appreciation for them into words. Thank you, Genevieve.

We have advanced by leaps to the Pacific, and left many a lesser
Oregon and California unexplored behind us.

H.D. Thoreau, *The Maine Woods*

The men are unspoilt by comfort yet they are free of the necessity
of having to exploit nature. They enter into nature rather as a
swimmer, who has no need to cross it, enters a river. They play in
the current: In it and yet not of it. What prevents them from being
swept away are time-honoured rules to which they adhere
without question. The rules all concern ways of treating or
handling specific objects or situations – guns, boots, bags, dogs,
trees, deer, etc. Thus the force of nature (either from within or
from without) is never allowed to accumulate; the rules always
establish calm, as locks do in a river. Such men feel like gods
because they have the impression of imposing an aesthetic order
upon nature merely by the timing and style of their formal
interventions.

John Berger, *G*

Your brother is young enough to believe that the past still exists,
he said. That the injustices within it await his remedy. Perhaps
you believe this also?
I don't have a opinion. I'm just down here about some horses.

Cormac McCarthy, *The Crossing*

1 Introduction:
A Walk at Herring Cove

On a cool September day in 1994, Ann Keith and her husband Dan led me down to Herring Cove, just west of Alma, New Brunswick. Ann is a fit, lively woman; it is hard to believe she spent two summers at cottages here sixty years ago, hoping the Bay of Fundy air would cure her whooping cough. She showed me relics from those times. Coins from Norwegian sailors. A grainy image of the stevedores' bunkhouse perched at the water's edge. A 1934 letter asking her mother to send the cat. We clambered on the treed banks like children or archaeologists, re-creating the community of cottages. Apple trees grew within and obscured the foundation of Captain Rolfe's home, there. That's the concrete base of Judge Jonah's cabin. The four-seater would have been right here. This is the old road.

The cottages are gone and the trees have taken over because a seventy-square-mile area around Herring Cove became Fundy National Park in 1947. As a park, the land was expected to be natural, so the property rights of individual owners were negated through expropriation, and the land was conferred to the people of Canada. The Canadian National Parks Branch – now called Parks Canada[1] – allowed the area to return to nature. In the late 1960s, the Parks Branch began to take an interest in the cultural history present in the lands they oversaw. Park historians began to document the "human histories" of the parks.[2] This awkward phrase hints at the problem inherent in these official publications: in each, the park's existence seems predetermined. All human settlement prior to park establishment seems de facto park property. As in the parks' geological and

The road to Herring Cove. Photo taken by R.W. Cautley in his 1930 inspection of potential New Brunswick national park sites. National Archives of Canada, PA187873.

biological surveys, the human histories made artifacts of their subjects but ignored the cultural forces that shaped the park at establishment and during the park's life.

National parks are about both nature (which we may define simply as all that is nonhuman) and culture (all that is human); a history of national parks that does not address both is incomplete. To create a national park is to favour the natural over the cultural, even if only temporarily and within a fixed location. Human plans for the land are (theoretically) superseded so that the land can (theoretically) thrive unhindered. Nature's efforts to return and block out the signs of past settlement, as at Herring Cove, seem almost a validation of the park: this is indeed a natural place. But although culture, like the cottages' foundations, is difficult to see, it is nonetheless still present. Where there had been old buildings left by past residents there are now more permanent structures for more transient visitors: campgrounds have replaced cottages, modern washrooms have replaced four-seaters. And more than this, the naturalness of the park is itself a product of cultural decisions. People chose this land to be a national park for a variety of aesthetic, economic, and political reasons, and every subsequent decision also bore cultural weight. We cannot see national parks as natural without understanding that it is our culture that has made them so and declares them so.

Of course, national parks are fashioned by nature, too. In Fundy National Park, fields were allowed to grow back into forests, but the forests grew on their own. Perhaps the best way of showing that parks are natural creations is by thinking counterfactually, by asking how Fundy would be different if located on Ellesmere Island. Or one hundred yards east of Alma, rather than west. Historians are bound to lose their way once they begin to think like this; describing the nature in a national park and what this has meant to its history is a difficult task. But it is a hike worth taking, if only because it is this nature that the Parks Branch wished to save (and change), and this nature that is the product of what they saved (and changed).

In national parks, the cultural and the natural merge, as they do everywhere else. But parks are particularly interesting because they are places where humans believe they have made nature paramount. As such, parks can help us see how nature has been viewed by people. How was this land chosen as a potential park, and by whom? What vision did those who made this choice have for the new park? What human activities were permitted to interfere with nature in the park? And how did this balance between nature and culture change during the life of the park?

What follows is a study of how the Canadian National Parks Branch struck that balance in the establishment and management of Cape Breton Highlands National Park (created in 1936), Prince Edward Island National Park (1936), Fundy National Park (1947), and Terra Nova National Park (1957). These are noteworthy national parks in that they were the first created in Atlantic Canada, and the only ones created in Canada between 1930 and 1968. For both these reasons, their histories can teach us a great deal. First, they can show how aesthetic judgments about different kinds of nature can affect the treatment of that nature. Second, they can show how the treatment of nature changes over time, in part because of changing aesthetic preferences.

In responding to the provincial demands and federal exigencies for Eastern parks in the 1930s and onwards, the National Parks Branch was forced to adapt its existing aesthetic to entirely new circumstances. Until then, the national park ideal was the sublime scenery found at Western parks such as Banff and Jasper, with canyons, waterfalls, and, above all, mountains. Atlantic Canada did not possess such scenic extremes, nor did it possess in quantity or variety the large animals that visitors to the Western parks enjoyed. Moreover, most parts of Atlantic Canada had been settled for generations, and thus the region was believed not to possess a requisite sense of wildness. Though the Parks Branch tried to apply a traditional model to the Cape Breton Highlands and Prince Edward Island parks, staff's feelings about the region's natural inadequacies ensured that the new parks would be, to their minds, imperfect imitations. Thus these first two Atlantic Canadian parks were designed as resorts with some pretension to gentility, but also as destinations for mass tourism. At

Prince Edward Island National Park in particular, so far from the Branch's image of what a park should be, tourism development was much more intrusive on the park landscape than would have been tolerated elsewhere.

The distinction made between "high" mountain parks and "low" coastal parks offers a fitting topographical application to one of my very central contentions: many of the Parks Branch's decisions were made in an attempt to maintain the parks as symbols of high culture. Beginning at Banff in 1885, national parks had always been resorts catering to wealthy tourists from Canada and elsewhere. Parks were sold as epitomizing fine attributes such as a taste for beauty, a love of nature, and national pride. In Pierre Bourdieu's *Distinction: A Social Critique of the Judgment of Taste*, the author argues that such aesthetic consumption, whether of art or landscape, serves the same function as any other sort of consumption: it helps to legitimate social differences. People learn to enjoy those objects, entertainments, and views which they believe will reinforce their social position.[3] In the case presented here, national parks promised to infuse their visitors with cultural capital. By their very attendance, tourists proved themselves well cultured. As the caretaker of the parks, the Parks Branch was in the position to judge what natural and cultural features did and did not conform to the park aesthetic. In turn, it was in the position to be thought a high-culture institution itself.

However, as Lawrence Levine explains in *Highbrow/Lowbrow: The Emergence of Cultural Hierarchy in America*, the boundaries between cultural groups are always shifting, so that those who wish to remain of a certain standing must continually adapt to new situations and adopt new tastes and preferences.[4] The Parks Branch could not rely on a timeless, permanent, single model as an ideal. To maintain its system's prestige – and its own – the agency had to adapt constantly to changing times. It was very difficult for the Parks Branch to accomplish such aesthetic shifts, however, because the park mandate demanded that parks be maintained unchanged. For this very reason, studying four parks at their establishment is quite worthwhile: it was here, in the early days of the park, that many of the developmental decisions were made, defining each park as one of a certain era and setting its future in stone. Fundy National Park, created in the late 1940s, was not designed as a resort like the first two Maritime parks only a decade earlier, but more like a suburb representing the promise of society-wide affluence. The headquarters area at Fundy with its many amenities was a response to the period's belief in recreational democracy. In the following years, the Branch noted the germination

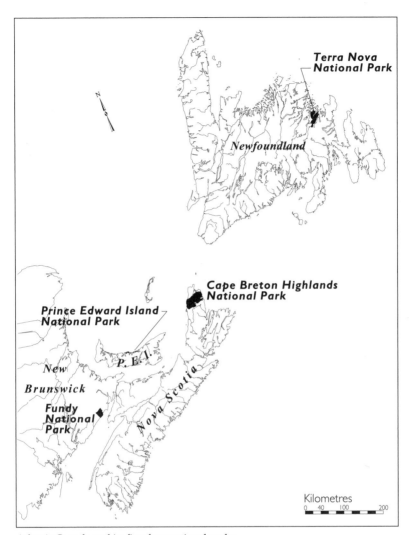

Atlantic Canada and its first four national parks.

of a back-to-nature movement in North America. Staff again recognized the opportunity to equate parks with high-culture ideals, this time by dividing visitors not directly in class terms but in more intellectual ones: by an ability or inability to appreciate parks for their natural rather than their cultural features. Terra Nova National Park, created in 1957, was therefore far less developed by the Parks Branch than Fundy had been, and was instead more of a wilderness-oriented national park.

I selected these four parks as my focus in large part, then, in hopes of documenting the changing notion of what the Canadian National Parks Branch thought a park should be from 1930 to 1970. But focusing on Cape Breton Highlands, Prince Edward Island, Fundy, and Terra Nova parks serves other purposes as well. First, it fills in several historiographical gaps. Most historical work on Canadian parks focuses on their infancy, from the establishment of Banff in 1885 to the 1930 passage of the National Parks Act. Studying the first four Atlantic parks permits a view of the more recent past, and of the expansion and maturity of the park system. In the same vein, the preponderance of historical research has been on Western Canadian parks. An account of the expansion to the East helps explain how making the park system truly national affected the system's policies and philosophy. Second, an examination of these parks can contribute to a better understanding of Canadian ecological science in this period. Parks were constantly referred to as "laboratories" and used for "experiments" in the latest wildlife, fish, and vegetation management policies. Finally, these were the first Canadian national parks to involve the removal of significant numbers of landowners. Studying the rationales for expropriation made by park staff helps to answer the questions of what "preservation" was meant to preserve, and why some "use" was considered more acceptable than others.

Practical considerations defined the limit of the project. I chose not to deal with all Atlantic Canadian national parks or to carry discussion of the four parks under study through to the present, in both cases because there were far fewer archival records available for the post-1970 period. The choice not to view all Canadian parks from 1930 to 1970 was more difficult, as it prevents the project from dealing comprehensively with topics such as predator control and townsite planning, which were important on the national scene but not the Atlantic one. But to have examined parks more broadly would have eliminated the possibility of studying a few more thoroughly at the local level. This seemed a greater loss; an integral part of policy formulation and implementation in Canada is the tension between the local, provincial, and federal levels. And while predator control and townsite planning are largely absent from this account, the greater issues they address – the manipulation of nature and the relationship between parks, the public, and business – are present in other guises.

I may be accused of ignoring an earlier expropriation of these park lands: the one imposed by European peoples on the Mi'kmaq (and Maliseet) in the Maritimes and the Beothuk in Newfoundland. However, this book is not meant to be a history of these sites from time immemorial but a history of their years within the national park system.

The remainder of this chapter outlines the three theoretical and geographical planes on which this work simultaneously travels. One plane deals with the interaction of culture and nature in nineteenth- and twentieth- century North America. A special concern will be to find a language that can adequately describe nature and discuss its role in history as an independent entity entitled to consideration for its own sake. Specifically, I will examine the ability of environmental history as a discipline to do so. A second plane is concerned with the institutional character of the Canadian National Parks Branch. The Parks Branch was forced to accommodate federal, provincial, and local interests while maintaining its own conviction of what parks were to be. Its history of seeking a balance between park preservation and park use is a real-life rendition of the theoretical struggle between nature and culture. A third plane will tell the stories of the Cape Breton Highlands, Prince Edward Island, Fundy, and Terra Nova National Parks themselves. This is perhaps the most important plane to me, for it is here that the ironies of our simultaneous love and exploitation of nature shine through. We need to understand small places like Herring Cove – its name a small fish and a small body of water – if we are to understand how culture and nature interact.

OF NATURE AND CULTURE

The most respected of North American national park histories, Alfred Runte's 1979 *National Parks: The American Experience*, is prefaced with the promise, "What follows, then, is an interpretive history; people, events, and legislation are treated only as they pertain to the *idea* of national parks."[5] Eleven years later, Runte followed up with *Yosemite: The Embattled Wilderness*, a book with a theme, style, and method similar to those of his earlier work. Yet here he announces, "What follows is an environmental history. People, buildings, and traditions are treated only as they pertain to evolving philosophies of park management and use."[6] Of interest are not the passages' clear similarities but rather their primary difference: why has an "interpretive" history become an "environmental" one? As Runte might say, what follows?

Environmental history was born out of the rise of environmentalism in North America in the 1960s and 1970s, and the corresponding rise of scholarly interest in the questions raised by the movement. The few historians who began to study the relationship of humans to nature soon found themselves part of a longer historical tradition. Thus Roderick Nash could, in 1970, write a review essay entitled "The State of Environmental History," simultaneously naming a new discipline and suggesting its prior existence. Older works were incorporated into an environmental history canon: George Perkins Marsh's 1864 *Man and Nature* (on anthropogenic changes to nature), Lucien Febvre's 1932 *A Geographical Introduction to History* (on human geography), Samuel Hays's 1959 *Conservation and the Gospel of Efficiency* (on environmental politics), and Clarence Glacken's 1967 *Traces on the Rhodian Shore* (on ideas about the environment) all qualified.[7] In this context, Alfred Runte's inclination toward environmental history is understandable. His work since the 1970s on the national park idea – what might be considered intellectual or institutional history, and what he calls interpretive history – had become environmental history by 1990. One might also conclude that Runte had more incentive to be defined as an environmental historian in 1990, because by then the field was becoming established, even trendy.[8]

In *A River Runs through It*, Norman Maclean writes that his book was rejected by one publisher with the comment, "These stories have trees in them."[9] If environmental history is simply history with trees in it, one has to wonder what purpose it serves. We can best understand environmental history today as experiencing an adolescence much as women's history did more than a decade ago. Arising from a political movement of the 1960s and 1970s, women's history considered women as a worthy subject of inquiry. The new history's mandate was to return its subject to its rightful place in history. In time, however, the weakness of this position became apparent. The very foundation of women's history – that women's lives have been socially and historically, rather than biologically, marginalized – suggested that women's history needed to be from that point forward forever parenthesized as "women's" history. As Joy Parr states, "'Tell me about woman' always to some degree meant 'Tell me about someone who will be recognizable to me as a woman.'"[10] The result was a move away from a mainly descriptive women's history to one that sought to know how gender identities were formed. Before studying what women (and men) "did," gender historians sought to know how they came to be, pushed and pulled by the forces of race, class, ethnicity, and sexuality.

Environmental history has followed much the same path, to a point. Many of its practitioners felt that highlighting nature's place in the history of human affairs was a reasonably straightforward enterprise. In all of Runte's work, for example, the history of national parks seems to be an exercise in connecting the political and intellectual dots that led Americans to an awakening love of nature. Just as women's history envisioned filling the historical canvas by painting women in, environmental history hinted that an even more total history would be achieved once nature was part of history. The first

Canadian environmental history reader, published in 1995, noted in the introduction that "Because environmental history calls for analyses that integrate processes such as class, gender, ethnicity, and setting, this approach has the limitless ambition of an all-encompassing historical understanding"[11] – environmental history would be the last new history. To be sure, many environmental historians have attempted to give the field a more firm grounding. The most influential of these, Donald Worster, described the "three clusters of issues" that environmental history seeks to address: 1) nature itself as it has existed over time, 2) the socioeconomic realm in which nature and culture interact, and 3) the intellectual realm in which humans supply nature with meaning. While Worster's model has met with some disagreement, it has a simple and coherent structure familiar to other environmental historians.[12]

Surveying the field of environmental history in a 1994 article, Dan Flores writes approvingly, "In its brief three decades, modern environmental history has made a name for itself primarily as a field that has offered up stimulating studies of environmentalism as a sociopolitical movement, of intellectual ideas about nature, and specific environmental events of historical importance."[13] This is fainter praise than he intends. If environmental history uses nature simply as a topic but seeks its explanations elsewhere, such as in political or intellectual history, it will necessarily be subsumed by those larger studies.[14] I may, like Runte, change my business cards and become an environmental historian, but this seems an incomplete conversion if I only believe that nature is a place where history occurs and do not also believe that nature is somehow involved in how history occurs. Environmental historians have not been very successful in expressing nature's role. Perhaps we have been slow to understand the implication of the discipline's foundation: that nature is historically contingent. This should lead us to the realization that "Tell me about nature" always to some degree means "Tell me about something that will be recognizable to me as nature." Environmental historians must follow the example of women's historians and stop taking their subject for granted. They must ask what nature *is* and what nature *does*.

"Nature," says William Cronon, "is the place where we are not."[15] This definition possesses the two necessary ingredients of a useful introduction to the idea of nature. First, it offers a simple, working notion of nature as all that is nonhuman. Second, it refutes the simplicity of this notion, by ironically intimating that we arbitrarily distance the human species from the rest of the planet. In reality, of course, the idea of nature is more complicated. We breathe in nature, stand on nature, depend on nature for our survival at every moment. And in return, we as a species change nature, in everything from providing sanctuary for bacteria to building dams that minutely shift the earth's rotation. The idea of a culture/nature dichotomy is an imperfect, anthropocentric one – which is not surprising, since we are humans (as naturalist John Livingston notes, his dogs are guilty of canimorphizing).[16] We need the term "nature" if we are to begin discussion of the relationship between humans and the nonhuman world without constantly reverting to quotation marks.[17]

But even having defined nature in such a way, it is much more difficult to describe it. What language is to be used? For instance, to describe Herring Cove, should the words of park engineers, geologists, biologists, or limnologists be preferred over those of expropriated landowners? Should the focus be on the visible landscape, on larger environmental indicator species, or on the microbiological and climatological levels as well? The answer certainly is that there will be occasion to depend on each, selectively, and to depend on the different descriptions that result to tease out a general description of what Herring Cove is. Any description that results will still be far from complete. As Cronon notes, "Whereas fields like women's history and African-American history have sought to recover the 'lost' voices of 'ordinary people' by letting their 'subjects speak for themselves,' we can never hope to discover quite so certain or autonomous a voice for the natural actors that participate in our own narratives."[18]

This might seem to stop the project before it begins. But environmental history may at this point benefit from a recent offshoot of literary theory called ecocriticism, which studies nature in literature.[19] Ecocritics accept that an inability to translate nature accurately to language is a fundamental reality of writing about nature. This should not impose paralysis, however, but rather modesty and accommodation. Ecocritic Lawrence Buell writes, "one has to imagine. One has to invent, to extrapolate, to fabricate. Not in order to create an alternative reality but to see what without the aid of the imagination isn't likely to be seen at all."[20] After all, the suggestion that descriptions of nature are social creations does not take away in the slightest from the suggestion that there is a nature with its own independent reality. The environmental historian's job is to observe the aesthetic prejudices and personal interests that tint each text, the better to understand the ways that humans represent nature; once this filter is recognized, the job then entails seeking to create a closer approximation of nature. This does not demand that all environmental historians be discourse analysts, but it does demand that we show a concern for understanding how nature has been described by others and is to be described in our own work. Above all, it means accepting that each description of nature is a mediation between language and real nature. As Buell rather unartfully notes, "Should this or that literary expression of gratitude at one's return to nature be taken as responding to nature, or disguising a human interest ... or simply as affirming the tradition of nature affirmations? The answer to such questions is always 'both.' "[21]

Finding a way to describe what nature is goes a long way toward finding a way to describe what nature does. It is a truism in environmental history that nature is more than just a stage on which history takes place; it is a participant. But if so, how important a participant? Some historians look longingly, I think, to the environmental determinist works of writers such as Frederick Jackson Turner and Walter Prescott Webb. To these American writers, nature dominated culture: the immense presence and harsh environment of the West stripped settlers clean of the vestiges of European culture, allowing them to grow into Americans, pursuing happiness, bearing arms, and so forth.[22] More recently, historians of landscape and landscape art have stood this notion on its head and argued that throughout history culture has dominated nature. In this view, nature is – in every way we look at and respond to it – a cultural construction. Moreover, it has not been a democratizing force; it has traditionally been an instrument that has served conservative, even fascist interests.[23] To these writers nature is so entwined with culture that nature as an entity unto itself seems to disappear. In this spirit, Simon Schama writes in *Landscape and Memory* that "Landscapes are culture before they are nature, "[24] as if associations have an existence prior to and more real than that which is being associated. Writers like Turner and Schama have given credit to nature as shaping history, but not in ways that environmental historians should consider progress: they transform nature from a setting to a prop, and often a weapon at that.

It is more helpful to see nature as an actor, but not necessarily the star. Rather, it is an actor relating to other actors in different ways, to different degrees, at different times. For example, in saying that park staff shaped Fundy National Park in response to the natural features they found there, we really mean they acted in response to their responses to those features. There is always a dynamic involved. We should, as Lawrence Buell writes, on the one hand "acknowledge that reported contacts with particular settings are intertextually, intersocially constructed" and on the other "acknowledge that the nonbuilt environment is one of the variables that influence culture, text, and personality."[25] This should be a sufficient degree of environmental determinism for environmental historians. If historians of class and gender can accept that neither is the sole creator of identity, surely environmental historians can accept that both humans and nature contribute to the relationship between the two.

Having said all that, I think readers will be surprised to find I do not always follow my own counsel to incorporate what nature is and what nature does. This book is often a more traditional account of environmental policy-making. Canadian national parks are some of the most studied places on the planet, with Parks Canada researchers having written a multitude of reports on geological, biological, and climatological conditions. I did not want to write a mere compendium of these studies, and so have chosen not to tread this ground already walked. My focus instead is on Parks Canada's own history, a topic the agency has covered less well.

Then why offer this model for environmental history in the first place? Three reasons. First, to document the conceptual juices in which this work and its author have been marinating over the past few years. Second, since this is one of the first Canadian monographs to position itself explicitly within the field of environmental history, it seemed worthwhile to address directly the state of the discipline. (In the words of nineteenth-century proto–environmental historian George Perkins Marsh, "it is hard to 'get the floor' in the world's great debating society, and when a speaker who has anything to say once finds access to the public ear, he must make the most of his opportunity, without inquiring too nicely whether his observations are 'in order.' "[26]) And third, because I believe the model *is* well represented in this book. Though what follows is at times environmental history in the older, Runtian sense of exploring humans' relationship to nature, it is often also environmental history in the sense of seeing nature as a continuing force in history. Throughout, I will call attention to how the nature in Cape Breton Highlands, Prince Edward Island, Fundy, and Terra Nova national parks is described and how this reflects both nature's real and independent existence. Nature will not be the foundational chapter never to be seen again, as it is in too many history books, but neither will it be the sole dominating character. It will, as in our own lives, be a constant presence, sometimes speaking, sometimes quiet.

OF PRESERVATION AND USE

The one-sentence mandate of Canada's 1930 National Parks Act is as follows: "The parks are hereby dedicated to the people of Canada for their benefit, education and enjoyment, subject to the provisions of this act and the regulations, and such parks shall be maintained and made use of so as to leave them unimpaired for the enjoyment of future generations."[27] What begins with "benefit" is brought up short by "unimpaired." That is, the national parks are to be used by people as well as preserved from the harm that people bring. The dual nature of this mandate is entirely appropriate, however. National parks are areas of land considered wild, free from the hand of man – that is, they are natural. Yet they are also created by and for people – that is, they are cultural. It is only logical, then, that the tug of war between preservation (the maintenance of the natural component) and use (the relinquishing of the natural to human demands) has been the constant, unresolved problem at the heart of park history.

One might think that the contest between park use and preservation would be a natural subject for Canadian historians. It is easy to visualize it as a metaphoric battle between nature and culture, evolved nature lovers vs. earth-paving greedheads, or eco-fascists vs. wise-use advocates, depending on your political stripe. But this has not been the case. Robert Craig Brown's 1969 essay "The Doctrine of Usefulness" is still the most influential study of Canadian national park history, and the only one with which most Canadian historians are probably familiar. Brown contends that at the birth of the national parks in 1885, there was no dispute between preservation and use, because "there is little evidence to suggest that national parks policy originated in any conviction about preserving the 'wilderness' on either aesthetic or other grounds."[28] Instead, John A. Macdonald's government set Banff aside to make it a tourist resort, and to exploit the trees, rivers, and mines on this reserved land. Macdonald stated at one point that "the Government thought it was of great importance that all this section of country should be brought at once into usefulness."[29] Brown uses this quotation to great effect, giving name to what he calls the doctrine of usefulness: the belief that parks were set aside so that the fullest use of their resources could be made by the federal government and private enterprise. Brown concludes that "the original parks policy of Canada was not a departure from but rather a continuation of the general resource policy that grew out of the National Policy of the Macdonald Government."[30]

Brown's thesis has met with no opposition in the past thirty years, perhaps especially since so much development has gone on in the parks from their beginning. Spray Lakes in Banff was dammed for hydroelectric power, coal was mined in Jasper until after the First World War, zinc and silver were taken from Yoho until after the Second World War, and logging in Wood Buffalo was only recently discontinued. The credibility of Brown's argument was reinforced by US historian Alfred Runte's claim that in the American case, Congress only chose sites to be national parks when it was certain that the area had no natural resources of commercial value. Runte's "worthless lands" thesis reads like a pessimistic spin of Brown's "doctrine of usefulness."[31]

W.F. Lothian's encyclopaedic four-volume *A History of Canada's National Parks* published in the mid-1970s did not respond to Brown's doctrine of usefulness, or deal more generally with preservation and use.[32] Lothian's work is that most cursed of scholarly works: the commissioned history. It is avowedly partial, yet because of the author's half-century of service with the Parks Branch, it is authoritative and comprehensive enough to steer other scholars away. As Lothian himself came to acknowledge, his masterwork is rather dry and institutional, providing little insight into the people involved in the Parks Branch and the ideal which they were attempting to enforce.[33]

Perhaps Brown's explanation for the approach taken toward the parks in their early days could only be formulated when the wisdom of that path was in question. Preservationist impulses grew stronger within the park system and society at large in the 1960s. In 1964, park policy was established that made preservation the parks' "most fundamental and important obligation."[34] A revised policy statement in 1979 endorsed even stronger preservationist views, and in 1988, amendment of the National Park Act made the maintenance of ecological integrity the Parks Branch's prime directive. Recent historical work on Canadian parks reflects such growing preservationism, holding to Robert Craig Brown's view of park history while being more critical of its implications. Leslie Bella begins her *Parks for Profit* (the book's title is its abstract) with the rather ahistorical declaration, "National parks are supposed to be about preservation."[35] She succeeds in proving that, in fact, this has never been the case. Parks, she writes, "have not been removed from economic development, but have been the focus of that development."[36] One cannot help but feel in reading Bella's book that parks, which she loves in theory, have been at best compromises with capitalism. Kevin McNamee's recent overview history of the park system, "From Wild Places to Endangered Spaces," accepts that economic initiatives steered the parks for much of their history, but suggests that public concern for the environment in the 1960s fortunately helped set parks on a higher road.[37] In such histories, the doctrine of usefulness is itself useful in a narrative sense: it allows the parks to begin as instruments of capitalism until redeemed by the grassroots efforts of environmentalists.

Brown's doctrine-of-usefulness thesis has been accepted rather uncritically as the perspective from which to view Canadian park history. But it is an incomplete analysis in four ways: first, it applies the term "usefulness" in an extremely misleading way; second, it ignores preservationist impulses present at the establishment of the first park; third, it does not speak to the beliefs or policies of the parks staff themselves; and fourth, it is valid as an organizing principle, if at all, for only a very brief time. As a result, the constant struggle between preservation and use in the parks has gone largely unexplored. By exploring Brown's article in greater detail, we can show that from the creation of Banff in 1885 there has never been just one doctrine directing the national park system.

To begin, Brown makes very slippery use of "usefulness" throughout his essay. Brown refers to "usefulness" continually, as if it were the Macdonald government's mantra, spoken time and again in debate to indicate plans to fully exploit the Banff area. Yet the word is spoken only *once* in the twenty-five pages of debate on the Banff bill, in the passage by Macdonald cited above. By keeping the term in quotation marks, Brown benefits from the reader's mistaken assumption that he is quoting its nineteenth-century use, when in fact he is clarifying (?) that this is a principle he is naming. (However, he compares the frequency of "utilization" in 1911 debates to usefulness, "the 1887 term," and elsewhere writes, "This is what Macdonald and his colleagues had in mind when they spoke of 'usefulness.' ")[38] And in the single instance in which Macdonald did speak of usefulness, Brown ends the quote in mid-sentence. Macdonald said, "Then it is of some importance – the Government thought it was of great importance – that all this section of the country should be brought at once into usefulness, that people should be encouraged to come there, that hotels should be built, that bath-houses should be erected for sanitary purposes, and in order to prevent squatters going in, the reservation was made."[39] Usefulness here referred specifically to making Banff a resort. All other development was secondary. It was accepted that mining and lumbering would be allowed to continue in the region because Canadians of the day accepted that nature was less important than prosperity. But to say that the Macdonald government condoned resource extraction in the park is different from saying it created the park for the sake of extraction.

Indeed, the House debates suggest that members believed parks were to be about preservation, and endorsed this goal as a good thing. The resort, after all, was to be government-run to prevent the capitalist excesses that had ruined the hot springs resort areas in Arkansas and Virginia. And whereas Brown contends that there was no interest in "preserving the 'wilderness' on either aesthetic or other grounds,"[40] it was agreed that not just the hot springs but also the surrounding land within view must be preserved. Macdonald noted in the sentence preceding his declaration on usefulness that it was considered an urgent matter, once the idea of a park was established, "that a reserve should be made at once, and that as much attention as possible should be paid to the protection of the timber in the general line of the park."[41] This is preservation for aesthetic reasons, but it is preservation nonetheless. Macdonald knew that resource exploitation reflected poorly on a national park and, if present, must be hidden away. Opposition members understood the concept of preservation well enough to claim that national parks could not condone resource development at all, offering the United States' Yellowstone and Yosemite as their proof. One member vowed to side with government to keep timber licences out of the area.[42] Another bluntly stated, "If you intend to keep it as a park, you must shut out trade, traffic, and mining."[43] The idea of parks for preservation was not alien to the members of the House.

If we are nonetheless to accept that the doctrine of usefulness was the dominant paradigm at the beginning of park history, we may still doubt the extent of its influence. In his article, Brown jumps from 1887 to 1911, and states that the doctrine had by then "taken on a somewhat greater degree of sophistication," meaning that by the later date it was recognized that resources should be efficiently conserved.[44] He does not refer specifically to parks at all in this period, so his thesis vis-à-vis park history is unproven beyond 1887. Also, because park records were unavailable to Brown in 1969, his model was formulated entirely from the pages of Hansard. But since we know that even government members knew parks were to be about preservation, should we not wonder how the staff responsible for the parks after 1887 saw their job? Did they follow a doctrine of usefulness, though they were not responsible for resource extraction, or did their efforts to sustain the natural areas of the parks incline them toward a doctrine of preservation? Did they feel, for example, that they had to defend the park areas from some uses?

Ultimately, the national park system cannot be interpreted in terms of only a doctrine of usefulness or a doctrine of preservation, because the Parks Branch has been expected to fulfil the mandates of both use and preservation simultaneously. As well, use and preservation do not necessarily contradict one another: in the period covered here, preservation gained in importance even as park use rose dramatically. A more valuable indicator is the changing level of intervention that the Parks Branch permitted and itself caused in the name of both use and preservation. In the 1930s and during the Second World War, intervention in parks was limited, because the Branch had a shortage of funds for development and showed little interest in managing preservation. From the mid-1940s to the late 1950s, funding increased rapidly for both preservation and use, and the science of ecology which guided park preservation grew increasingly activist; as a result, the Parks Branch was much more interventionist in this period. Around 1960, the birth of a North American back-to-nature movement on the one hand created rapidly rising park attendance, and subsequent pressures on the park system; on the other hand, it led to calls for less direct human intrusion on the parks' nature. Responding to demands for more intervention for park use and less intervention for park preservation, the Parks Branch in the early 1960s would have to clarify how it planned in the future to fulfil its troublesome dual mandate.

There can be no "victory" for either preservation or use: a park system demands both. Unrestrained use would make the park no different from places outside its borders, and the park as an idea would be meaningless. Likewise, unrestrained preservation would demand the exclusion of persons, a policy not only politically untenable but ecologically contrived, in that it would arbitrarily leave out one species to preserve a nature that had already been shaped by that species. Park policy demands greater latitude, so that "benefit" and "unimpaired" can continue to share space in the same sentence.

OF FOUR PARKS

Beyond an interest in humans' relationships to nature, and a desire to understand how national parks work, a fundamental reason for writing this book has been to learn the stories of these four particular parks – in particular, the stories of how relations between park staff and local citizens have evolved. Canadians may remember the firestorm of opposition unleashed by the expropriation of land for New Brunswick's Kouchibouguac National Park in the 1970s. Rioting, an unlawful occupation of the park, and a decade-long threat of violence convinced Parks Canada to change the way it acquired land for parks.[45] In comparison, the expropriations of the first four Atlantic Canada parks were docile affairs. The Parks Branch chose land it thought appropriate for a park, the provinces expropriated the land, and landowners settled. But this does not mean that these landowners were treated more fairly than those at Kouchibouguac; just the opposite. Until around 1970, expropriation in Canada was weighted heavily in the favour of federal and provincial governments. Since it was governments who needed expropriation carried out, the legislation they passed best served their needs, at the expense of landowners' needs and rights. Rather than pay "market value" for properties or the price of establishing a similar property elsewhere, the governments offered the potentially much lower "value to the owner." They also tended to inform landowners inadequately, set compensation as they saw fit, and even set up their own boards to arbitrate dissent. Whereas Great Britain had implemented a more progressive expropriation system in 1919, Canada seemed in no hurry to follow suit. Expropriation was "the neglected Cinderella of Canadian public law," in the words of one legal scholar.[46]

Most of those people who lost their homes and their land to make room for Cape Breton Highlands, Prince Edward Island, Fundy, and Terra Nova national parks responded with fatalism rather than opposition. Canadians of the day generally accepted (though resented) that governments could do as they pleased. This was perhaps especially true in rural Atlantic Canada, mired as it was in patronage-driven politics. Dispossessed landowners felt they had no choice but to leave, and they worried that by holding out they would ultimately receive less money, so they accepted the government's offer. They often moved to a community right next door to the new park. The parks were born in bad feelings, then, which lingered. After 1945, unprecedented infrastructure development in Canada led to an unprecedented amount of expropriation – the building of Ontario's Highway 401 involved twenty-one thousand properties, for instance. Canadians increasingly came to see the expropriation system as arbitrary, unclear, and unfair, and their sympathy for victims of expropriation grew. Those people who had lost land to parks grew more convinced that their rights had been trampled, even if time had given the park more claim to ownership. Add to this local displeasure with park hiring, preservationist, and development policies, and the resulting stories of these four parks are of constant tension between parks and locals.

I was first drawn to these stories when examining the creation of Prince Edward Island National Park as part of a Master's thesis on PEI tourism.[47] Smaller, equally compelling stories kept appearing: stories of cows trespassing on the golf course in solemn but messy protest over the expropriation of land, or stories of the Parks Branch painting the Anne of Green Gables house white with green gables to make this real house, which was said to be the model for the fictional one, more closely resemble the fictional one. In other words, I was attracted to the ironies that were showing up. In fact, if a book has a defining trope, this one's is irony. Perhaps this is not surprising, because as Linda Hutcheon has suggested, Canadians feel comfortable with irony. It offers them the chance of addressing speakers politely but with a touch of sly confrontation.[48] And as William Cronon states, "like most modern historians," environmental historians "have a special fondness for stories that convey a sense of irony, because irony best expresses our sense of the multivalent complexity of the world."[49]

Irony has, I hope, been visible in the previous two sections of this chapter, in which culture/nature and use/preservation have been named as important operating principles with which to work, and then shown to be false dichotomies. The conceptual separation of human culture from nature is perhaps understandable in that we are human, but it falsely suggests that we are distanced from biological processes. This, of course, is exactly the opposite of environmental history's intent. Similarly, polarizing preservation and use misses how they tend to draw on one another. Preservation has historically been an active task involving land reclamation and species reintroduction, and use for nationalistic or economic purposes has often been the reason for nature preservation. Referring to the irony in such conceptual matters seems harmless. But what of the ironies that are found in people's lives, that are present in the stories of the four parks to be studied?

There are good reasons to have reservations about irony. It can be too easy – pointing out the peccadilloes of others with the clarity offered by hindsight. It can be too light – minimizing the importance of events with humour. In effect, it can be condescending – implicitly suggesting our own superiority by playing on the mistakes and weaknesses of the past and its people. Such condescension has been particularly dangerous in discussions about our relationship to nature. Too much environmental thought of the past few decades has proclaimed the foolishness of humans' behaviour to nature, while reverentially mouthing a preference for the company of nature. As a result, environmentalism has been perceived as elitist and uninterested in finding solutions suitable for people as well as nature. Much of the writing has been dogmatically anti-modern (see any number of books on escaping to the wild and building your dream cabin), anti-Western civilization (such as Frederick Turner's 1980 *Beyond Geography: The Western Spirit against the Wilderness* or Calvin Luther Martin's 1992 *In the Spirit of the Earth*), and even anti-human (see portions of Edward Abbey's later writings, and Earth First!'s manuals for eco-saboteurs). Environmental thought, in other words, has somehow become simultaneously escapist and apocalyptic. Not surprisingly, against this extremism there has been an equally extreme backlash, with the publication of books such as Gregg Easterbrook's *A Moment on the Earth* that seek an "ecorealistic" alternative to environmentalism, promising that we can have our Western civilization and eat it, too.

And yet, irony persists. In the expropriation of farmland for a national park, the fields turned into a golf course (and the farmer taking up golf). In preserving a place as representative of the best of Canadian nature because it resembles the Scottish highlands. In building a causeway to permit tourists to visit an offshore island, and in doing so changing the tidal behaviour and causing the island to erode away. To use irony fairly in the writing of history, one must temper it with honest affection. The ironies that will be described in the stories of Cape Breton Highlands, Prince Edward Island, Fundy, and Terra Nova National Parks will, I hope, express my affection for the people whose stories these are. And not only an "easy" affection for those who lost land to the national parks, but also an affection for park staff faced with the weighty task of balancing nature and culture, preservation and use.

This book is divided into two parts. Part 1, "In Search of Eastern Beauty," traces the establishment of the first four Atlantic Canada national parks. Chapter 2 introduces the National Parks Branch under Commissioner James Harkin, from the agency's creation in 1911 through to the mid-1930s. Chapters 3 through 6 discuss the selection, establishment, expropriation, and development of the four national parks in question. Part 2, "A Pious Hope," shifts to thematic chapters involving all four parks from 1935 to 1965, and the changing philosophy of the Parks Branch during this time. Chapter 7 studies issues of use, including tourism (with special reference to class and discrimination), business concessions, and park inviolability. Chapter 8 studies issues of preservation concerning wildlife, fish, and vegetation. Chapter 9 examines the relationship between the national parks and their surrounding communities.

Part One

In Search of Eastern Beauty

Nova Scotia Should Have Its Fair Share

"Nova Scotia Should Have Its Fair Share," by Donald McRitchie, Halifax *Herald* 30 March 1935. Republished with permission of The Halifax Herald Limited.

2 James Harkin and the National Parks Branch

In June of 1911, the minister of the interior, Frank Oliver, called his private secretary, James Bernard ("Bunny") Harkin, into his office and offered him a new job. The government was setting up a separate branch to oversee national parks, and Harkin was asked to be its commissioner. In his memoirs, Harkin would write, "Overcome by surprise I could only say that I doubted my ability since I knew nothing about the parks or what would be expected of me. 'All the better,' he [Oliver] said, in his laconic way, 'You won't be hampered by preconceived ideas and you can find out.'"[1] The park system was now in the care of a man with no experience in parks, no belief in what they should be about. Harkin would later claim that his naïveté even helped him to accept the post, "thinking that the care of a few beautiful places would be almost too cosy and delightful a task."[2]

This begins the Canadian National Parks Branch creation myth, the traditional story offered by park historians.[3] The young bureaucrat journeyed west to visit the national parks under his control, and was transformed by what he saw there. He became a convert to conservationism and soon blossomed into Canada's most influential nature lover, slowly steering the parks away from the doctrine of usefulness and toward a doctrine of preservation. He would oversee the Parks Branch for a quarter-century, leading it as it grew from an Ottawa staff of seven to more than eighty, and from six parks covering four thousand square miles to seventeen covering thirteen thousand. His work was unfinished when he retired in 1936, though, and it would not be until decades later that the preservationist doctrine he helped formulate would gain a sort of ascendancy in the parks.

When histories of parks and conservation in Canada began to be written in the late 1960s, Harkin was an obvious candidate for founding father. In *Working for Wildlife*, Janet Foster writes that Harkin "articulated the most complete philosophy of wildlands preservation." He was fifty years ahead of Americans in appreciating wilderness at a time when most Canadians were fifty years behind. W.F. Lothian saw Harkin as "an ardent conservationist," J.I. Nicol called him an "idealist," and Roderick Nash noted that he "had a clear conception of the aesthetic and spiritual value of wilderness."[4] These writers forgive Harkin for also being a great booster for the economic advantages of park tourism, for filling the parks with hotels, roads, and golf courses; this is interpreted as pragmatism. To further the parks agenda Harkin had to make concessions to the Philistines who thought purely in economic terms. Harkin's biographers have fed off one another to the point that Gavin Henderson in a 1994 article entitled "The Father of Canadian National Parks" offers precisely no new information on the commissioner, yet goes furthest in naming him an environmentalist. Henderson writes that despite Harkin's concessions to tourism, "as we know now, he was committed to a broader perspective."[5]

In fact, we know nothing of the sort. Historians' eagerness to see Harkin's drive for dollars as a strategy rather than a philosophy tells us more about them than about him. Underlining Foster's, Lothian's, McNamee's, and Henderson's work on James Harkin is the knowledge that he was the first commissioner of the Parks Branch and he had a lengthy term at a time when the parks were relatively free of political interference, when the future policies of the park system were being conceived. In other words, he *must* be important, and since the parks have done such an admirable job of preserving Canadian wilderness, he must have been a great preservationist. Therefore, Harkin becomes a hero, if only by default. That he began with no knowledge of parks makes his story all the more remarkable: national parks must truly have the power to convert people to nature.

Harkin's biographers have based their assessment of him on a very thin stack of sources. We know that he was born in Vankleek Hill, Ontario, in 1875 and was educated there and at Marquette, Michigan. He became a newspaperman at the *Ottawa Journal* in 1892, leaving in 1901 to become secretary to succeeding ministers of the interior. He was appointed to the new Parks Branch in 1911 and was its commissioner until 1936, when he retired. He died in 1955. We know little else about Harkin's career, and the sources that we would expect to be illuminating are only frustrating.[6] His archival papers do nothing to clarify his role in the formation of national parks philosophy. Rather than containing a record of his achievements in park creation and maintenance, wildlife preservation, highway construction, historic site commemoration, and so on, the papers consist almost entirely of his notes and correspondence on Vilhjalmur Stefansson's planned 1921 expedition to the Arctic. According to his long-time secretary, Harkin always considered his role as the federal government's envoy to Stefansson "The greatest thing he did for Canada."[7] Does this mean that he did not see his park work as particularly significant, either because it failed to interest him or because he believed that the credit for it should go to others? There is no way of knowing. In my own research covering just a small fraction of the total park files, I came across references to Harkin taking a three-month sick leave in 1905 (at age thirty), being away from the office due to sickness in 1927, 1931, and 1934, and having an operation in 1936.[8] Did Harkin have a recurring illness or disability, and did it affect his ability to run the Branch (or did it lead him to find nature medicinal, a notion he then promoted in park literature)? Or was Harkin perhaps a virile but clumsy mountain-climber? We do not know enough about Harkin, and as a result cannot know enough about the Parks Branch.

With so little biographical material available, writers have turned to Harkin's written work to flesh him out. They quote heavily from a few articles, the posthumously published extracts from a planned memoir, and his annual reports as commissioner. In doing so, his biographers have made several sizeable assumptions. They have accepted uncritically that Harkin wrote all that is credited to him; that his conservationist sentiments are sincere but his economic justifications are strategic; and that what he writes is honest and unmotivated and thus correlates exactly with how he would behave. There are valid reasons for making these decisions: they simplify the story, making Harkin more heroic, which in turn puts a face to the Branch's work. But such narrative choices have consequences. By making James Harkin the focus of all the Parks Branch's successes, writers have made him a man ahead of his time rather than a product of it. The Parks Branch becomes a mere backdrop to Harkin's work, and necessarily an ineffectual organization. After all, Harkin's philosophy could only be considered progressive in relation to what would have been had he not been around.

It is easy to sympathize with historians who, in simplifying and personalizing the story, have focused on Harkin at the Parks Branch's expense; I too face the difficulty of bringing together an unruly set of facts and opinions and making them cohere. This chapter must set the scene for the establishment of parks in Atlantic Canada that are the subject of the remainder of the work. To do so, it must touch on the Canadian National Parks Branch philosophy, the aesthetic conventions of national parks, the rise of mass tourism to parks in the 1920s and 1930s, and the early efforts to truly nationalize the park system. The question is, how to tell the story.

This chapter begins with Harkin because I accept the premise that Harkin's tenure as commissioner of parks must have been important if for no other reason than that it was the first and went on for a quarter-century. In this long, stable period, many of the Branch policies were designed, tested, and corrected. But rather than search out a way to prove Harkin's personal importance, it seems more sensible to study him as representative of the Parks Branch as a whole. Arriving with no expertise about parks, he relied on his staff to help draft policy, and so served as a conduit for the philosophy germinating within the Branch. He can be seen as a personification of the organization's history – an epitome rather than an anomaly. (In the same fashion, the Parks Branch's emerging philosophy was itself a product of its times. Discussion of Harkin and the Branch's beliefs on nature and its value to society can therefore also serve as an introduction to broader Canadian thought on these topics.) True, this also is a storytelling form, but it is one that is not only more faithful to the information we have but demands less of information that we do not have. For example, gaps in our knowledge of Harkin would not be filled in with assumptions about his conservationism; rather, they would be used to discuss the difficulty of uncovering Branch history. In discussing Harkin's career and writings, it may be said that they parallel the parks' broader history in three ways. First, though his history is understood in broad strokes, there are important gaps that have not been filled. Second, like any good Canadian, he was enamoured of whatever was going on in the United States. Finally, his philosophy has been oversimplified for the sake of more efficient storytelling. His writings suggest a complex interest in both use and preservation, as well as a belief that these two were not necessarily contradictory.

THE COMMERCIAL AND THE HUMANITARIAN

Though Harkin came to the Parks Branch with an admitted lack of experience in parks, he immediately began to write knowledgeably on the financial benefits of parks and the values they promoted. This has been interpreted as a rapid conversion to wilderness advocacy, but it is much more sensible to suppose that he was relying on his staff to help draft policy. A search through the Parks Branch papers shows that often what is attributed to James Harkin was in fact first drafted by others, especially his assistant, F.H.H. Williamson, and his expert on wildlife policy, Hoyes Lloyd.[9] It is quite possible, of course, that Harkin was still very much involved, either in telling his staff what he wanted written or redrafting what was given him. But we know that the commissioner's need for advice was especially acute. What are credited as Harkin's beliefs are more likely to be the beliefs of the Parks Branch as a whole than his biographers have suggested. (For the sake of simplicity, in the coming pages I will credit Harkin with sole authorship of reports published under his name, rather than continually note that these were likely joint efforts.)[10]

Harkin also relied a great deal on information obtained from the American National Park Service and American sources in general. The Canadian park system had always been close to the American one: the wording of the Rocky Mountain Park Act that officially created Banff in 1887 even mimicked the wording on the act that had created Yellowstone in 1872. Whereas one would be hard pressed to find any mention of Great Britain in Canadian park correspondence,[11] it was natural that Canadian administrators felt much in common with their southern neighbours: the two countries had similar cultural histories, geologies, biologies, and climates – and similar parks. By the early 1910s, the Canadian park system was arguably just as strong as the American one, and considered to be such a rival – drawing American tourists to scenery they could just as well see in their own country – that the US parks turned down a proposal to share publicity campaigns.[12] Only in 1916, when the US government formed the National Park Service and doubled the park budget, was it clear that that country had the better-developed park system.[13] But from the very first days of his appointment James Harkin had believed that the Americans knew how things should be done, and that the Canadian park system should emulate theirs. He corresponded regularly with the National Park Service, the National Park Association, and the American Civic Association on everything from predator policy to legal aesthetic rights.[14] The Branch's annual reports quoted American conservationists such as J. Horace MacFarland, Stephen Mather, Teddy Roosevelt, and Henry Thoreau with the understanding that the two countries' situations were similar. After citing MacFarland in the 1914 report, Harkin added, "His remarks, of course, referred to American parks but change the word 'American' to 'Canadian' and the concluding portion of his address crystallizes a thought of equal application to Canada."[15] The commissioner's reports were especially fond of quoting John Muir, America's most prominent conservationist and founder of the Sierra Club, in speeches, articles, and reports. Historians Henderson, Foster, and McNamee make much of this, as if restating another's philosophy is equivalent to stating one's own. But Harkin may have quoted Muir not only because he agreed with his beliefs, but also because he knew his audience was familiar with the American, and because to quote him was to become identified with him (which Harkin's biographers make clear has been the case).

My point is not that Canadian national park philosophy was insincere, only that it was not unmotivated. The Parks Branch was trying to solidify its own meaning as an organization, and to secure funding at the same time; a discussion of its stated beliefs should recognize both realities. This is especially important in understanding Harkin's defence of both use and preservation in the parks. The usual interpretation is that the parks were first created as part of the doctrine of usefulness, a natural extension of Sir John A. Macdonald's National Policy. Under Harkin's leadership, though, use itself became a use – a tactic to bring attention and money to the parks. Years later Harkin wrote in explanation, "How could the hard-headed members of the House of Commons be persuaded to increase parks' appropriations? It is an axiom that no society will pay for something it does not value."[16] The solution was to be found in tourism. Harkin wrote to American states and European countries for tourism revenue statistics and used these to convince the Canadian government that the parks were worth tens of millions each year. Often using economic multipliers multiplied by multipliers, he was able to offer exuberant statements on the economic value of scenery. Harkin calculated that wheat fields were worth only $4.91 per acre to Canada, but scenery (which presumably does not include wheat fields) was worth $13.88.[17] Park historians have accepted without argument Harkin's defence, made decades after the fact, that "while we were forced in the beginning to stress the economic value of national parks we realized that there were other values far more important which would be recognized in time." These were results that he elsewhere calls "of a higher order – results that serve the individual as to the welfare of his body, the activity and efficiency of his mind, and the beauty and harmony of his soul."[18] When Harkin speaks of such physical, spiritual, social, and national benefits of parks, he is taken at face value.[19]

But Harkin's writing suggests that the secular and the sacred justifications for parks were both sincere in their time. A 1914 memo classified these as "Commercial" and "Humanitarian" reasons for parks, noting, "The commercial while very important is nevertheless subordinate to the humanitarian." The commercial is then discussed for the next thirteen pages.[20] The commissioner's annual reports always begin with the economic value of parks, and always in the appropriate business terms. Tourists, Harkin wrote, were "eager to spend money on trips to see outstanding natural beauty. Therefore the Canadian National Parks may be said to be in the business of selling scenery."[21] Elsewhere, it is noted that a great advantage of tourism was that "When we sell scenery, no matter how large our sales, our capital stock remains undiminished. We have the same scenery to sell over and over again."[22] This idea of selling scenery not only gave economic justification for the parks' existence, but also validated Canadian scenery's international stature.

Considerable space in annual reports was devoted to discussion of the humanitarian advantages of parks, particularly in the first years of Harkin's tenure. Parks made people better: physically and mentally healthier, closer to God, and more aesthetically refined. These benefits were often intertwined, and none was given preference (the boosterism is such in the commissioner's reports that at different times each benefit was declared the most important). But the space devoted to each did change. In the first years, the commissioner especially stressed the value of parks for recreation. "[T]he commercial side of National Parks is only an incident, though indeed a very important one ... National parks are in reality national recreation grounds," Harkin wrote.[23] Periodically, people needed to escape the stresses of their modern lives and enjoy the benefits of the simple life. Harkin theorized that all great cultures throughout history had allowed for the exercising of the "play spirit," a rejuvenation through recreation. Though this was a physical rejuvenation, the park visitor did not necessarily have to exercise: just breathe fresh air, take in the sun, and regard nature. With the First World War, the physical demands on Canadian bodies became more important, and so the sort of rejuvenation that the parks were offering changed. In 1916, Harkin wrote, "National Parks exist for the purpose of providing for all the people of Canada facilities for acquiring that virile and efficient manhood so noticeable in Canadian military training camps."[24] By the following year, with Canadians trench-deep into an exhausting war, Harkin promised the parks would "materially assist in remedying whatever the war may do to Canada's human assets."[25]

Whereas before the war Harkin referred to rejuvenation away from the daily stresses of life, in the postwar years these stresses now seemed intolerable. "To-day," Harkin stated, "the strenuous life of civilization compels man to live under stress and tension; men by sheer will power and concentration hold themselves down for long hours to desks and machines."[26] Urbanism deserved special blame.

Cities brought out the animal in man by removing him from his natural environment. Harkin claimed that "the data collected by scientists within the last few years has shown beyond question that life in our modern cities tends almost universally to a deterioration in type and that vitality ... under modern city conditions is constantly being dissipated."[27] F.H.H. Williamson, Harkin's assistant, warned that unless something was done slums would take over Canada as they had "New York, Chicago, Philadelphia, St. Louis, Pittsburgh, Detroit, and the cities of the old world." He noted that "a pig in its natural state is a cleanly animal but pen it up and it will wallow in the filth; a dog gets cross and vicious when shut up in a house."[28] Neurasthenia – nervous exhaustion – was a product of the modern world, but could be undone by a return to nature, and a renouncing of the life that had made the patient ill.[29] A visit to a national park would be sure to leave one restored, fortified, and inspired. How did this work? No one was sure – "What the secret of her magic elixir is our scientists have not yet discovered"[30] – but Harkin knew that it was deep within our biology. As he reflected, on the subject of "man" in general, "The impulses which were stored in his physical cells through countless centuries of human existence re-awaken and he experiences a strange pleasure in reverting to the primitive." He also argued that "The older, the more basic, the more primitive the brain patterns used in our hours of relaxation, the more complete our rest and enjoyment."[31] Harkin's favourite analogy was to a wild strawberry plant. Plant it in your garden at home and it would lose "that wonderful tang which was its soul." But take it back to the wild and it would quickly recover its zest.[32]

All this may seem like a radical critique of industrial life: if modernity was so toxic, should it not be cleaned up? But parks allowed civilization to continue as is, with the worker escaping periodically to nature to be recharged. The rejuvenated nature-goer was always meant to return and become a productive member of society. Having parks near cities, "where thousands of workers could spend the week-end camping and fishing[,] would result in human dividends worth many times the capital invested."[33] Elsewhere, Harkin states, "the ideal on which National Parks are administered is the production of dividends for Canada – dividends in gold and dividends in human units."[34] "Human units" was a term often used, presumably to prove that the parks could be shown to have computable, statistically verifiable value. Here was a moment where humanitarian and commercial interests merged: in helping individuals, parks helped society. Harkin's biographers see him as using the commercial to justify the humanitarian; a cynic could suppose that he used the humanitarian to justify the commercial. In the absence of evidence that one outweighed the other, it is best to accept that the Parks Branch saw the bountiful parks as meeting both objectives equally.

In the commissioner's annual reports, every mention of human units seems counterbalanced by a declaration of the parks' mystical properties. Harkin knew better than to try to explain or define these; it was enough to know that in nature people were elevated. People would find, it was said, "mystical agencies of healing and rejuvenescence for body, mind, and soul, peculiar agencies that can be found nowhere else."[35] "In these silent wildernesses there are "holy places.'"[36] The way the reports referred to nature – or Nature – was, consciously or not, the same way others might write about God. When Harkin remarked, "Perhaps in broad terms the ultimate purpose served by national parks is to draw people towards Nature, to give them a better understanding of Nature and finally to make them realize that it is from Nature alone they can get things which they need," we can see how easily "God" could replace "Nature."[37] Analogies of forests to cathedrals and mountains to temples helped reinforce the notion that God could be found in nature – and even hinted that God *was* nature.[38]

Harkin's writing glides so easily from one justification for national parks to another – referring to them as both holy places and common stock on a single page[39] – because the use and preservation of parks were understood to be related. Parks needed to be sold to Canadians, but precisely because of what they were and what they promised to be for generations. In one article, Harkin tells the story of meeting "a typical matter of fact businessman" who spoke warmly of a trip to Banff a dozen years before to recover his health. He had fallen in love with the place and spent much of his time there wandering alone. One morning he came upon "a well-dressed, prosperous looking man," and they took to talking. On discussing the park's beauty, the other said, "Beautiful! Why, sir, I had to come to Banff to learn that there was a God."[40] Here is a little bit of everything that the parks were to be, all in one anecdote: mystical, attractive to the tourist, healing, beautiful, and invigorating for the hard-working businessman.

THE UNJUST DISTRIBUTION OF BEAUTY

But what was characteristic of national parks that made them a tonic for the sick, a reprieve for the weary, a destination for the American? The commissioner's reports rarely bothered to say, since what made the parks so distinctive was so obvious. The parks had meaning and value because they were beautiful, and they were beautiful because they were wild and natural, and they were wild and natural because they were mountainous. National parks had mountains, and the valleys, crags, and waterfalls that went with them.[41] Harkin's first annual report includes twenty photos of the national parks. In eighteen, mountains are shown, and in fifteen of these they are the picture's focus. Harkin offered pages of testimonials on the benefits of mountains in his second report, but gave very little of this sort of evidence in subsequent years. It was enough to say that the parks constituted "the best scenery and the best recreational areas" in the country, and were of "a high standard of scenic beauty."[42] That the first national parks – Banff, Yoho, Glacier, Jasper, and Waterton – were all in the mountains of Alberta and British Columbia could not help but suggest where "the best scenery" was located.

To understand this love of mountains, it is necessary to understand Romanticism, the predominant nature aesthetic in the nineteenth – and twentieth – centuries. Romantics privileged feeling over reason, and believed that the purest expression of feelings could be found through an intense observation of natural scenery. Through nature one could come to a greater spiritual knowledge, be energized in body and mind, and re-establish a wholeness of humanity that had been dissipated in civilization (all, notably, things which Harkin believed parks accomplished). In their quest Romantics carried as beacons two important aesthetic categories, the "picturesque" and the "sublime."[43] The picturesque referred to distinct, irregular landscapes often bearing signs of human past, whether ruins of classical antiquity or peasant huts. North America lacked evidence of this sort of past occupation (I am ignoring, as did the Romantics, relics of past First Nations occupation), so the picturesque never attained the influence it enjoyed in Europe. Instead, Romantics here focused their nature appreciation on the sublime. Though commonly used to describe a landscape, the sublime is more accurately a reaction to that landscape: a desired feeling, both pleasurable and terrifying, of being overcome by nature. Though the sublime as a nature aesthetic had travelled far since its introduction by Edmund Burke in 1757, it retained its original sense of an appreciation of what would, in other circumstances, be dangerous. Burke wrote, "When danger or pain press too nearly, they are incapable of giving any delight, and are simply terrible; but at certain distances, and with certain modifications, they may be, and they are delightful."[44] By approaching large, wild nature that bespoke no human presence, the viewer could hope to glimpse a spiritual infinity through a geological one. Although any number of natural formations, from oceans to forests to waterfalls, could conceivably be sublime, it was primarily mountains that captured the Romantic imagination.

Harkin's references to the "sublime grandeur" of the parks and to the mountain parks' "primeval solitudes and sublime heights" make it easy for us to detect his appreciation of sublimity.[45] But the sublime was much more than just a handy term to use when describing parks. It was the principal aesthetic used in deciding what landscape could constitute a national park. (City parks, on the other hand, were picturesque. This is understandable: towns are more likely to be built on relatively flat, arable land than on the side of a mountain. Still, it made for a nice "aesthetic division of labour.") Nineteenth-century North Americans found sublime nature, at places such as Niagara Falls and the Grand Canyon, proof of their continent's greatness. When Niagara in particular fell victim to crass commercialization, there were calls to preserve the sublime sites being continually discovered in the West.[46] As historian Stanford Demars writes in the American context, "From the beginning national parks in the United States were established more for their sublime than for their picturesque scenery. Indeed, for many years it was the monumental and the grotesque in nature that were perceived as being worthy of national park status."[47] This idea that cultural insecurity was a source of nature appreciation seems as applicable to Canada as to the United States. Banff was notable because of its mountain sublimity, and it was set aside to preserve this sublimity from unwanted – that is, uncontrolled – commercialization.[48]

The concept of sublimity lived on, though changed, in the early twentieth century. It had become increasingly difficult to work up to a bracing confrontation with nature, knowing that there were a hundred tourists around you doing the same thing, knowing that thousands had gone before you, and knowing above all that the efforts you had made to reach this point called into question the authenticity of your emotional response. Tourism had domesticated sublimity. In the 1910s, Harkin's reports still speak of the mountain parks in the language of the Romantic sublime, but not a sublime of trembling self-belittlement in the face of Nature. It is instead an inclusive sublime that reminds man of a world outside the industrial, yet connects man to nature rather than separating him from it. Searching once more for a way to explain what parks could do, Harkin once wrote, "national parks as they are may be likened to a great power house in remote mountains which carry light, heat, and energy to far-away cities."[49] That this emblematic park was in the mountains comes as no surprise.

But the light from the mountains did not always reach the cities. Though taxpayers helped pay for the parks of Western Canada, over three thousand miles separated the parks from the majority of Canadians. Those who did visit were those who could afford the trains, the hotels, and the time that a trip to the mountains of Western Canada demanded. During the 1910s this reality was either ignored or shrugged off, as when Harkin noted that Canadians needed to make going to nature a habit, even though "for geographical reasons all the people of Canada cannot visit the national parks."[50] The Parks Branch argued that parks, even when unavailable to many, still served as examples to cities, encouraging them to establish small parks within their limits and larger ones on their outskirts.[51] In 1914 Harkin himself called for parks "measured in square miles" to be established near cities, places "which may not provide as spectacular scenery as the Mountain Parks of the West" but which would offer a place for city dwellers to commune with nature.[52] He did not suggest, however, that these would be national parks, presumably since they would not be large enough – and could not be sublime enough – to qualify.

The Parks Branch increasingly became aware that the lack of Eastern parks betrayed the spirit of its publicized philosophy. How could people wearied by modern life be rejuvenated if those most desperate for help could not possibly reach it? F.H.H. Williamson gave the most spirited discussion of this problem in a memo to Harkin. At present, he wrote, the parks were of benefit almost solely to "the moneyed man or the middle classes." But it was the poor who most needed fresh air and the outdoor life, because their present situation was the least tolerable. Williamson believed that "The lowest type of people undoubtedly could not obtain the highest enjoyment from the mountains and the woods such as that derived by an artist or professor of botany, but their animal nature would demand exercise and this in turn would evolve to clearer perceptions and livelier faculties, resulting after a time in love of Nature and its antithesis – abhorrence of slumdom." The solution, of course, was that parks "should be dotted all over the country in the vicinity of centres of population."[53] And yet even Williamson stopped short of saying they should be considered national parks.

In the 1910s and 1920s, a few new national parks were established in places outside the mountains of Alberta and British Columbia. Elk Island, just east of Edmonton, was established in 1913 in a forest reserve. Ontario's Point Pelee, established in 1918, was an important site for bird migrations. St Lawrence Islands and Georgian Bay Islands national parks were reserved in Ontario in 1914 and 1929 respectively, for the purpose of saving some land for the public in popular tourist areas. But none of these parks was created for the express purpose of nationalizing the park system, or because its landscape offered cities the commercial and humanitarian benefits that national parks provided. Elk Island was meant to preserve buffalo and elk, and any humans visiting there were secondary. The Ontario parks were very small pieces of land, made up almost exclusively of easily transferred government property. None of these was chosen because it embodied or broadened the national park ideal; they were all exceptions rather than exceptional. This is not to suggest that these were not "real" parks, but only that they did not force the Parks Branch to re-evaluate what a national park was. The ideal was still typified in mountain parks such as Banff and Jasper.

Instead of the Parks Branch bringing parks to the cities, after the First World War cars increasingly brought the cities to the park. Attendance at Canadian national parks climbed from 151,000 in 1921, to 250,000 in 1924, to 391,000 in 1926.[54] More and more of these tourists were of moderate means, able to travel cheaply by car and to accommodate themselves by camping. "The automobile," Harkin trumpeted, "has brought about a wider, and more democratic use of the parks."[55] Cars also meant that a greater percentage of visitors were Canadians, there to see what the fuss was about. Recognizing that the majority of park visitors were not Europeans or Americans but Canadians, the Parks Branch redefined its goal, from servicing "recreationists" or "workers" to servicing "Canadians." Harkin's annual reports began to refer to Canadians' right to the best of their country's nature, to the parks bearing "the nation's stamp of approval."[56] Parks were now seen to make better citizens. "Already the national parks are arousing a new love and pride of country. ... Like great works of art, they are enriching the emotions and stimulating the imaginations of many and so helping to build up that finer cultural background which is necessary if Canada is to be a great nation among the nations of the world."[57]

With growing respect for its clientele's Canadianness and the benefits the parks could bring Canadians, the Parks Branch for the first time showed signs of wanting to make the national park system truly national. Harkin's 1922 report called for national parks within reach of Canadian cities, and did so again in 1923. "The great benefits accruing from the National Parks make it seem more and more desirable that these should be established more generally throughout Canada." There were wilderness areas in Ontario and Quebec that could be purchased. Harkin even noted, "It is also very desirable that areas should be set aside in the Maritime Provinces at an early date including some part of the beautiful sea coast and the original forest if any area where this remains can be secured."[58]

And yet, the Parks Branch did not follow up on this bold prescription for more than a decade. The next reference to Maritime parks in the commissioner's reports would be 1936, when the Cape Breton Highlands and Prince Edward Island national parks were being established, having been proposed, surveyed, and selected without mention in the reports. Though the Parks Branch expressed a willingness to expand the park system to the east, it did nothing in the 1920s to make this idea a reality.

The seemingly insurmountable obstacle was the lack of land that was both available and suitable. The early park system had been fortunate in that, even after the Western provinces joined Confederation, the federal government held claim to the region's natural resources, and so parks could be set aside quite easily. In fact, the first Rocky Mountain parks were created in the fear that the lands would soon be used, and that the opportunity of saving them for future generations would be lost. No such land was available in the East, other than the pockets of provincially controlled Crown lands. The three small parks created in Ontario in the 1910s and 1920s came from native and admiralty lands that were transferred to Parks Branch control. As early as 1915, a few politicians and businessmen in the Maritimes began to ask the federal government for national parks, but the response was always disheartening. Harkin replied to one petitioner, "it does not appear to be possible to find enough suitable land which still remains the property of the Provincial Government." "I am afraid it is out of the question to even think of the purchase of private lands," he wrote another. On one occasion he responded, "The moment the creation of a park depends upon the purchase of large areas of land, that moment we get into very deep water because it would be exceedingly difficult to get any Government to propose an expenditure of large sums of money for such purpose."[59] Creating national parks had always been difficult enough in the sparsely populated West; reclaiming land in settled Eastern Canada seemed politically and financially untenable.

The question of suitability was more tricky. Though Harkin professed to be in favour of a Maritime national park, he wrote, "I have some doubts as to whether any of the eastern provinces have Crown lands that would be suitable. Prince Edward Island has not. I think it is almost equally true with respect to Nova Scotia. Even if New Brunswick may have some suitable areas, there is always the probability that the province would not be willing to cede such lands to the Dominion."[60] Harkin does not explain what he meant by "suitable," and the Parks Branch never expressly pointed out which landscape was and was not worthy of being a national park. However, three requirements were always sought in choosing a national park site. It must first be sufficiently large. Harkin guessed that two hundred square miles was roughly the minimum requirement.[61] Second, it should be "virgin wilderness,"[62] untouched since European contact (First Nations people apparently did not deflower wilderness). These two conditions should serve to remind us how filled with cultural meaning national parks are, how parks had to fulfil a Romantic ideal of big and unspoiled nature. Third, parks must of course be beautiful, but more than this, beautiful in the national park sense: with the sort of sublime mountain scenery not to be found east of the Rockies. It remained to be seen whether, if asked to evaluate a site for a park in the East, the Parks Branch would seek "suitable" Western-style scenery (a hopeless task, making the nationalization of the park system an impossibility) or the closest "suitable" imitation, the most mountainous and sublime land that the East had to offer.

Not surprisingly, the Parks Branch saw no reason to clarify its aesthetic philosophy. If questioned about a Nova Scotia park, it would have been bad form for Harkin to say the province's beauty was wanting. Better to question the availability of a large, unspoiled parcel of land. The most compelling evidence that the Parks Branch staff felt Eastern scenery did not measure up is circumstantial: when their reports trumpeted the values of parks to people, when they regretted that more Canadians could not make it to the parks, they did not take the logical next step of calling for more national parks to be created closer to the majority of Canadians.

If the Parks Branch tiptoed around the suitability of Maritime scenery for national parks, Maritimers were more likely to meet the matter head-on. The region's natural image was of hilly lands, scrubby coniferous forests, and brooding seascapes. E.W. Robinson, a Nova Scotia member of Parliament, told the Commons during discussion of a park, "In the first place, the scenery throughout the Maritime provinces all belongs in the same category. There are not in this respect, three provinces; there is one province. We have not mighty mountains; we have no great rivers. Everything is on a small scale."[63] And much of the land was not only populated, but taken over for agriculture, making for even less sublime landscapes. In Prince Edward Island, provincial secretary Arthur Newbery – himself a designer of city parks – wrote a national park advocate there that the idea was impossible, "the whole Province being really one immense cultivated garden rather than a Park, and no one spot seeming better than another. We lack grand mountain scenery, bold cliffs, dense forests, extensive valleys, great water falls, cascades and rapids, etc., some or most of which are essential in the formation of a National Park."[64] For both Robinson and Newbery, the very term "national park" called to mind a specific style of landscape that the Maritime provinces did not possess.

MAKING THE PARK SYSTEM NATIONAL

Even the Maritimes park booster Casey Baldwin, who would be instrumental in pushing for the creation of Cape Breton Highlands National Park, bore the Parks Branch no ill will for ignoring the region for so long. "The reason, in the past, for this apparent discrimination is perfectly fair," he said, "and we have nothing but praise and commendation for this policy." But Baldwin, speaking in 1934, now felt that discrimination against the East for whatever reasons was no longer tolerable. The Depression had hit the country and made the nationalization of government programs essential. As well, the value of tourism was better understood, and so, many Maritimers wanted more tourists. Parks were now as much about tourism as about landscape, Baldwin said, so it was only fair that the Maritime provinces should be given national parks of their own.[65]

Ironically, it was in the hopes of preventing the park system from becoming more about tourism than about landscape that, beginning in the late 1920s, the Parks Branch adjusted its philosophy to promote the idea of the system being truly national. Federal and provincial politicians had worked in concert to create Prince Albert National Park in Saskatchewan and Riding Mountain National Park in Manitoba during this period, to be the foci for recreation and tourism for those provinces. The Parks Branch had been left with little say in whether either park would be established or what sites would be selected.[66] Having spent a decade preaching the economic and humanitarian benefits of parks, the bureau was now finding that its very success threatened its control of the park system. It needed to define the standard for parks – a standard it would have the authority to maintain – in a way flexible enough to deal with the growing calls for parks throughout the Dominion. It did so by reformulating its purpose, arguing that its job was to make national aesthetic judgments: to find, set aside, and maintain places that were both the most beautiful and most typical representations of different parts of Canada. (Most beautiful and most typical: this always reminds me of Garrison Keillor's description of Lake Wobegon, where all the kids are above average.) This allowed the Parks Branch graciously to concede the impracticality of having a Canadian park system with only Western parks, while cementing its own right to decide what was deemed parkworthy.[67]

Harkin could now contend that parks were "typical of the early conditions of a province rather than for the protection of some particular outstanding physical feature."[68] More than this, his letters claimed that Maritime parks had always been part of his vision: "I know that there is nothing more fascinating to the inlander than the sea. I have always felt that in years to come when Canada has an immense population there will be thousands of people each year from the interior who will want to spend some time near the sea. ... I have dreamed for many years that sooner or later we might be able to find areas in the Maritimes which would include ample sea shore and which would be incorporated in the National Parks."[69] Harkin even stated that "one of my ambitions, unrealized as yet, has been the creation of one National Park in each of our Provinces."[70] The fact is that if Harkin or anyone in the Parks Branch had wanted Maritime parks, or the system made national, they said nothing of it until faced with the threat that such policies might be thrust upon them.

Importing ideas from the United States helped bring the Maritime parks closer to reality. Maritime boosters looked to Maine's thriving tourist industry as something for the region to aspire to. The state was so popular a destination that James Harkin used it as a tourism success story in his first annual report, though noting that "Maine's Adirondacks cannot be compared to Canada's national parks."[71] Then in 1919 the US National Park Service relieved the Eastern seaboard of feelings of topographical inadequacy by establishing Acadia National Park in Maine. This site, though perhaps without the grandeur of Rocky Mountain parks like Yellowstone, was as rugged and dramatic as anything on the East Coast, and it had an ocean to boot. To both the Parks Branch and Maritime park advocates, Acadia served as an example that the East Coast did have parkworthy scenery.[72]

At the same time, the Canadian Parks Branch borrowed from its US counterpart the method by which Maritime park land could be obtained. Until the late 1920s, Harkin and his department responded to requests for new national parks around the country by saying they could not consider purchasing the necessary land. But in 1929, Harkin wrote Horace Albright, assistant to the director of the National Park Service, asking him how the land for parks such as Acadia had been obtained. Albright explained that the Park Service did not buy the land. Instead, it investigated proposed sites, and if the report was favourable it recommended the park's establishment, subject to the acquisition of the land by either the state or, as at Acadia, private interests.[73] Coincidentally, R.W. Cautley, the chief surveyor in the Canadian Department of Interior, wrote Harkin several months later, also outlining the American policy and suggesting it be implemented in Canada. Not only would this new policy rid the department of the responsibility of funding land purchases, but also, "The above would seem to be an equitable arrangement which has this great advantage, namely, that it goes far to prevent any inter-State friction in regard to the expenditure of Federal money within a particular State."[74] Impressed by the simplicity and economy of the American policy, the Canadian Parks Branch decided to imitate it. From 1929 on, rather than turning requests for parks in Eastern Canada down flat, Harkin explained that although it was the Parks Branch which would evaluate whether a site was or was not suitable for a national park, it would ultimately be up to the province to make the land available. So from the example of an American national park, Maritimers began to appreciate the possibility of parks in their own region; from the American park system, the Canadian Parks Branch learned how it could make such parks a reality.

In 1930, it became politically expedient for the Canadian government to announce that a park for every province was its official policy. It was seeking support for the passage of the National Parks Act, which was to replace the 1911 Dominion Forest Reserves and Parks Act. The old law, though setting up the Parks Branch, had lumped parks in with reserves, suggesting that they were, in the words of historian C.J. Taylor, "fundamentally resource reserves, allowing for the controlled exploitation of a range of resources, such as minerals, timber and water as well as scenery."[75] Although amendments to the 1911 act had somewhat separated reserves from parks, the status of resources on park land was still unclear. Harkin's Branch fought for the new law to enshrine the principle of inviolability, whereby parks were declared to be maintained intact forever, free from development or encroachment.[76] To foster national acceptance of the new act, the Liberal government proclaimed time and again during the winter of 1930 its intent truly to "nationalize" the park system.[77] As minister of the interior, Charles Stewart declared during debate on the Parks Act, "It is the policy of the government to develop a national park in each province provided the province makes available for this purpose, free of charge to the Dominion and free of encumbrance, a compact area of national parks standard."[78]

Within the space of only a few years, the idea of making the national park system national had moved from an implausibility to a probability. The turn-around had occurred incrementally, with political motivations, fairness, American ideas, and the desire to share in park benefits all playing a part. Together, these were enough to convince politicians and at least some of the public that the East could conceivably have national parks, without damaging the integrity of the park system. Never, though, was there a sense that the way of seeing Eastern landscape had changed, or that those who were considering Eastern parks were even looking at its landscape. The Maritime landscape of 1930 was viewed much in the same way as it had been in 1920.

EXTERNAL PRESSURES

There would not be a national park in the Maritimes for another few years, as events delayed the nationalization of the park system once more. The arrival of the Great Depression in 1929 meant that there was little chance of finding funds for park creation. Harkin told the premier of Nova Scotia, G.S. Harrington, in 1931 that "at the present time all proposals for new parks or to enlarge our existing parks are temporarily suspended as all available funds are being utilized to meet more pressing demands" and offered the same answer when Harrington asked again in 1933.[79] As Harkin's new minister, Thomas Murphy, stated in Parliament after being besieged with requests for new parks, "The spirit is willing but the pocket book is weak."[80]

This may seem entirely sensible. Yet the US faced the same Depression, and in the first half of the 1930s spent more on park system development than ever before. Since parks can always be considered luxuries, we need to look beyond economics to understand why some governments have embraced parks and others have not. Another Canadian administration might have decided that during the Depression was exactly the right time to engage the unemployed in constructing parks across the country. But the Conservative government of R.B. Bennett, which had been elected in 1930, was intensely anti-parks, and especially opposed to J.B. Harkin. This was in part because Harkin had been appointed twenty years before by Clifford Sifton, and was believed to be the Liberals' man.[81] More important, R.B. Bennett was the member of Parliament for Calgary West, which included the community of Banff. He had locked horns with Harkin a number of times in the past over what he saw as unconstitutional federal interference in the lives of Banff's citizens, and his prime minister's papers are filled with correspondence slamming the Parks Branch.[82] This feud undoubtedly reinforced Bennett's political philosophy that the state too often intruded in public life.[83] During the Bennett years, the Department of the Interior was gutted and the Parks Branch lost thirty-two employees.[84] According to historian C.J. Taylor, Bennett himself regularly phoned Harkin and asked him to resign.[85] Not surprisingly, nothing was accomplished during the Bennett years toward extending the park system.

For the provinces, the Depression made national parks all the more attractive. The Maritimes were increasingly dissatisfied with paying for a service for which they received no material benefit, and which the great majority of their citizens could not enjoy. On top of this, the park system was given a sizeable unemployment relief project in the early 1930s, which of course primarily benefited Westerners.[86] With little understanding and less compassion, the minister of the interior noted, "The east is beginning to get a little jealous of the west with its great national parks."[87] During the early and mid-1930s, clamouring for national parks grew nationwide and politicians had little choice but to join in.[88] For instance, J.L. Ilsley, a Nova Scotia MP, told Parliament that he was personally opposed to a park, fearful that it would only attract picnickers who would set fire to the woods. Still, he could not refrain from adding, "Of course if there is going to be a national park I would suggest that my constituency is the most beautiful in the province."[89] Politicians used the call for a national park as an excuse to wax romantic on the wonders of their constituency. Even if their words were completely ignored in the House, at least their local paper might take note and compliment them for representing the region so well.[90]

Interest in tourism was awakening at the federal level in the early 1930s. The Depression brought a sharp decline in traffic from the United States, and with that decline came the realization that a laissez-faire tourism policy was inadequate. A new hands-on strategy for tourism was spearheaded, ironically enough, by a Bennett appointee to the Senate: W.H. Dennis. The publisher of the Halifax *Herald*, Dennis had a wealth of experience in promoting the Maritimes. In his inaugural speech as a senator in 1934, he called for a Special Committee on Tourist Traffic. The committee that resulted was a watershed in what could be called tourist appreciation in Canada, in that it brought together experts on publicity, transportation, tourist accommodations, and national parks to study how to rationalize and make national the tourism in Canada. With Dennis as chair and Senators W.E. Foster of Saint John, New Brunswick, and Creelman MacArthur of Summerside, PEI, actively involved, the Maritime provinces were not only represented but also received an inordinate amount of attention.[91]

At the forefront of the committee's interest was extending the national park system. J.B. Harkin was the first witness to be heard, on the morning of the first day, and he outlined what needed to be done to bring a national park to a province that did not have one. Nova Scotia businessmen Casey Baldwin and George E. Graham called for Maritime national parks, referring to Maine's success in attracting tourists. The Committee's final report noted that since the establishment of Banff, Canadians had poured over $22 million into national parks "almost exclusively in one part of the Dominion." Therefore, it was resolved that the national park system should "be extended, as a truly national policy, to embrace all the provinces."[92] This was a committee report that was not forgotten: its recommendation for the creation of a Canadian Travel Bureau was implemented within a month, and, as suggested, Leo Dolan became its first head. The following year, a federal-provincial conference seconded the Committee's recommendation for Eastern Canadian national parks. The worst years of the Depression had apparently made more Canadian politicians receptive to the economic benefits of national parks.

The National Parks Branch was also entering a period of improved opportunity for park creation. Following the victory of William Lyon Mackenzie King's Liberals in 1935, the Department of the Interior was merged with the Department of Mines, the Department of Immigration and Colonization, and the Department of Indian Affairs under one minister, and in the following year became a single Department of Mines and Resources. Though James Harkin would have held the same position as before, he would now have an additional level of ministry organization above him. Rather than accept what must have seemed a demotion, he chose to retire, with F.H.H. Williamson replacing him. The Parks Branch's profile, so small during the Bennett years, appeared to be shrinking again. But the new minister of mines and resources, Thomas A. Crerar, soon demonstrated a willingness to build up the Parks Branch budget, especially for immediate work relief projects. As park historian C.J. Taylor has noted, "Unlike ministers of the Interior, Crerar regarded the whole of Canada, not just the west, as his domain, and under his administration the parks bureau began to pay attention to the east in a way that it had not done before."[93]

By 1935, then, it would seem that all the stars were in alignment, and the Maritimes would have one or several national parks. Provincial governments saw federal funding, the federal government saw an extension of national interests, tourism boosters saw a larger tourist trade, and the Parks Branch saw the fairness of granting to Maritimers what it had granted to Westerners – and the power that could be achieved by doing so. But this did not change the fact that parks were expected to meet certain aesthetic criteria. Since 1885 the Canadian national park system had developed with a single idea of what was suitable national park land. Staff in the Parks Branch were indoctrinated with the belief that a park's beauty demanded virgin territory, huge parcels of land, and mountains. It was only a dozen years since Harkin had first mentioned in an annual report that "It is also very desirable that areas should be set aside" in the Maritimes – and his reports had not said another word about it. Did the aesthetic standard of what constituted a national park change just because it was decided that it should? R.W. Cautley, who had spent over thirty years surveying in parks in the West but who would be sent East to investigate sites in Nova Scotia and New Brunswick, recognized this dilemma in a 1930 report and suggested that it went beyond the Parks Branch:

It may be said that it is unfair to compare the mountains and lakes of New Brunswick with scenic features of a similar kind in other Provinces of Canada, but it must be remembered that the object of my report is to select a site for a National Park of Canada – not a Provincial Park – and that the two main objects of a National Park are: – (a) To set apart an area which shall truly represent the best of each distinctive type of Canadian scenery. (b) To attract tourists from other Provinces of Canada and from all over the world. It must also be considered that if I refrained from making these comparisons no power on earth could restrain the tourists from making them.[94]

Cautley may very well have been right; tourists in the 1930s might not have found what became Fundy National Park as beautiful as Banff, and they may not today. This is of interest, but not as much as Cautley's own feelings, since he went on to help select two of the Maritime parks ostensibly in spite of those feelings. In the above quotation he seems to define his duty as finding "the best of each distinctive type of Canadian scenery" wherever it might be; finding, for example, the best hill in Canada – which would be in the West, of course – rather than the best scenery of each region. In selecting parks in Eastern Canada, did Cautley and the rest of the Parks Branch staff seek a Western-style beauty in the East and accept these new parks as pale imitations; did they seek out a new aesthetic, one that corresponded to the geology and the biology of the East; or did they see their task as basically hopeless, and treat aesthetics as secondary, perhaps to tourism? Any of these decisions would not only affect what the Maritime national parks would be like, but would rebound to affect the park system as a whole.

3 Sublimity by the Sea: Establishing Cape Breton Highlands National Park, ca. 1936

In the fall of 1934, R.W. Cautley, chief surveyor for the National Parks Branch, travelled to Nova Scotia to scout out locations for a potential national park site. R.B. Bennett was still prime minister, and there were no immediate plans to create a Nova Scotia park, but the federal department bowed to the provincial government's request that a suitable future site be found. Cautley drove down from Ottawa in his own car, one specially suited to fit his six-foot-six frame. Commissioner James Harkin had given him a checklist of qualities that any park would be expected to possess: accessibility, potential for development, opportunity for game preservation, and as little extractable natural resources and human settlement as possible. But Harkin wrote, "Primarily, and before all other considerations, it is desired that the site shall be the best possible example of the beauty and character of Nova Scotia scenery."[1] Harkin did not elaborate; he knew there was no need to. Cautley, a civil engineer, was like everyone else. He would know beautiful, typical Nova Scotian scenery when he saw it.

A site in Cape Breton had been talked of for years, but Cautley had claimed, sight unseen, that it would not be suitable. "Now I have never seen the country," Cautley wrote in 1929, "but from such evidence as can be obtained from a study of the information, photographs and maps at our disposal ... it does not possess those high qualities of scenic attraction that are absolutely essential to the success of a National Park."[2] Like Harkin with his instructions, Cautley was vague but insistent. From what he knew, northern Cape Breton was not sublime enough to be home to a national park. But at the request of the Nova Scotia government, Cautley was now visiting three locations throughout the province, including one in Cape Breton's northern peninsula. With two provincial representatives in tow he spent five days exploring the Cabot Trail, and then spent another week crisscrossing northern Cape Breton alone.

To his own surprise, Cautley was very impressed. In his official report to Harkin that December, he recommended a national park for northern Cape Breton and wrote that "the scenic values of the site are outstanding."[3] He then spent four and one-half pages trying to explain why the area was beautiful enough to be a park; he was not entirely successful. Again and again Cautley returned to the stock terms: "beautiful" five times, "picturesque" five times, "scenic" and "scenery" fourteen times. Cautley never used the word "sublime," perhaps associating it with an effeminate aestheticism. But his description of his appreciation for the Cape Breton scenery can only be called the discovery of a coastal sublime. Travelling along the craggy coast, one looked down at the churning water from an impressive height, and thus had the opportunity to experience the terror, the spectacle of infinity, the sense of insignificance that were all associated with the sublime. In this sense, though the Cape Breton site was not truly mountainous, it permitted a comparison with the Western parks. It also opened up the possibility that other Maritime landscapes, in their contrast between land and sea, offered a type of worthwhile national park scenery.

Whereas the two leading boosters for a Cape Breton national park, local politician Casey Baldwin and head of the Cape Breton Tourism Association Samuel Challoner, had envisioned it in the northern interior, leaving the coastline for the small fishing settlements that surrounded it, Cautley insisted that the coastline was essential.[4] The interior was in itself, he wrote, "singularly devoid of scenic attraction. There are no large lakes within it and very few small ones. There are no 'mountain ranges' or 'peaks'. It is only as one approaches the coast that the original plateau has been so cut up by the erosion of many extraordinary steep mountain torrents as to become a picturesque mountain terrain, with serrated sky line and distinctive peaks. The great scenic value of the site is the rugged coast itself with its mountain background. The interior or plateau country is only valuable as a Park asset from a game preservation point of view."[5] The coastline had the special benefit of the Cabot Trail, the recently completed road that threaded along the coast of the northern peninsula. Cautley wrote, "The point that it is desired to make quite clear at the outset is that the merits of the Cape Breton Park site rest on the coast line scenery and the Cabot Trail as a means of see-

At Cap Rouge. Photo taken by R.W. Cautley in his 1934 inspection of potential Nova Scotia national park sites. National Archives of Canada, PA121449.

ing it."[6] As a man who built roads for a living and as a tourist to Cape Breton himself, Cautley saw the park site from the perspective of an automobile driver.[7] He had long worked on national park roads in the Rocky Mountains, and understood the aesthetic appeal that mountainous landscape and travelling through it had for drivers. Also, he himself enjoyed the park site from behind the wheel of his car, and knew that his pleasure would not be unique. Cautley's report constantly reverts to the tone of a tourist brochure: "the tourist is faced by," "the tourist should visit," "no tourist should plan his trip," and so on.[8] In his four and one-half page description of the park, he refers to "tourists" and "tourism" ten times. The Cabot Trail, swooping down to the sea and then rising to the highlands by grades of as much as twenty degrees, helped make the hills hillier and the sea more dramatic by permitting the tourist to enjoy both simultaneously. Cautley understood that the Trail not only made the area more accessible but also made it more beautiful. For these reasons he wrote, "the Cape Breton site is almost entirely an automobile route and must be judged accordingly."[9]

Cautley's vision for what would become Cape Breton Highlands National Park closely approximated the finished product. The process of creating the park in the mid- and late 1930s was in great part about developing the Trail, and about expropriating from those people who had land and homes around it. The land and landscape of the region's interior were of little significance to the park because they were tucked away behind the highlands, out of view for those driving on the Trail, the one way of seeing the park.

Until the opening of the Cabot Trail in 1932, the entire northern peninsula was unknown to outsiders. Few travellers ventured as far as Cheticamp, the western entrance of the park today, or Ingonish, the eastern one. The roads were poor and there was no significant settlement to travel to, so there was little opportunity or reason to visit. Most nineteeth-century tourists to Cape Breton bypassed the region, as did most travel writers.[10] Even in Beckles Willson's 1912 *Nova Scotia: The Province That Has Been Passed By* the region was passed by. Those who referred to the area spoke in terms of its inaccessibility. An 1883 guidebook, relying on the second-hand information of a "gentleman visitor," described the land above Ingonish as "wild and grand, romantic and picturesque, though long since associated with marine disasters." Crossing the peninsula was an adventure, and the author bubbled that "To appreciate the grandeur of the scenery and enjoy its benefits it must be visited, with the companionship of one map and one or two *Guides!*"[11]

In 1925, the province began to open up the northern peninsula by constructing the Cabot Trail. Minister of highways A.S. MacMillan in later life recalled that the idea of the trail came to him in a dream which showed that a road would not only serve people of the northern peninsula cut off by land for parts of the year but also bring tourism to the area.[12] (It would also enrich MacMillan's own construction company, as we shall see.) And when the Cabot Trail opened in 1932, northern Cape Breton did indeed became a destination for the adventurous tourist. The Trail was steep, narrow, and quite treacherous, to the point that Nova Scotia's Relief Map Directory warned, "In our opinion, inexperienced or timid drivers should not take this drive; and under no circumstances should the drive be taken unless your brakes are working perfectly, and your car working well in low gear." Some drivers took this as a challenge.[13] The appeal of northern Cape Breton was as much the drama of seeing the scenery as the scenery itself.

Though R.W. Cautley himself recommended the northern region on the basis of its coastal drive, he believed that the drive alone was not enough – and no doubt thought it would in fact discourage some travellers. He therefore proposed that a parcel of land in the Bras d'Or area, forty miles to the south, also be included. Travellers had long found the region to be a diamond in the rough – even the crusty American writer Charles Dudley Warner had been moved to call it "the most beautiful salt-water lake I have ever seen, and more beautiful than we had imagined a body of salt water could be."[14] Cautley himself had previously recommended the Bras d'Or as the most

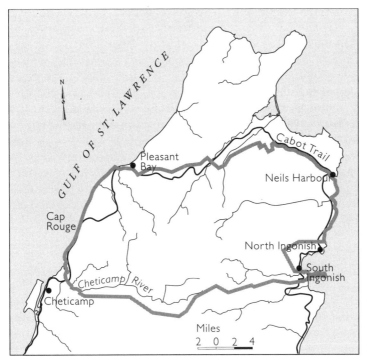

Cape Breton Highlands National Park (Boundaries are ca. 1955)

likely place for a Nova Scotia park. And now he saw that its genteel, more picturesque nature would nicely offset the wildness of the Cabot Trail. The difficulty was that Bras d'Or was already populated and popular, and thus land there would be expensive to buy. The Nova Scotia government had been so disinclined to take over the area that Cautley had not been asked to visit the region. Harkin had suggested it himself.[15] The enthusiasm shared by Harkin and Cautley for the Bras d'Or site indicates their determination to make the Nova Scotia park, wherever it was, the centre for provincial tourism.

After reading Cautley's report, Harkin had serious doubts that the Cape Breton site "measures up to National Parks standards of scenic predominance." He quoted Cautley's own words that the highland interior which would constitute most of the park was singularly "devoid of scenic attraction," "a Parks liability rather than a Parks asset."[16] However, Bras d'Or was another matter. In conversation, Cautley "freely admits that any reasonably-sized area in the Bras d'Or country would, in his opinion, make a better National Park than the Cape Breton site." The area constitutes "the most beautiful and internationally remarkable features of Nova Scotian scenery." Most important, according to Harkin the Bras d'Or possessed "extraordinary facilities for recreational development"[17] which the northern site did not.[18] It is not clear whether Harkin was referring to existing manmade facilities like docks, or to natural features like beaches, but in either case his emphasis was on making the area a centre of tourist development, the fairground to the Cabot Trail's rollercoaster.

What is missing from the Parks Branch's correspondence is even a pretence of interest in a Nova Scotia park site for the preservation of its flora and fauna, or even its scenery. There was no talk of reclaiming the interior's natural form, or even of helping to develop a mass tourism attuned with nature. Harkin and the Parks Branch clearly were focused not on preserving Nova Scotia's past but on developing Nova Scotians' future. The main concern was how to do so effectively, and that involved not just scenery tourism but satisfying all parts of the tourist's experiences. Harkin wrote, "I have grave fears that if we name the northern portion of Cape Breton Island a National Park and then find ourselves with nothing more to sell than a highway, even though a considerable portion of it has real scenic value, we may do more harm than good from a tourist standpoint. We cannot afford to have any visitors go away feeling that what we called a National Park was not of a quality which they had reason to expect."[19] Despite his own misgivings, Harkin knew that the park was to be created, and that if it was a failure it would sully the names of national parks nationwide.

In 1935 the park plans temporarily stalled. The federal election of W.L.M. King's Liberal government and its decision to merge the Department of the Interior with three other departments under one ministry slowed any park progress for a time.[20] Only Thomas Crerar's enthusiasm for nationalizing what had previously been the Western-centred ministry gave hope for the Parks Branch's future. But Harkin soon found that the minister's enthusiasm had its drawbacks. In the spring of 1936, with the new federal Liberal government ready to move forward with a Nova Scotia park, Crerar and his deputy minister met with Nova Scotia politicians to discuss the site, with neither Harkin nor any of his staff present. Though the provincial representatives agreed to ensure that a piece of Bras d'Or Lake would be turned over to national park use, Crerar on his own chose to make no mention of the Bras d'Or in his official letter to Premier Angus L. Macdonald announcing that the park could go forward.[21] Harkin considered this unilateral decision an undermining of the Parks

Branch's authority on park matters, not to mention an utterly unnecessary concession. The Bras d'Or was "absolutely indispensable if the Department expects this Branch to make a success of the Nova Scotia National Park from a tourist standpoint," he wrote the deputy minister. It was "a first-class central point for holding the touring public and gradually distributing them to the various attractive points in the island." Harkin was especially concerned that what to his mind was an incomplete park would not only be of limited use to the province, but would also weaken the national park name. "It must be kept in mind," he wrote, "that while the name 'National Park' has a very great selling value, especially in the United States, even a National Park must deliver the goods. I feel that we cannot afford to have the name 'National Park' used in any area where the result may be to seriously damage the prestige of all our National Parks."[22] Harkin's arguments went unheeded, and in April 1936, Crerar sent Macdonald an agreement creating a park in the northern section of Cape Breton alone. By mid-June, Parliament was ready to pass a bill establishing Nova Scotia and Prince Edward Island national parks.

Sixty-one-year-old James Harkin chose not to continue the fight for national parks. He had survived the Tory years of the early 1930s, only to be given a smaller role within the new Liberal government. Harkin took early retirement in November of 1936, and was replaced by his long-time aide F.H.H. Williamson. Neither Williamson nor any future head of the parks agency would have the prestige and power that Harkin had enjoyed, not because of a lack of will or personality but because decisions were increasingly made over their heads, by more senior bureaucrats in the ministry or by the minister himself.

COMMUNITIES

When Samuel Challoner had proposed a national park for Cape Breton in the 1920s, he had pointedly suggested that it be "exclusive of the farms and villages along the coasts."[23] The park was to be beneficial, not detrimental, to the people of the northern peninsula. James Harkin had instructed Cautley before his survey that "the proportion of alienated land ... will have an important bearing on the practicability of any site selected,"[24] though his main concern was with the cost of expropriation. Having surveyed the site, Cautley felt that excluding all settlement would be impossible. His vision of the park was dominated by the Cabot Trail, and for this, land along the populated coast was essential. Yet Cautley's choices of which settlements must be submerged – choices that would disrupt communities and shape the park forever – indicate his interpretation of what life in rural Nova Scotia should be like.

Beginning in the southeast at Indian Brook, twelve miles below Ingonish, Cautley suggested that a number of communities should be left outside the park. Around Indian Brook, "some of the present owners have fairly well-developed farms,"[25] so they should not be displaced, and at Ingonish, "There is a considerable population scattered throughout."[26] Neils Harbour and White Point were likewise discounted, plus all of Victoria County north of the Aspy River (including Bay St Lawrence, Cape North, and Dingwell) because of its "considerable population."[27] But at Pleasant Bay and Cap Rouge on the west coast, Cautley's thinking shifted considerably. After citing Pleasant Bay's Scottish past and describing its present settlement of almost three hundred people, he suggested it be incorporated into the park as is. "It is out of the question," he wrote, "to expatriate all the inhabitants of such an old established settlement." The farms of Pleasant Bay would be "really an asset to the scenic grandeur of the region, since they provide a relief for the sombre character of the scenery."[28] It would be easier to control the community as a park townsite than as an independent settlement just outside the park boundary. Pleasant Bay should stay. This was an unusual suggestion for Cautley to make, since the Parks Branch understood from its experience with Western parks that townsites tended to be sources of constant aggravation. Also, these Western townsites were only tolerated because they supplied necessary services to park visitors, which a community like Pleasant Bay probably could not. Nonetheless, Cautley lobbied hard for the little community to be allowed to survive.

Cautley's treatment of Cap Rouge, a small French-speaking settlement along the trail from Pleasant Bay to Cheticamp, stands in sharp contrast. He overlooked its existence altogether. The surveyor stated only that there were at most a dozen poor fishermen's "cottages" along the whole road north of Cheticamp. He referred to such homes as "isolated fishing stations," places where fishermen eked out a scant living from the sea and from the few crops and animals they raised. In reality, Cap Rouge had some thirty families and was not much less substantial than Pleasant Bay (there were, for example, 77 Cap Rouge voters in the 1937 election, and 138 in Pleasant Bay).[29] Unlike full-time farmers, the fishing families of Cap Rouge did not have large houses and barns. But they did have, in the words of travel writer Gordon Brinley, "good crops and well painted houses."[30] Cautley could not have missed Cap Rouge.

If Cautley had argued that the community of Cap Rouge did not deserve the same consideration as Pleasant Bay, his thinking could be considered pragmatic. Cap Rouge was smaller, and would be easier to uproot. And it was at a point on the Trail where the coast rose quickly into the uplands, so there was no room for the highway to bypass settlement. (Still, he could have suggested the houses stay, as a folk spectacle for park visitors.) But by ignoring Cap Rouge's existence as a community altogether, the surveyor was showing prejudice. He neither understood nor approved of how the inhabitants of Cap Rouge lived. The fishing community did not show the material signs of permanence which Pleasant Bay, with its big farms, did. Cap Rouge had one foot in the sea. Cautley decided that moving the shore fishermen would be to their own advantage. He wrote, "It is difficult to understand why shore fishermen have settled in some of the places where they are found ... [I]n this case, the impropriety of moving people away from the locus of their occupation does not exist, as in the case of Pleasant Bay for instance, because the thriving settlement of Cheticamp and the only good harbour for boats on the entire west coast is within a few miles and affords much better opportunities for either fishing or farming than where they now are."[31]

No other member of the National Parks Branch or the two levels of government visited Cape Breton to confirm Cautley's findings, so when the federal-provincial agreement to create the park was drafted in 1936, it contained all the recommendations the surveyor had made. Most notably, residents of Pleasant Bay – and Pleasant Bay alone – would be permitted to remain and lease back their land from the federal owners. Only when other park staff began to arrive did the arbitrariness of Cautley's recommendations become apparent. The new architect of the park was the acting superintendent, James Smart. The son of a turn-of-the-century deputy minister of interior, Smart was a forester from Manitoba who had helped get the new Riding Mountain National Park successfully started.[32] After he had inspected the Branch's new territory, the first thing he did was give Harkin nine reasons not to make Pleasant Bay a townsite. Instead, "We should insist on the removal of the settlers and demolish their buildings." Residents would demand schools, hospitals, telephones, and all other possible amenities to be provided by the park. Residents' houses would not conform to the park's high standards. Most bothersome, "A great majority of the settlers, especially the younger generation, will always look to the Park for employment." They would even "demand ... the right to engage in businesses other than their supposed occupations, such as fishermen or small farmers." Park residents, Smart concluded, would be a "running sore" to the department.[33]

At Harkin's request, James Smart also investigated Ingonish as a substitute for the Bras d'Or as a centre of tourist development. He was especially impressed with a jutting piece of land called Middlehead owned at that time by Julia Corson, widow of an American million-aire. He wrote, "This is an area of outstanding beauty and includes one of the finest ocean beaches in Nova Scotia." The Corson house was in rather bad repair, but it was somewhat majestic and might usefully be turned into a park hotel, on the order of the hotels at western parks like Banff and Waterton Lakes. Nearby was opportunity for the main development, a golf course, and "all other features usually included."[34]

F.H.H. Williamson, assistant director and soon to be Harkin's re-placement, visited the new park and agreed with Smart's assess-ments. James Smart had arrived in June, and by November, after discussion with provincial representatives and with very little fuss, he had reshaped the park. It was decided that Pleasant Bay would not be bought up and torn down, but neither would it be included in the park. The park boundary would skate around the little community. Cap Rouge would not be so lucky. The seventy-square-mile tip of the park which had run all the way to the north coast on the Inverness side of the boundary of Victoria and Inverness counties was removed because it was deemed redundant, and because at the moment the park was larger than had been permitted in Nova Scotia's 1935 en-abling legislation.[35] In return for saving a considerable amount of money by these deletions, the province agreed to pay for the pur-chase of an additional thirty-eight square miles around Ingonish, in-cluding Middlehead.[36] The new national park had a headquarters.

In 1936, an Act of Canada had defined a national park in Nova Scotia, and in 1937 an amendment to that act significantly altered its boundaries. What was and what was not park was still subject to change. This suggests, I think, to what degree the new national park was not about the preservation of nature. Land was first and foremost the medium of park creation, and whatever land – by a combination of aesthetic, political, economic, and environmental factors – was finally made park land was what the Parks Branch would in the fu-ture give meaning to. Once one starts thinking about land in this abstract fashion, it is easier to understand why to the park staff the people living on the land were also abstractions, and why even as the park's shape changed in 1936 the people most directly involved were not informed. Only once the land was expropriated and the people removed could the Parks Branch begin to think of the Cape Breton site as a national park, its nature worthy of expropriation.

PUTTING THEM OUT

Angus Leblanc of Cheticamp, formerly of Cap Rouge, told me that when mention was made of a national park for the area, no one was quite clear what it was to be: they had nothing to compare it to. "Cheticamp was as far as many had gone, you see ... some had been to Sydney to work." He paused. "– knew there would be a road through. That's all." I was about to ask another question, when he added, "– figured there would be a fence around. Elephants, giraffes maybe for people to see."[37] He chuckled at himself.

Considering that the new park was about to reshape land ownership and land use in Cape Breton's northern peninsula, it is remarkable how little was known about it by those most directly affected. Locals knew that Casey Baldwin had been lobbying for such a park; he had given speeches to Boards of Trade, written letters to editors, and sponsored an essay contest on "The Advantages of a National Park in Cape Breton." But these provided only hopeful descriptions for the park. When in the spring of 1936 the Nova Scotian government actually authorized the expropriation of land and deeded this land to Canada, even the press knew little or nothing. The local *Victoria-Inverness Bulletin* does not seem to have heard about the park officially, while the Sydney *Post-Record* wrongly declared that it was to comprise "the entire north end of Cape Breton."[38] A Canadian Press story mistakenly described the park as a disappointingly small project of twenty-five square miles (about four hundred square miles short) and noted that those with improved property would be allowed to remain there (another error: this was to be true only for Pleasant Bay residents). Even the vigilantly anti-Liberal Halifax *Herald* ran this last piece without question or criticism. Considering the public interest generated when Cautley inspected potential sites, this lack of knowledge is surprising. The provincial government may not have wanted to draw attention to the upcoming expropriation, but one would think it would have wanted credit for making the park a reality. And while newspapers may have taken the official creation of the park as an unexciting paper transfer, one would think they would have seen its significance both to the province and to those losing land.

Landowners knew nothing about the new park, and the Parks Branch was wary about changing this. In an earlier memo, Cautley had warned that in creating a national park in an older, long-populated province, "there is always a danger of creating hostility and feelings of injustice if old-established rights of residence are disturbed."[39] When he wrote his 1934 report, he again suggested that involving small Cape Breton communities was bound to be troublesome: "If the land was being purchased for a park, on the understanding that the purchase involved expatriation, it would probably prove to be a more difficult matter to get them to sell willingly at all."[40] Though expropriation was a provincial concern, the Parks Branch knew it must walk gingerly, that it would have forever to deal with any bad feelings created by the expropriation process. For this reason the Branch, which had originally planned to move ahead quickly with development in 1936, shifted its funds toward work that would not interfere with the original owners.[41] The Engineering Division began rebuilding the Cabot Trail to national park standards. This job was as politically expedient as it was necessary, in that it demanded about a hundred men to be hired from the area and did not involve development on what had been private land. Staff on site were assured by their superiors, however, that when needed they had every right to make use of all park land. Residents believed they still owned the land, so they found this puzzling. In a letter to the *Victoria-Inverness Bulletin*, landowner Hector Moore, while stressing he was not against "the famous National Park," complained, "The Engineers and Surveyors, in fact, any employee of the Park, appear to be allowed to trespass on any private property, and take a parcel of land of said property, without even consulting the owner."[42]

To the Nova Scotia government, keeping the landowners uninformed was nothing more than procrastination. They had already deeded the land to Canada, and were resigned to settling with owners sooner or later. It is also clear, however, that the government felt that by planning to resolve settlements with landholders slowly and piecemeal, they were doing the federal government a favour. The province's deputy attorney-general, Fred Mathers, said as much to the Parks Branch's chief engineer. He explained that under the Nova Scotia Expropriation Act, the province could take possession of a property (or threaten to do so) if the landowner refused to settle. Therefore, Mathers noted, "The Dominion authorities would be far better off without the transfer being formally completed," when the Parks Branch would assume the difficult responsibility of eviction.[43] It was to the two governments' mutual advantage that the Parks Branch move forward with visible development on land that had been deeded to it, while worried, uninformed past landowners looked on and became increasingly motivated to settle with the province. Residents did not learn until 1937 that the new park would mean the expropriation of settled land, not just the reservation of the highland plateau as Casey Baldwin had preached. Nor would anyone learn that the park boundaries had shifted to include Ingonish and omit Pleasant Bay and the northern tip of Inverness County until August of 1938, almost two years after the province and the Parks Branch had agreed to this.[44]

With the true state of the park so unclear in people's minds, it is not surprising that the first negotiators sent out by Nova Scotia in 1937 had no success reaching settlements; landowners would simply not deal with them. The Department of Highways then handed the responsibility for expropriation over to the Department of Land and Forests. Provincial forester Wilfred Creighton was called in to reach settlements. It was, he later wrote, a difficult task for a young man: "The work was interesting, sometimes funny, often frustrating and more often distasteful."[45] Creighton found that he would have to deal with about three hundred properties in all, including about seventy homes mostly at Cap Rouge or around Ingonish. He also soon learned that northern Cape Bretoners expected a price of about $6 per acre for woodland, with farmland running from $20 to $100 per acre.[46] Creighton would travel around making offers to everyone and hoping for a few settlements, then would leave the park area for a few weeks, go back and do it all over again.

Today, Creighton believes that the great majority of property owners got settlements above the market norm. He tells the story of a woodland property at Pleasant Bay of 110 acres. Its owner, a young Gaelic-speaking man, accepted $700 for it, but Creighton discovered that the deed had been signed a week after the official park expropriation. Creighton felt they should seek out the previous owner to sign a release. When they did so, the previous owner felt he should receive half the settlement, $350. Creighton writes, "The young man took a dim view of this suggestion. Eventually, the old gentleman did sign, but with considerable reluctance. As we drove away, the young man kept muttering to himself in Gaelic. Finally I said, 'That old man was unreasonable expecting to get half the money. You don't owe him anything, do you? You paid him for the property?' A soft Highland voice replied, 'Ach, yes, every cent, $150.'"[47] Creighton's story is not just a fable on the wiliness of the Scots, it is his evidence that the young man got a very good, above-market price, that expropriation was fair. And certainly those selling only woodland did profit by the expropriation. But for those who had been living in the park, whose families in many cases had been there for generations, it was impossible to interpret the process in strictly financial terms. Creighton himself remembers, "I minded seeing people put out of their homes. ... Oh, I hated ... – I was in favour of the park, the park was good. But I thought it was cruel to put them out."[48]

For those who moved, the memories are vivid. Wilf Aucoin was five when his family's land in Cap Rouge was expropriated. He remembers their barn cut into pieces and hauled on oil drums down the coast to Cheticamp, and his house torn apart and salvaged. His family saw the expropriation as a blessing, however. They no longer had to walk ten miles to church, and they had greater opportunity for jobs. Aucoin's father, Christopher, became one of the first employees in the park, and stayed there for thirty-three years. Some people, Aucoin knows, were upset about moving, about losing their favourite hunting places, but "these are small things when you look at the whole thing."[49]

Others felt very differently. Maurice and Emma Donovan of Ingonish Beach told me that their house at Clyburn Brook was "on Number 9" – that is, on the ninth hole of the park's golf course today. They moved their house to Clyburn Brook in 1937, not knowing that the province had already decided to add the Ingonish section to the national park. The soil was beautiful, the Clyburn ran through their land, and with partridge, deer, rabbits, strawberries, fish, and their own farm goods, they felt self-sufficient. As Maurice Donovan tells it, the first word of the park came when a Sydney lawyer overseeing the settlements, Smith McIvor, came up to them and announced, "As of today, the park owns your property." Donovan said, "I haven't sold it yet." McIvor replied, "You don't sell it, it is expropriated for public use." The Donovans asked for $50,000 for their thirty acres cleared and thirty-four acres of woodland – an impossibly high price – and were offered $1500. When negotiations stalled, Maurice Donovan got a job helping build the golf course, even as it wound around his house. The couple accepted the government offer in the end. They realize that their story has certain tragic overtones to it, and recall it romantically. Donovan states, "I wish I had a camera the day we left there. Every child who was old enough to lead an animal had an animal on a rope, when we cleared off me and my wife and the babies, everyone was crying, leading the animals over the mountains. I said that evening, I wonder what the Acadians thought of their leaving the day they were expelled from Nova Scotia."[50]

Some Cap Rougians made the same connection. Angus Leblanc was fifteen when his family heard about the expropriation. His family had seventeen members. They had a real farm going, with four hundred acres, a big barn, some land cleared, eight cows, two horses, forty-five sheep, and a few pigs. He remembers catching rides to school on the back grille or spare tire of tourists' cars. When word first came of expropriation, there were meetings at the Cap Rouge schoolhouse about resisting, but nothing came of it: a few of the older people, fearful of facing the government, settled quickly and resistance rapidly dissolved. Leblanc's family ended up selling for $2700. His family never voted Liberal again, never forgave the government. Leblanc especially resented that the jobs they were told would come with the park never materialized.[51]

Gordon Doucette of Ingonish remembers returning from the war in 1945 to find his mother surrounded by the park. His father was in a hospital in Antigonish with tuberculosis, and his mother had not wanted to move. "The method that they used," he states, "when I came back from overseas, the park had put a fence around the property, and had a gate on it, and the gate was locked, and my mother was locked into it."[52] Park staff were using a Doucette field to hold stray livestock, and the barn to store park vehicles. For years, the Doucettes stayed where they were. They did not move out until 1950, when they were "worn down" and realized that their old home would need renovating if they were to stay there.

The most recalcitrant of owners was Julia Corson, who lived at the jutting end of Middlehead in Ingonish. Whereas the park staff felt free to work on other owners' land, in Corson's case the project's engineer did not feel quite up to the task. Corson sought $125,000 for her property and when the province balked she took her case to Crerar and Prime Minister King.[53] She inundated them with letters, even sending King a poem written by her niece:

You who say you have purchased for tourists
   This bay and the mountains around
Though all manner of eminent jurists
   Make it yours to the moles underground
Yet we keep more than all you are claiming
   You will never find what we found
For the pulse of the places you're naming
   Has beat in our blood for so long
That their rhythm can never be bound
   To sell to your throng.[54]

Like most others, though, Julia Corson tired of the fight, and in time settled for a price far below her original one.

Common to the stories of expropriation at Cape Breton Highlands National Park is a feeling of inevitability. Eva Deveau told me that her father fought the park at first, "but the government is stronger than anybody." A number of people told me that people back then had a greater respect for authority, a feeling that what the government says goes. Nevertheless, in Cheticamp there is a generally held belief that Pleasant Bay was taken out of the park because of the strength of resistance there. As Eva Deveau summarizes it, "I think the English people from Pleasant Bay, they didn't want to move and they're still there, so I think if the French people had done the same thing they would have stayed."[55] For the Acadians of the area the story has become a lesson in their own lack of solidarity and will. In reality, Pleasant Bay was omitted from the park before knowledge of the boundaries was well known, so it is very doubtful that any alleged resistance was ever a factor. Pleasant Bay was omitted because Cautley's plan to make it a park townsite was unwieldy from the beginning, and the Parks Branch, on the advice of James Smart, realized that it would be simpler just to move the park boundary to outside the village. No lesson can be learned from this about how to deal with government.

The single biggest winner in the expropriation process was the biggest landholder, the Oxford Paper Company, which did not even own the property it was losing. In 1920, this Maine-based company had bought "the Big Lease," a property of over one half million acres in northern Cape Breton originally leased from the Nova Scotia government for ninety-nine years for $6000 per year.[56] Throughout the 1920s the company worked the property until it had cut all wood that could be retrieved economically and until the Depression made exporting any pulpwood to Maine economically unfeasible. The company held on to the Big Lease as a reserve, with the hope that in time it could be cut again. The Nova Scotia government believed that this inactivity promised opportunity: the large, unused forests would be a fine core to a park, and would presumably be easily obtained.[57] As the *Victoria-Inverness Bulletin* wrote when tipped off that the park was going through, the proposed site would take in "what is commonly called 'The Big Lease'. The acquiring of this property would not cost a great deal as most of it is Crown lands."[58] In all, this part of the Big Lease being considered for the park consisted of over 175,000 acres. Though Cautley had said that the highland interior was "only valuable as a Park asset from a game preservation point of view," it did give the park size and ruggedness, which were considered essential.[59]

To the government's surprise, the company had no intention of giving up its lease for a pittance. Oxford Paper made a total claim on the land for about $2.3 million, citing its value as a pulpwood reserve and mentioning that it had value for other purposes – including as a park. In a 1940 arbitration hearing, the company defended the land's worth while government witnesses insisted that land they were wanting to make a national park was ugly, impassable, and useless for forestry. The interior was too wet and hillocky, its trees exposed, wind-stunted, and branchy. Forester Wilfred Creighton, having become an expert on evaluating Cape Breton land, testified that one of the Oxford Paper Company's own men had told him, "Mister, I have cruised a lot of country; some of it was good and some of it was bad, but of all the country that I ever saw, this is the worst." Creighton agreed: "Acre for acre, it is the worst piece of land that I know, not only in Nova Scotia but anywhere in eastern Canada."[60] Though both men were speaking of the area in terms of its value as forest, it would be surprising if they had found what was so economically ugly to be aesthetically beautiful. After a six-week court case, Oxford Paper won a settlement that with costs, interest, and reimbursement totalled just over $520,000. Both sides voiced satisfaction with the judgment,[61] but the Oxford Paper Company must have been especially content. It had made over one half million dollars on land it could not afford to use, land that may not even have been usable, land it was leasing at the time for only $6000 per year.[62] The Nova Scotia government paid almost twice as much for reclaiming this one lease as it did for purchasing all three hundred other properties throughout the park.

MAKING A PARK

In his 1934 instructions on examining potential Nova Scotia park sites, James Harkin reminded R.W. Cautley that he was to find a spot that would be good for wildlife: "The preservation of all game is a cardinal principle of National Parks administration." In the next sentence, however, the parks director noted that this alone was not enough, that "there is a great deal of country which would make a first class game preserve but which would not have any value as a National Park."[63] This was the thinking that ruled the creation of the Cape Breton Highlands National Park after the land was turned over to Canada in 1936. It was understood that the business of making a park was a two-part affair. First, the land must become a reserve, with all its wildlife and other natural resources free from harm. Second, the reserve must become a park, not only by offering nature a chance to thrive but by offering visitors the facilities needed to enjoy it. A park was not a park until it offered first-class roads, accommodations, and recreational possibilities. From 1937 to 1939, the National Parks Branch worked at this two-part process in northern Cape Breton, transforming the land into a reserve and the reserve into a park.

    I will examine in detail the Parks Branch's policies for the management and preservation of wildlife, fish, and vegetation in chapter 8. It is sufficient here to say that the Branch's initial concern was to learn about the natural features that it had acquired, and then to ensure that local citizens understood that past land and resource uses no longer applied and would not be tolerated. In his 1934 report, Cautley noted that the area was home to whitetail deer, and could be restocked with caribou and moose.[64] No other animals were mentioned. Cautley was enthusiastic about the park's fishing potential, particularly trout fishing on the Cheticamp River.[65] In regard to both fish and wildlife, the Parks Branch was concerned that locals would violate the reserve. Locals, it was said, were responsible for overfishing the Cheticamp River.[66] Cautley believed that they were also responsible for killing off most of the area's wildlife and that "It is probable that the sons of the men who exterminated game in the northern end of Cape Breton are just as keen sportsmen as their fathers were and that the protection of imported game would be both difficult and expensive."[67] In reality, market hunters and sportsmen had probably been more responsible for local extinction of some species. Nevertheless, it is quite true that Cape Bretoners felt free to take deer, partridge, and rabbit from the highlands for their family's use. When the park was formed, three of the six-man staff were game wardens, responsible as much for publicizing the park's existence as for actually enforcing regulations. The first wardens were all locals and they seem to have accepted some poaching by those who had lost land or a favourite hunting spot.[68] Around Cheticamp, warden John Roche is remembered for taking a patrol every day at the same time. After he had walked through, hunters knew it was safe to return to the park.[69]

In the protection of fish and wildlife, the new park sought to regulate and change the customs of the long-time residents of the region. The park staff were to preserve the nature they had been granted, regardless of how this nature had been used in the past. Nevertheless, for the good of the park, it was understood that the park staff themselves must shape the nature of the park. Whereas the Parks Branch demanded vigilance against traditional resource uses, it was accepted that, to develop the park facilities, long-term gain demanded some short-term pain. The difference was that the park's enforced changes would be one time only and would be to the ultimate benefit of park visitors.

The most important development in the new park was the improvement of the Cabot Trail. From Cautley's report, the Branch had learned that though it was to be the park's signature attraction, it was at present just too dangerous. What was needed was to keep the present road as much as possible but shave off the steepest grades on the worst sections, especially on the western coast of the park. Cautley accepted that this would be terrifically expensive, but, he wrote, "my recommendation of the Cape Breton site for a National Park must depend entirely on whether the Dominion Government is willing to construct a good road from Cheticamp to Pleasant Bay."[70] The provincial government was thrilled, because the park meant they were free of completing and maintaining a very expensive highway. Nova Scotia was already in the process of a massive hard-surfacing program, with about a thousand miles paved between 1934 and 1939.

No one could possibly be more happy about the new park than the Hon. A.S. MacMillan, minister of highways, chair of the provincial Tourist Commission, architect of the Cabot Trail … and head of the Fundy Construction Company, hired by the federal government to do the roadwork on the trail. Symptomatic of the close ties between government and business in 1930s Nova Scotia, MacMillan's clear conflict of interest raised no eyebrows.[71]

When the park land was deeded to Canada in the summer of 1936, the Trail became the focus of developmental work. The Trail was the main conduit for tourist travel, and it was understood that tourists could not arrive in great numbers until the road was in good shape. Also, the Trail fulfilled the promise of giving jobs to locals. When work on it was at its peak in the summers of 1937 and 1938, there were over a hundred men hired, the vast majority from communities surrounding the park, and many of these were people who had land expropriated.[72] Road construction provided other employment, as a writer to a local newspaper claimed: "National park is the talk of the day. Our young men are anxious to get some work which will be a great help to their fathers. … Farmers should get together and get the supplying of eggs, produce and meat for the different camps and it would help the situation, fishermen the fish trade and all in all would be quite an item."[73] In 1937, over six and a half miles of the Trail were built or reconstructed, and in 1938 another ten and a half miles were completed or near completion. In both years and in 1939, bridges, culverts, and guard rails were also worked on.[74]

Road-building was a central reason for the province to want a park in the first place, but it was not considered justification on its own for the park. Upon learning that the Parks Branch had little planned for the park in 1937 other than road-building, Cape Breton member of Parliament D.A. Cameron complained to Crerar that the park seemed to be nothing but a road, and "if the development of this park is to be along those lines – speaking for myself, and I am sure I am speaking for the residents of Nova Scotia generally – I would much rather that the park had never been undertaken."[75] What was needed was a much larger outlay of funds for the construction of accommodation and recreation facilities. The provincial authorities made much the same complaints. Calling the 1938 federal grant "not much of a showing," highways minister A.S. MacMillan threatened that although he was expected to move forward with expropriation, "I do not feel justified in making this expenditure in view of the small amount that your Government has expended, and are proposing to expend this year."[76]

The National Parks Branch was uncharacteristically silent in the face of these complaints, probably because it agreed that the federal government should move development forward. Harkin himself had told Cautley early on, "It was also very important to consider what natural attractions exist or what future developments can be made with a view to attracting visitors and tourists, such as fishing, boating, swimming, motoring, golf and trail riding."[77] Cautley saw places for all of these in his dreamed-of park, and even envisioned a bathhouse, similar to that of Banff hot springs, "with a warmed salt-water plunge" which would be "an attraction to a special class of tourists suffering from arthritis and ailments benefited by salt-water bathing."[78] Originally, Bras d'Or was to be home to such development, but when Ingonish was made the centre of the park, all planned development moved there. When the Parks Branch was finally deeded control of this area in July of 1938, development plans began immediately.[79]

The Parks Branch had no central and considered plan for the development of Ingonish; the work was done piecemeal. First, a location for the golf course had to be selected. Golf courses were seen as absolutely essential to national parks ever since James Harkin, himself an ardent golfer, had commissioned Canadian golf architect Stanley Thompson to build courses at Jasper and Banff in the late 1920s.[80] Banff's was a huge engineering project involving the removal of huge quantities of rock. At a cost of $500,000, it was at the time thought to be the most expensive course ever built.[81] The Parks Branch so trusted Thompson's expertise that it allowed him to decide where the Cape Breton Highlands course should be. He chose the land around Clyburn Brook and on the Middlehead promontory, saying, "This area provides craggy cliff land, sandy shore and wooded valley land traversed by a river. I know of no place on the whole Atlantic Sea Board where such a variety of terrain can be found in so limited an area."[82] When the Middlehead area was turned over to the Dominion in the summer of 1938, Thompson began work on the first nine holes with a crew of fifty local men, draining the land, grading and topsoiling it, preparing the seed bed, and planting and fertilizing the new grass. Work progressed so steadily that in an October meeting with the executive of Mines and Resources, Thompson suggested that they push forward with the next nine holes. Rather than continuing the course along the sea as had been proposed, Thompson recommended development up the Clyburn. "The Clyburn Valley," he insisted, "traversed by the river with towering hills on each side, affords gorgeous scenery and character, and would add a great deal of variety to the play, as well as shelter from the raw Ocean breeze. Its ruggedness is in keeping with the whole Park development. This route happens to be on land already deeded to the park."[83] All present agreed, and work progressed on the course. A year later all eighteen holes were complete.

In his meeting with the Department of Mines and Resources executive, Stanley Thompson had noted that one reason the course should move forward so quickly was to provide employment for those whose land had been expropriated, but who had not yet settled.[84] This may have been genuinely charitable, though it made for the awkward situation that men like Gordon Doucette and Maurice Donovan took jobs on land that they still considered their own, while awaiting a fair settlement from the province. However, it was also "good business" because it appeased people in need of appeasing. The energy with which the golf course was completed, as with the work on the Cabot Trail, encouraged local landowners to believe that the park truly would bring jobs and prosperity. But this same energy meant that all major development opportunities were used up between 1937 and 1939. Once park development was finished, the park staff's primary job was to keep the park unchanged – a mandate that demanded far less employment. But by then settlement with landowners would be complete. Not surprisingly, the most common complaint about the park I heard from interview subjects and read within the archival records was that prosperity was a short-term affair, and long-term jobs promised to locals never appeared.[85]

Once the Trail and golf course were built, the Parks Branch felt that its responsibility in park development was largely complete. The only remaining job was attracting investors for accommodations. Though this will be discussed in detail later, it is sufficient here to say that the Branch hoped that investors would come forward to set up a grand hotel at Middlehead in the national park tradition. Parks Branch director R.A. Gibson had accepted that the golf course should move forward swiftly, in part because it was the best possible inducement for "attracting to the Park people with sufficient capital who are willing to develop the type of tourist accommodation that we are trying to secure to satisfy high class tourist business."[86] Though the Parks Branch used its own network to seek out potential investors, especially the Canadian National Railway (who already ran a number of park hotels) and local hotel companies, it was unsuccessful. There was little interest, especially when the outbreak of war in 1939 promised that tourism would be sluggish for the indeterminate future. In the winter of 1940, the Nova Scotia government, fearful that another summer would see tourists cruising around the Trail without stopping, offered to set up a hotel on the old Corson property at Middlehead. They asked for the land to be deeded back to the province, and were surprised when the Parks Branch instead offered them a licence to run the hotel, as if they were any other lessee. In 1941 the provincial authorities set up a bungalow camp and had spruced up the Corson house, calling it Keltic Lodge.[87] Though tourists liked it and it was a hit with Nova Scotians, the new hotel was a drain on the province's finances and the government continually encouraged the Parks Branch to assume control over it. By the later years of the war, the province refused the unasked for and unwanted responsibility of keeping Keltic Lodge open.

HAME NOO

Making the Cape Breton park was a process of some steps. R.W. Cautley discovered a highly scenic road that could be a show-piece to some of the best landscape that Nova Scotia had to offer. The park he envisioned had only approximately defined borders, so the federal and provincial governments worked to finalize the site's exact dimensions. The past owners of lands and property within the boundaries were bought out. The Parks Branch enforced new restrictive land uses for the area, while developing some sections of it to an unprecedented degree. The new park consisted of a collection of communities and a still unknown interior region which together had never been thought of as a unit. The final stage in turning this land into a park was to give it a single meaning, to make the park establishment and the park itself seem natural. This was done by promoting the park – emblematic of the best in Canadian and Nova Scotian scenery – as an intrinsically Scottish place.

In his article "Tartanism Triumphant," Ian McKay describes how Nova Scotia came to see its essence as tied to Scottishness, particularly under the premiership of Angus L. Macdonald beginning in 1933.[88] At a time when the province was coming to see tourism as a key to its economic future, Macdonald gave tourism promotion a focus by spotlighting Nova Scotia's Scottish heritage and its present-day independent, Gaelic-speaking Highland stock. Through Macdonald's support, Nova Scotia got a Gaelic College, the Gathering of the Clans, a piper to welcome tourists at the border, and many other trappings of Scottishness. Such effort signalled that there was a Scottish presence to be preserved in Nova Scotia as well as created its own physical evidence of Scottishness: in time, it could be expected that people would forget that these things were evidence of Scottishness once removed.

It would not be surprising, then, to learn that Cape Breton Highlands National Park was an extension of this thinking, that Macdonald pushed for the park to show that Nova Scotia was Scottish right down to the ground. But there is no evidence that Macdonald supported this proposed park site over others, or that he had paramount influence in associating it with Scotland. Instead, the park became Scottish because many Nova Scotians – and, true, foremost among them was Angus L. – considered this its obvious theme, and because the National Parks Branch worked actively to make it so.

Northern Cape Breton's rocky landscape had long been considered reminiscent of Scotland, and surely the Gaelic one heard there helped foster this association. In 1868, J.G. Bourinot wrote that "Those who have travelled Scotland cannot fail to notice the striking resemblance that the scenery of this part of Cape Breton bears to the Highlands."[89] But this in itself does not tell us much. The land could be interpreted in other ways. A 1935 magazine article on the proposed coming of the park described the region's landscape as "not unlike that of Sweden" and cited an American visitor who said that one view was "not excelled by the famed San Joachim Valley, California." In prose run amok, a later writer gushed that Ingonish "runs the gamut from the gloom of a Norwegian fjord to the windswept sunshine of Bermuda."[90] That Northern Nova Scotia was generally associated with Scotland in the 1930s does not mean that this was the only way its landscape could be understood.

But some residents did consider the area so intrinsically Scottish as to think it patently obvious that this should be the theme of the park. They also felt that this would be both patriotically and economically profitable. At a 1934 meeting of the Baddeck Board of Trade, a woman proposed that if "the guides in the park could speak Gaelic it would be an attraction to the Tourists. The president [John M. Campbell] then assured her that when the Park was established the guides would be dressed in kilt and speak gaelic, while the caribou would be trained to play the bag-pipes if that would attract more tourists to Cape Breton."[91] The summer of 1934 also saw the late Dalhousie professor Donald MacIntosh bequeath a hundred acres in Pleasant Bay to the Crown to construct "a small park" centred on a lone shieling, a mountain hut like those found on the Isle of Skye.[92] These instances are not evidence that the park was chosen because of its Scottishness, but suggest that once it was chosen, its Scottishness was sure to be pronounced.

When the park became a reality, the National Parks Branch worked to associate it with Scotland. The 1936 departmental report compared the area's landscape to that of Scotland. The Branch certainly knew that the Nova Scotia government wished this theme to be pressed. Describing a meeting with Angus L. in the summer of 1936, Smart noted that the premier "was very anxious that the Scottish suggestion should be carried out, both in the name of the Park and the general theme or style in our development."[93] Smart personally disapproved of this, stating "in my opinion, the characteristics of these people are not today typical of the Highland Scotch" and thus the Parks Branch should "go slowly in giving the impression that the Cape Breton Scots are representative of the historic Highlanders."[94] (It may be worth mentioning that Smart was married to a lass from Edinburgh.) But his concern for authenticity was not shared. R.A. Gibson wrote Williamson in 1938 that their minister, Thomas Crerar, mentioned "the desirability of making Cape Breton Highlands National Park a highland park, putting our park officers in kilts."[95] Some employees would be Gaelic-speaking, and in time heather and woolly Highland cattle would be introduced to Ingonish, helping tourists imagine themselves across the Atlantic. The new provincially run hotel was christened Keltic Lodge and the park golf course – its very existence a seeming homage to Scotland – would allow players to travel a map of Scottish associations, on holes dubbed "Tam O'Shanter," "Muckle Mouth Meg," "Hame Noo," and so forth.[96] And in 1947, the lone shieling was unveiled, fulfilling Professor MacIntosh's bequest. The Parks Branch had provided tourists with a single, simple way of understanding the four hundred square miles of the park.

Only James Smart, in charge of the park's early development, attempted to ensure that the French culture of the park region was represented in the park theme. Angus L. Macdonald had sent Ottawa a list of suggestions for the park's name, most of them Scottish in nature. Interestingly, though, his first choice was "Isle Royale," "the old French name for the Island," which he felt gave it "the charm of antiquity" and "conveys the idea of the sea, which is, to my mind, one idea that we should keep before the minds of people in connection with advertising the park." Macdonald liked "The Highlands" as a name but thought "Highlands National Park" maybe "a little cumbersome" and that "Cape Breton Park" was unsuitable because it had "no particular historical or geographical meaning."[97] It was Smart who defended "Cape Breton Highlands National Park" expressly because it served as a reminder of both the Scottish and French pasts: "the idea being to associate the French background in the use of 'Cape Breton' and, of course, 'Highlands' would give the Scottish suggestion." Smart pointed out that though much of the region was now Scottish, the French still congregated around Cheticamp and that Ingonish had once been French.[98] Smart also tried to make the new built environment suggest the park's French heritage. While buildings at the Ingonish end were to be "the Crofter style of Scottish building with rough masonry walls and thatched roof," Smart asked that those of a proposed Cheticamp-end bungalow camp "take on the style of habitant buildings similar to those in Old Quebec and, in fact, a few of the older buildings which are seen in the Acadian settlements in Cape Breton and the Gaspe country."[99]

Smart deserves credit for attempting to be inclusive, but his efforts demonstrate how strange the whole preservation of ethnic associations really was. The Parks Branch could tolerate the associations with people but not the people themselves. It did not wish to preserve the cultural remnants of the people being expropriated, but sought to introduce idealized versions of their culture for the amusement of tourists. Symbolic memory was to replace actual memory. And if the Parks Branch was not commemorating the people losing land but rather the people of the park region, why did they not give the French associations preference, since Cheticamp had two and a half times the population of Ingonish? And why was the Ingonish end considered Scottish anyway, when the population of both Ingonish and Neils Harbour – the two communities of significance in the eastern end of the park – was overwhelmingly (almost 95 per cent) not Scottish, not French, but English?[100] The tide of tartanism swept such questions away, and even Smart's efforts to retain vestiges of Frenchness in the park were forgotten. By the official opening of Cape Breton Highlands National Park in 1941 (complete with kilts and bagpipes), the National Parks Branch had transformed a smattering of French-, Scottish-, and English-Canadian communities encircling a largely unknown interior into a single unit with a single owner and a single theme. Its success in defining this piece of the most spectacular, the most typical of Nova Scotia landscapes seemed complete.

But either humans are never entirely satisfied with another's attempt to make sense of nature for them or, more likely, it is beyond human capability to give nature a single meaning. In 1948, the American poet Elizabeth Bishop visited the northern peninsula, and in her poem "Cape Breton" describes a quiet Sunday walk along the Cabot Trail (a part near, albeit not actually within, the park's final boundaries):

The road appears to have been abandoned.
Whatever the landscape had of meaning appears to have been abandoned,
unless the road is holding it back, in the interior,
where we cannot see,
where deep lakes are reputed to be
and disused trails and mountains of rock
and miles of burnt forests standing in gray scratches
like the admirable scriptures made on stones by stones –
and those regions now have little to say for themselves
except in thousands of light song-sparrow songs floating upward[101]

Should we take this passage as elegiac or pleased? To Bishop, the trail has denaturalized the nature near it: that which is within sight of the road, a view made available by engineers, she no longer thinks of as nature. Yet the road acts as a fence, keeping nature in. There is still a nature in the interior that humans have accidentally left largely untouched. The Parks Branch had worked diligently to shape how visitors would understand Cape Breton Highlands National Park, but it necessarily did so in a way that seemed appropriate to the time. The best example of this was that it built the park around the idea of a scenic drive – an idea that demanded not only a society that possessed cars but one that saw driving as an entertainment in itself. In coming decades, other cultural ideas about nature and recreation would gain ascendancy. In turn, the idea of what this national park was, and what all national parks should be, would change.

4 The Greening of Green Gables: Establishing Prince Edward Island National Park, ca. 1936

Green Gables did not have green gables. When the National Parks Branch took over the land which would become Prince Edward Island National Park in 1936, and with it the farmhouse that had inspired Lucy Maud Montgomery to write the children's classic *Anne of Green Gables*, there was no paint on the house, just whitewash, and no trim of any colour. Inspecting the recent acquisition, parks surveyor R.W. Cautley wrote his superiors that "While the exterior does not actually need painting, the present colour scheme is not altogether suitable, and it would accordingly be desirable to repaint the building at an early date in order to emphasize the gables of the house, which should, of course, be green."[1] Of course.

This story, with its easy-to-remember irony, can serve as a healthy reminder that parks are cultural constructions. But this is a sort of irony that can satisfy too easily, blinding us to ironies beneath the paint. For one thing, to preserve Green Gables the Parks Branch had to improve it structurally, spending as much on its renovation as had been spent to buy it, and more than what it had cost to build. The principle of park inviolability meant that the house needed to be permanently in prime condition, and the principle of national standards meant that it had to be very attractive; in other words, it received a treatment an ordinary Prince Edward Island home (like itself) would not be given. More than this, Green Gables' preservation was itself grossly unrepresentative. Throughout the rest of the park, the Parks Branch tore down the houses and barns of families who had worked and lived there for generations, sparing Green Gables only because of its association with a fictional story.

And the ironies run deeper still, into the land itself. After the land became a park, developments within it were shaped more by the memory of its human past than by any sense of its natural history. With some effort one can notice that Green Gables was a late-nineteenth-century farmhouse, in this exact location because a small stream flowed nearby. Today the stream seems insignificant even as scenery. Look closely at the golf course and it is a rough map of farmers' fields, duffers the most profitable of crops. The previous land use is often visible, if not immediately noticeable, but beneath that are even earlier landscapes no longer retrievable. The white birch, fir, and spruce that have taken over fields in the park are not the trees chopped down by European settlers to clear the land for agriculture. Those sugar maples, yellow birch, tamarack, and cedar are no longer here to supply seed.[2] The Parks Branch is allowing the fields to grow back, but it cannot restore the area to an "original" condition. Look at Prince Edward Island National Park too long and the sense is that all nature is nothing but history, time masquerading as space.

That is not the sort of impression one is supposed to get from a national park. Certainly Prince Edward Island National Park is well trampled by humans both past and present. It is just a thin strip of Prince Edward Island's north shore, most of it farmers' fields expropriated in the 1930s. As one drives along the highway that runs through the centre of the strip, the park's boundaries – the Gulf of St Lawrence on one side and fields and cottage development on the other – are always visible. At less than ten square miles it is one of Canada's smallest national parks, yet it is also one of the most heavily attended. Its beaches, dune systems, and views are beautiful and not difficult for the Parks Branch to sell; more difficult is selling it as pristine and ecologically significant. As a Parks Canada staff member said after learning that I was researching Prince Edward Island National Park, "Well, it's not *really* a national park, is it?"

But it really *is*, as much as any other. National parks are never found remnants of untouched, self-contained nature. They are unnaturally bordered plots of land selected by people for a variety of reasons, one being the perceived quality of their nature. Rather than denigrating PEI National Park's status, we would do better to understand how it came to be a park when in terms of size, wildness, and sublimity it was so unlike the national park ideal. This chapter will explore what properties made the park area attractive to the Parks Branch in the 1930s, and what kind of park it was expected to be. Most important, it will show that the park's small size, lack of sublimity, and long European settlement and use affected the Parks Branch's judgment of it, which in turn affected how it was developed.

Man on beach looking over New London Bay. Photo taken by F.H.H. Williamson or
W.D. Cromarty during 1936 inspection of potential Prince Edward Island national park
sites. National Archives of Canada, PA203583.

Unlike Cape Breton Highlands National Park, which with its high
coastal vistas was accepted as a traditional Western park in miniature, Prince Edward Island National Park was always seen as something of an anachronism, not really a national park at all. Such
thinking was self-fulfilling.

POLITICAL PLANS
AND PUBLIC ASPIRATIONS

There were very few calls for a national park in Prince Edward Island before the mid-1930s. A provincial MP, Donald MacKinnon,
did write to James Harkin in 1923 that "The larger freaks of nature
are in the public eye to-day, but I think you can help the Province
into the limelight."[3] In testimony to the notion that Eastern Canada
had more history than nature, the federal ministry promised a new
historic site instead.[4] This was appropriate, both in that the Parks
Branch oversaw these sites and in that Harkin had introduced them
specifically for areas of the country not likely to have national park
scenery.[5] Seven years later, a park request from an MP and the president of the PEI Publicity Association received a much more positive response. Minister of the interior Charles Stewart wrote, "In
reply I may say that it is our hope that eventually there will be a National Park in each province of the Dominion and as far as I can see
there is no reason why Prince Edward Island should not be so
favoured, provided the local authorities are prepared to transfer to
the Dominion unencumbered land suitable for that purpose."[6] The

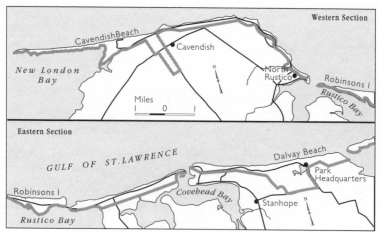

Prince Edward Island National Park (Boundaries are ca. 1955)

Depression cancelled any chance of establishing a park right away, however, and the impetus was lost. When in 1933 James Harkin listed for his superiors which provinces were looking for new parks, he noted that no overture on the subject had been made by Prince Edward Island.[7]

This would come, though, as the Depression made national parks much more attractive. The Island's economy was disastrously hit in the early 1930s. The province's gross value of production was halved between 1929 and 1932, and its two major industries, farming and fishing, especially suffered. The opposition Liberals were able to maintain that the Island's economy was being mishandled. With thousands of Islanders dependent on federal public works programs, the ability to deal effectively with the federal government was considered the greatest test of a provincial administration, and a test the Tories were said to be failing. The Liberals blamed the Conservatives for not keeping a promise that the province would not have to pay anything for its segment of the Trans-Canada Highway. Most damning of all for the provincial Conservatives was its association with the Conservative prime minister, R.B. Bennett, whose programs were proving totally unable to return Canada to prosperity. In the summer election of 1935, the Prince Edward Island Liberal party swept to power, carrying all thirty seats in the legislature. Without any formal opposition and asked to lead a province in economic chaos, the new government had the strongest possible mandate for change.

Considering this opportunity, it might seem surprising that the new premier, Thane Campbell, chose to make the call for a national park one of his first acts in office.[8] There had been no mention of a park in the election campaign and no reason to think that Campbell had wanted one. But at his first federal-provincial conference months after being elected, he had been surprised to learn that a park was within the province's reach. "At the present time," he told his legislature on return, "the national park idea at Ottawa seems to be directed, to a large extent, to the promotion of the tourism industry."[9] On 17 March 1936, his government passed an Order in Council asking the Canadian government to establish a national park. This was a public declaration of intent that the Nova Scotia government had not yet made, though the Cape Breton site had been inspected and all but selected two years before. The Parks Branch had known of Campbell's interest in a park, but his announcement tipped its hand. A day after the Island request, James Harkin wrote Nova Scotia premier Angus L. MacDonald that the federal government had decided to move ahead with Maritime national parks.[10]

That PEI could be granted a national park does not explain why it might want one. Certainly there was hope that a park would help the tourist industry as similar parks had in Western Canada. Prince Edward Island had slowly been developing tourism since the late nineteenth century, but lacked a focal attraction. Thane Campbell was no Angus L. MacDonald: he did not see tourism in visionary terms as an industry of the future, did not think that the province needed to sell itself in a single way. Campbell was a traditionalist who believed that good provincial governments provided good infrastructure – and great ones got the feds to pay for it. So it is not surprising that Campbell's first letter to the Canadian government about a park was directed to the minister of transportation, C.D. Howe. They had talked about a park at the federal-provincial conference, and now Campbell wrote, "We also anticipate that any site which will be chosen will in all probability be some distance from the hard-surfaced highways in the province, and we therefore anticipate requesting that in connection with the national park the Dominion Government should make provision for a hard-surfaced, dust free highway leading from our present pavement to the site of the park."[11] At such an early stage Campbell could not possibly know that the park would be a distance from the main highways unless he planned to make sure that it was. Ottawa saw through these machinations but tolerated them, if in return for some roadway it would get credit for a new national park. Executive Council president B.W. LePage told the Island legislature that "we were simply told by the authorities at Ottawa, that we would get nothing for our highways in lieu of the National Park. If we wanted a Park we could have it; if we did not, it would not be forced upon us."[12] This park's beginnings were more about paving than preservation.

This is not in itself unusual. British Columbia's Kootenay National Park had been created in 1920 as a federal condition for helping build a highway through the province, and the new Nova Scotia park was being located in Cape Breton in part to free the province from maintaining the worst of the Cabot Trail. But those sites still had the scenery and wildlife necessary to maintain the national park ideal. There were no preconceptions that this would be the case for a Prince Edward Island park. Premier Campbell assured his legislature that the cost of a new park would be trifling, and that only three to five hundred acres would need to be involved.[13] Thomas Crerar, the minister of mines and resources, said much the same in Parliament as a reason why Prince Edward Island was suddenly getting a park and New Brunswick, which had been seeking one since 1930, was not: "the expenditure for a park in Prince Edward Island will not be extensive, in the very nature of things."[14] Both governments seemed confident that a suitable location, agreeable to all sides, would be found.

The Campbell government already had a site it liked. Dalvay-by-the-Sea, or Dalvay House, seemed a splendid, sentimental choice for the national park. Nestled in an inlet on the Island's north shore, this opulent Victorian summer home had been built in 1896 by Alexander MacDonald, a director of Standard Oil from Cincinnati. For fifteen years the MacDonalds occupied the house and entertained in style, employing many people from the nearby community of Grand Tracadie as servants. Dalvay was a fitting reminder of an era in which travel was solely for the rich. And though it could hardly be considered sublime, it was impressive enough in scale, perhaps, to meet national park standards. What could be more fitting than to preserve a millionaire's cottage as the centre-piece of a public playground, a symbol of the democratization of twentieth-century tourism? And the province could easily purchase the property. Having changed hands several times, it was now owned by Lieutenant-Governor George DeBlois, who was willing to sell.

The provincial government could not commit itself to Dalvay, however. It had declared its interest in a park so early that there was public debate on where the park should be (unlike Nova Scotia, where there had been little public knowledge that a selection process was under way). Thane Campbell asked that the Parks Branch be brought in to inspect a number of sites, giving the selection process an air of scientific objectivity and relieving his government of blame for choosing one site over another. Crerar agreed to have inspectors travel throughout the province, but noted in confidence that it would probably be a formality. Though neither he nor anyone in the Parks Branch had seen it, "this Dalvay by the Sea premises will probably be the most acceptable." Crerar suggested that PEI should begin work on buying Dalvay; when the park was established, Ottawa would pay the province $25,000 for it.[15]

Prince Edward Islanders did not know this, and debated the relative beauty of their districts in the two Charlottetown papers, the *Guardian* and *Patriot*. Through the summer of 1936, letters to the editor argued whose grass was greenest, whose air was freshest, whose streams were clearest.[16] Quite common were denunciations of the north shore, which was rumoured already to be favoured. Writers complained that the winds off the Gulf of St Lawrence made the north side of the Island bleak, cool, and lifeless, and that the strong tides made for dangerous swimming, especially for children. At a public meeting to generate interest in a park in Bonshaw, an area on the Island's south side, it was said that the north shore had "pests such as snakes, sand fleas as large as chickens, and other annoyances." A letter in the newspaper two days later responded that to say such a thing was foolish – and besides, everyone knew Bonshaw was haunted.[17]

Periodically, this style of debate would be punctuated with a more basic question: what should a Prince Edward Island national park be like? A number of writers favoured locations around Charlottetown, believing that the Island's poor in particular would be better able to make use of it. The editor of the *Guardian* disagreed. "What is the National Park for but to draw tourists to a province?"[18] He felt this was reason enough to have a park, while others felt it was reason not to have it. "Pro Bono Publico" could not believe the government would spend money "to provide a playground, or perhaps something less innocent for tourists." Rather than encouraging "careless, speed mad motorists," better to spend on schoolyard fences to protect children from them.[19] Confusion over the reason for the park's existence even led to claims of conspiracy. "Inquirer" asked, "Is it a place where the area enclosed shall revert again to a state of nature. ...? Or is it for the benefit of the tourist and the sightseer or the seeker after health or pleasure, or for the convenience of those who would make wassail on the contents of the kegs borne shoreward by the waters of the Gulf?"[20] Some still asked whether Prince Edward Island even deserved a national park. One strand of letter writers believed the province had the wrong type of beauty. "Citizen" told the *Patriot*, "We have no wild mountainous country that would keep tourists more than an hour to find out what it's all about. There are no places where tourists can take their packs and get lost or tramp for days."[21] Another writer added that though the Island was beautiful, "We however have nothing more than scenic beauty and not much variety in that."[22] Others took the opposite tack, stating that since all of Prince Edward Island was beautiful it should all be made a park, or at least scattered districts throughout the province should benefit, rather than one site. The annual convention of the Women's Institute recommended that all the Island be considered park land, though what this would mean in practical terms was not clear. It was the kind of argument that appealed to unelected politicians. Conservative leader W.J.P. MacMillan proudly proclaimed, "During our time we didn't fiddle with the national park, which was not necessary because the whole Island is a national park."[23]

This broad range of ideas suggests a confusion over the meaning of national parks. But most letter writers understood the basics: parks must have the most beautiful scenery, they would be nature reserves, they would encourage tourism and recreation. Though most Islanders had likely not visited a national park, they would have known of Banff at least. What confused Islanders was how this thing called a national park would be translated to their province. That some should see their home as "park" – scenic, pastoral, and preserved from industrialization – and others see it as "not park" – lacking the size, wildlife, and sublimity of the Rocky Mountain parks – shows a local understanding that Prince Edward Island would necessarily have a new sort of park, depending on which characteristics the Parks Branch would choose to emphasize.

A TYPICAL SEASIDE RESORT

Within the Parks Branch there was as yet no plan for the Prince Edward Island park. Despite Harkin's musings about Maritime parks over a decade earlier, the promises of a park for every province in 1930, and the Senate Committee and federal-provincial conference recommendations in 1934 and 1935, there is no evidence of enthusiasm from within the Branch itself about nationalizing the system. The federal and provincial governments had agreed to a park for PEI without involvement from the Branch, and even the selection of Dalvay seemed a *fait accompli*. In correspondence during the establishment and development of Prince Edward Island National Park, park staff often referred to the advantages the province would enjoy from the new park but never to any benefits that would accrue to the Parks Branch or the park system.

The first major attempt to define the proposed park was made just days after it was announced that the Maritime parks would go ahead. F.H.H. Williamson, deputy commissioner of parks and assistant to James Harkin since the early 1910s, wrote a memo evaluating the park's prospects, even though it would be months before he would travel to the Island and see it for the first time. Williamson would not only soon be responsible for overseeing the creation of the PEI park but also, upon Harkin's retirement in less than a year, become head of the parks agency. So his sight-unseen analysis of the Island park reflects not only the old-school Branch opinion of Maritime nature but also the direction that would be taken with the first Maritime parks in their first years. Williamson wrote, "It will be impossible to apply the same National Park standards to the proposed park in PEI since its characteristics are almost totally different from any of the National Parks so far established in Canada. ... [I]t is diminutive in size; its scenic features are incomparable with the grandeur of the Rocky Mountain Parks or the extensive lakes, streams and forests of Prince Albert and Riding Mountain Parks; it is not a wilderness sanctuary for animal life; its attractions cannot include long motor drives or trail rides."[24] All this was beyond dispute (though in later years the Parks Branch would advertise the PEI park in terms of its scenery, its rare wildlife, its "motor drive"). Williamson believed that the park should rely on its strengths: "Naturally its development should centre around its principal feature – its extensive beaches with warm sea bathing. It is my opinion that we should not plan its development in terms of attracting only hundreds of tourists but many thousands a year. I see no reason why we cannot expect a quarter of a million visitors a year in a few years. ... I suppose for every person who wants wilderness loneliness there are a dozen who prefer to be in a crowd during their holidays."[25]

It may seem jarring that Williamson follows his frank assessment of a place's deficiencies with a claim that it could be a real crowd pleaser. He simply assumed that different landscapes naturally attracted different strata of society. Sublime nature was of interest to those who could draw out its emotive power – "class" – while nature like the Island's was of interest to the many satisfied with pleasant surroundings – "mass." Williamson saw his job as developing a place which would satisfy such mass taste while permanently maintaining as much beauty and class as possible under the circumstances. He wrote, "Under proper control I feel we can supply such desiderata, which are a healthy and altogether beneficial quality, and attract to the PEI park far more visitors than by simply more or less allowing people to use what Nature has provided in the way of beach and sea and letting it go at that."[26] Elsewhere, Williamson told Harkin, "It seems to me the best contribution the park can make in the interests of Prince Edward Island and the country generally is to develop it as a typical seaside resort, sans the obnoxious amusements."[27]

Here is where ideas within the Parks Branch of what constituted beauty had real consequences. Williamson believed that the Island's nature was insufficient to carry itself. The Branch's job, then, was to forego trying to preserve it as a traditional park and instead develop it in its most attractive incarnation, a resort. This was not a normative decision; it was an aesthetic judgment made through the filter of a quarter-century spent working within a park system which saw mountain landscapes as its ideal. Even prior to its creation, Prince Edward Island National Park was being treated differently because its nature did not conform to this ideal.

It was decided that Williamson would travel to Prince Edward Island with W.D. Cromarty, chief of the National Parks Architectural and Landscaping Service, in June of 1936 to examine potential park locations.[28] With Premier Campbell as their guide, Williamson and Cromarty examined twenty-two sites scattered throughout the province.[29] At many of the sites local boosters tagged along, including eleven members of the legislature, Summerside mayor Brewer Robinson, and Lieutenant-Governor DeBlois.

The report that Williamson and Cromarty offered Commissioner Harkin the following month had both expected and unexpected features. The inspectors agreed with the provincial government's recommendation of Dalvay and suggested that it be made a hotel in the tradition of the great park hotels at Banff, Yoho, and Waterton. But they also believed that the province should be urged to acquire a strip of coastline for the twenty-five miles west of Dalvay, through Stanhope, Brackley, Rustico, and Cavendish beaches all the way to New London. They proposed the inclusion of shoreline, a strip several hundred yards deep to build a park road and lay out developments, plus a buffer zone to discourage private tourist operators from setting up businesses too close to the park. The inspectors were especially impressed with Cavendish, stating that "the area offers the greatest opportunity for Park development, since it not only possesses a beautiful and extensive beach of firm, white sand but it has the added attraction of picturesque red rocks, worn into unusual formations by the sea." Cavendish contained "an unusually wide appeal both to those who appreciate beauty and to the younger element who desire variety in their holiday environment."[30]

Of particular interest to the inspectors was Cavendish's association with *Anne of Green Gables*. They wrote, "According to a large signboard erected in this district by the Prince Edward Island Travel Bureau, one of these lakes is the 'Lake of Shining Waters' portrayed in the Novel 'Anne of Green Gables' by Lucy Maud Montgomery and the house (Green Gables) stands in the area proposed to be included in the Park. ... These features have been preserved in their natural state and outside of the general interest of the house it is thought that the adjoining woodland and small trout stream merit inclusion in the Park area."[31] Williamson and Cromarty did not then draw the obvious conclusion: this was land very much in use. Cavendish was already on its way to being a destination for literary-minded tourists. Lucy Maud Montgomery told of seeing souvenir-mad visitors overrun Green Gables in the late 1920s.[32] The present owners of the house, Myrtle and Ernest Webb, were giving tours and had even built a second storey over the "ell" of the house specifically to accommodate more guests. In the surrounding district, more and more farming families were opening their doors to travellers, building cottages, and finding more of their energies directed toward tourism.[33]

Of course, even before tourism, the land at Cavendish and along the strip back to Dalvay had been in use by farmers for generations. By extending the proposed park far beyond Dalvay, with its single owner, the inspectors were setting the park up for confrontation with the hundred or so farmers and other private landowners. Williamson and Cromarty, though, made no mention of present land use in their report. It was not their concern, since the province would be responsible for appropriating the land and handing it over to Canada. And the inspectors probably had assurances from Premier Campbell, travelling with them, that the land could be made available.

As in Cautley's report of Cape Breton, Cromarty and Williamson's description of scenery does not satisfactorily explain their appreciation of the proposed area. The term they fell back on time and time again in their report to describe the scenery was "picturesque." The beaches (page 3), red rocks (4), sandbars (5), red cliffs (6), scenery (6), woods (9), Winter River (10), a drive by the cliffs (13) – all were picturesque. By this, the inspectors meant only that the scenes were pleasant – they were not using the term in its formal sense, to mean a pastoral setting with signs of past occupation. Of course, it was past occupation and the resulting clearance of land which gave Williamson and Cromarty the sightlines they needed to pronounce the scenery picturesque in the first place. Yet their report makes no mention of the land's long history of human use. Overall, Williamson and Cromarty's evaluation of the Island's North Shore landscape was not nearly as enthusiastic as Cautley's had been of Cape Breton. Whereas he was impressed with the visual contrast Cape Breton Highlands offered between land and sea, they seemed merely content that the PEI site possessed the requisite amounts of land and sea.

In proposing a park with both Cavendish and Dalvay as centres of development, Williamson and Cromarty were hoping to accommodate two different sorts of tourists. "The Dalvay house," Williamson had suggested in an earlier memo, "will undoubtedly be the concentration point for the more or less wealthy visitors and I think we should develop this property principally for elderly people."[34] Dalvay had been built by wealthy tourists: its size, design, and historic atmosphere (though it was but forty years old) would appeal to the tasteful and historically aware rich. Dalvay, it was said, "Would attract best types of people and when they get to go there, will attract others."[35]

Others were fully expected to come. The western, Cavendish arm of the park with its beaches and Green Gables would attract middle-class families. Unlike tourists at Dalvay, who would find relaxation and quiet, those who came to Cavendish would find entertainment and large crowds. After returning to Ottawa, Williamson wrote Harkin a list of some of the developments he believed would be needed to popularize the area:

(Cavendish end particularly for children with "Green Gables" stream reserved for young children's fishing and the woods for children's village, picnicing etc; toy yachting on "Lake of Shining Waters", children's canoeing, paddle boating etc.) "Green Gables" might be a children's rest house with museum, aquarium etc. ... [Need for] bowling greens and buildings for same; dancing, roller skating and carnival arena, etc. Concessions may be rented for bungalow camps, hotels, boarding houses, restaurants, soft drinks and ice cream, boats, canoes, wheeled bathing houses; beach donkies and ponies; beach nigger minstrels; bands; moving picture theatres and other shows. ...[36]

The important thing, of course, was that this all be tastefully and artfully done, maintaining the dignity of Canada's national parks. If so, it could be an example to all Prince Edward Island of how to facilitate tourist business, "sans the obnoxious amusements." The Parks Branch's plan for Dalvay and Cavendish is a wonderful example of what Lawrence Levine calls a spatial bifurcation of culture.[37] Though visitors at the two beaches would be enjoying almost identical landscapes, they would be given different means of experiencing them, depending on their class's perceived interests. Though the Parks Branch claimed to be responding to societal wishes, it was most certainly helping to reinforce these social differences.

The inspectors' report was made public in September. It must have been a surprise to the Prince Edward Island government. The province knew it did not have the kind of scenery that the park system was used to, and so had sponsored Dalvay on the assumption that, thanks to its cultural associations, this was the most majestic and sublime place the Island could provide. Cromarty and Williamson's report convinced the government that it had sorely misjudged national park standards. It would seem that the Parks Branch appreciated the seaside scenery of the North Shore generally, and was willing to develop it heavily, much more than had been assumed. A week after the report was made public, Thane Campbell's government approved its recommendations "insofar as the same recommends" a new park from New London to Brackley. The eastern end of the park, including Dalvay, was to be omitted.[38]

The Parks Branch's reaction illustrates the tightrope it was walking. Ottawa, caught wholly unprepared by the Charlottetown decision, responded with a flurry of telegrams. Not hiding his alarm, Williamson impressed upon the province just how critical Dalvay and its environs were. He spoke of Dalvay's wooded acres as wilderness writ small: "a sanctuary for wild life, the vegetation is unspoiled, the stream a potential paradise for fishermen and for canoeists, the woods ideal for hiking." But these were clearly secondary considerations. More important, in a national park consisting solely of Cavendish, he warned, "Its constricted size will not permit of accommodating a large number of visitors which is the only justification for the establishment of a National park on the north shore of the Island. When thousands of visitors congregate in the area … there will be congestion and all the evils of overcrowding. Such conditions will lower its status as a National park and react against the other National Parks which so far possess a high standard of quality which the public at present recognize and appreciate."[39] The threat of losing wilderness did not scare Williamson, since wilderness was not the park's justification. Rather, he feared that the new Prince Edward Island national park would become so successful – and therefore, it would seem, so vulgar – that it would change how people thought of national parks in general. Dalvay was a necessary run-off, a place where tourists with finer sensibilities could cling to the illusion that the park philosophy was unchanged. So could the Parks Branch. Cromarty was rushed to Prince Edward Island with an ultimatum for its government: no Dalvay, no park. Not surprisingly, Dalvay was reinstated immediately.

A ONE-SIDED PROPOSITION

The expropriation of land for the new Prince Edward Island national park was bound to have a different dynamic from that of Cape Breton Highlands National Park. In Nova Scotia, entire properties, entire communities were taken. On PEI, the government was lopping off a strip generally a few hundred yards wide along the shore across a number of rural communities. This made expropriation both simpler and more difficult for landowners, for the provincial government, and for the Parks Branch. On the one hand, many landowners did not have to move, so they were better able to abide their loss. On the other hand, many landowners did not have to move, so they would be constant presences in the future of the park.

The National Parks Branch was still new to relying on provincial governments for the purchase of planned park land. In Nova Scotia, though the expropriation process was not without problems, it was at least under the supervision of a government interested in having a park. Thane Campbell's government, in contrast, demonstrated no desire for any particular sort of park; any would do. Nowhere is this more evident than in its naming. Campbell suggested that an essay competition for Island students be organized to select a name, but when the finance minister, Charles Dunning, refused to pay for prizes, the province went no further. By the spring of 1937, with the park's establishment soon to be officially announced, naming it grew in importance. Crerar wrote asking Campbell what name he preferred, but did not hear back. Crerar's secretary offered a list of suggestions, from the plain "Prince Edward Island National Park" to the imaginative "Silversands National Park."[40] Again, no answer. After some deliberation within the Ministry of the Interior, it was decided that the best possible name was the easiest to remember: "Prince Edward Island National Park." There having been no input from the province on the matter, it was also the easiest to choose.[41]

The Campbell government's policy in regard to expropriation would be a mixture of neglect and compulsion. In June 1936 the Liberals passed "An Act Respecting the Establishment of a National Park," which provided the machinery for expropriation. The new law allowed the government to make an offer for a property, at which time the owner could accept or appeal. If the owner appealed, the government might raise its offer or stand firm, but in either case this was the end of the transaction and the land was turned over to the province. The landowner had no recourse to the courts. This legislation drew no public notice when it was passed, because the legislature was without an opposition and because the park still only existed on paper.

Once it was clear in the late summer of 1936 what the rough boundaries of the new park were to be, the provincial government took its first step toward buying the land. The Dalvay transaction was made speedily, the lieutenant-governor handing it over for $15,000.

The federal Parks Branch, the provincial government, and the Island Travel Bureau all agreed that making Dalvay a CN hotel would be the best use of it and give the park instant credibility. All agreed, in fact, except CN itself. CN president S.J. Hungerford felt that with only twenty-six bedrooms and five bathrooms, Dalvay was too small to be profitable.[42] The Parks Branch was so obviously unhappy with this response that Hungerford had his general manager visit Dalvay to come up with other ideas. He proposed a simple solution: demolish Dalvay and build a "wooden structure" with seventy-five rooms in its place.[43] The Parks Branch thought not, but had no idea what it would now do with the house.

Since Green Gables was to be a centre-piece for the park, its transfer had to be accomplished smoothly. So members of the Campbell cabinet along with Queen's county federal representatives, Senator John Sinclair and the finance minister, Charles Dunning, together paid a visit to Green Gables' owners, the Webbs. This show of force succeeded in convincing the Webbs that they had no choice but to settle. They would be given a fair price, they were assured, and Ernest Webb could stay on as park caretaker. The Webbs resigned themselves to their fate – they were just unclear what their new life would be like. Myrtle Webb wrote an apologetic letter to F.H.H. Williamson, asking, "can you imagine the shock the family received when the word came out in the press the Cavendish Area had been your choice and the Green Gables property was to be included. ... Could you give us just a little idea of what we can expect in the way of changes ...?"[44] A price of $6500 was agreed upon for the house and barn, a price that Premier Campbell considered somewhat steep, but acceptable considering the "historic and sentimental associations" of the property.[45] Green Gables – never possessing that name until the book was a bestseller, never Lucy Maud Montgomery's home, and never more than a real home to a fictional character – was always more valuable for its associations than for what it was.

No other property owners received the attention that the owners of Dalvay and Green Gables did. On 24 April 1937, the park land was officially taken over by the province and transferred to federal ownership. The current landowners had not yet been offered a price for their property, or even been informed that their land was needed. Parks surveyor Cautley recognized the potentially incendiary nature of the land expropriation. When setting up the park boundaries, he saw that the citizens were only vaguely aware of where the park was to be and what this would mean to their holdings. When asked by his superiors in the Parks Branch to set up permanent markers on the new park boundaries, he refused. "I am quite willing," he wrote, "to

step outside my ordinary line of duty ... to serve you and my Department, but am most certainly not willing to get mixed up with the
Province's Expropriation proceedings."[46] By the summer of 1937,
Cautley warned Williamson that the province was bungling the land
transfer. He recounted an incident typical of the problems he was
having: "On the 8th Instant I called upon Mrs. Dr. Bonnell at her request, she being at present in residence in their cottage on the
McCoubrey farm next to Green Gables. I found that she wanted to see
me in connection with the grazing of their one cow, and that she was
genuinely surprized when I explained as courteously as I could that
we now owned their cottage."[47] The indifferent manner in which
landowners were informed made such situations inevitable. The first
public notice of the land transfer came forty days after the land had
become Canadian property. Even then, the message was not directed
at specific owners. It instead gave a detailed, legalistic description of
the land to be expropriated, and concluded that "former owners of
these properties are entitled to claim compensation," the claims to be
sent to the deputy provincial treasurer.[48]

Upon pleas from the federal government, which hoped to begin development in the park, the Campbell administration moved forward
with settling claims in the summer of 1937. It appointed three commissioners – E.T. Higgs, Daniel MacDonald, and Robert MacKinley –
to put valuations on the expropriated properties. (Government members came to refer to the commissioners as "two farmers and a businessman," to show they were just common folk.) The Higgs
Commission travelled across the North Shore visiting landowners
that summer, and returned to the government with proposed prices.
Using a scale ranging from $65 per acre for extra good land, down to
$6 per acre for sand dunes, they set prices for eighty-four properties.[49]
The Executive Council, apparently finding the commission over-
generous, arbitrarily cut many of the recommended offers by exactly
10 per cent – reassessing, for example, Alof Stevenson's land from
$446.50 to $401.85 were dropped by as much as 32 per cent. Only seventeen of eighty-four recommendations were left unchanged.[50] Letters
were sent out to the landowners, with this final, non-negotiable offer
enclosed.

Only then, four months after the land had been turned over to
Canada, did a groundswell of opposition to the expropriation process
arise. The Charlottetown *Guardian*, the daily newspaper sympathetic
to the Conservative party, took the lead. In a letter to the editor, "One
of the Dispossessed" explained point by point how the landowners
were given unfair prices for their land, were unable to appeal the
matter in court, and were unable even to convince the government to

buy all their property if that was their wish. The writer concluded, "As I write this letter I can see from my kitchen window a bountiful field of waving grain on my farm, a field which next year and all the succeeding years will be barren because of the determination of the Campbell Government to establish what is now becoming known throughout the Province as 'Expropriation Park.'"[51] This was powerful imagery to an agricultural province. Through the fall of 1937, the expropriation issue was a hot topic in the editorials and letters of the *Guardian* and the Liberal organ, the *Patriot*. The *Patriot's* calm defences of the government's wisdom and frugality only fed the *Guardian's* rage; the latter claimed that to deny the landowners the right to appeal was a violation of the Magna Carta, and it entitled editorials "What Price Democracy?" and "Our Dictators in Action."[52]

Opposition in the press generated political action, in the formation of a Committee of Dispossessed Landowners. The group arrived in Charlottetown forty strong to march on Premier Campbell, with a petition of more than fifty names protesting "against the denial of our rights as British Subjects of access to the Courts of Law."[53] Unfortunately, the premier was out (the committee having failed to make an appointment) and by the time he returned he faced a much-dwindled delegation. Campbell rebuffed them, promising to add an appeal amendment if and only if the current expropriation process ground to a halt. To maintain unity within their ranks and to broaden awareness outside their communities, the dispossessed landowners held public meetings in York, Kensington, and Corran Ban. Nothing much came from these to change their political strategy, except to amend a resolution being sent to Ottawa by omitting the phrase "arbitrary, capricious, tyrannical and oppressive."[54] Widespread public opposition never blossomed, and as fall turned to winter, the national park issue disappeared from the daily newspapers.

Strangely enough, however, the Campbell government quietly did raise the settlements of twenty-seven different landowners during late 1937 and 1938. Were these good Liberals? The *Guardian* editor had no doubts: "The result has been a settlement, in some cases at least, not on the merits but on the basis of partisan pull."[55] It is nearly impossible to confirm this today, given the difficulty of knowing the landowners' politics and of evaluating their land. But it is worth comparing the names on the Committee of Dispossessed Landowners' petition with the final land settlement list. In all, there were eighty-six people compensated, and forty-one of those had signed the petition. Of these forty-one, ten (or 24 per cent) had their settlement raised. Of the forty-five compensated who had not signed the petition, seventeen (38 per cent) had their settlement raised. In other words, landowners were more likely to get more money if they had *not* complained. If we assume that petitioners were more likely to be people who opposed the government in general terms – i.e., were Conservatives – then maybe Liberals did have better luck when selling their land.[56]

The Conservative leader, R.B. Bennett, tried to address the expropriation matter in Parliament in the spring of 1938, stating, "I hope we are not to undertake the management of a national park which has been brought into being by depriving citizens of their rights under the law."[57] But Crerar, the minister of mines and resources, parried that the process for obtaining land was entirely a provincial matter, and the federal government had neither the power nor the right to step in and tell the province how this should be done. Though unsatisfied, Bennett let the matter drop.

In the all-Liberal legislature, the government made a unified defence of the expropriation process. Dougald MacKinnon said that the three smart men who made up the Higgs Commission would surely know more about farm values than would a judge. J.P. MacIntyre said much the same: "I know if I had land to dispose of for highway purposes, I would far sooner see two farmers and a business man coming than to take it before judges of the court." Of course, most of the Higgs Commission's recommended settlements were reduced, but this too could be justified. As Donald McKay stated, "I don't see why seven or eight men at the head of this Government are not just as fair and as honest as any judge would be in settling matters."[58] The last word here should probably go to Premier Thane Campbell, who, in defending the machinery by which expropriation was being handled, noted that there were no complaints at all until a full fifteen months after the law had been passed. This was a specious point: the law had not been applied until then. In any case, Campbell was happy to report that at this point sixty-nine of eighty-three claims had been settled, "by complete accord and satisfaction and agreement."[59]

Organized opposition to the national park withered away, with only a few landowners fighting private battles with the park in the coming years. But for a short time at least in 1937, there had been a unity among disgruntled owners that there was not at Cape Breton Highlands, and that there would not be at Fundy and Terra Nova. Why? First, the *Guardian* played an important role in coalescing opposition to the park through its editorials, and giving ordinary citizens the chance to vent their anger. In the cases of the other parks studied here, there was no paper with a readership broad enough to reach all those affected and yet local enough to treat this issue with weight. Second, the very harshness of the expropriation method meant that landowners could respond to it in terms of universal rights, not just personal grievance. Expropriations in the other three parks may have been severe, but they did not forbid appeal. Finally, landowner opposition was galvanized by the protests of a few particularly vocal tourist operators, particularly Katherine Wyand and Jeremiah Simpson. Perhaps because landowners saw government in this instance as a competitor for tourism dollars rather than a regulatory body, they were less willing to accept it as an immovable force than were landowners at the other parks.

But in the end, the forces that pulled owners toward expropriation were stronger than those that stood in its way. Many Islanders looked forward to the national park, and some property owners on the north shore no doubt welcomed the opportunity to sell some land and hoped that their remaining land would increase in value. Those who opposed expropriation watched more and more of their neighbours receiving settlements, and felt greater pressure to sell, probably fearing that the government would offer them less the longer they waited.[60] Also, as early as 1937, tourists were beginning to arrive in the new park and proclaiming their own brand of squatters' rights.[61] And there were logistical problems in organizing opposition. The distance between communities meant that many landowners did not have the opportunity to speak to each other or attend meetings. More generally, resistance had to unite communities that had never before had reason to think of themselves as a unit. As Fred Horne writes, "This very likely marked the first time the whole area had a common, unique experience."[62]

Most important, landowners never got the majority of Islanders on their side. The people did not want to believe that their elected officials would grab land unfairly – or, worse, pay too much for it. As a writer to the *Patriot* suggested, "the general public will give credit to Premier Campbell for looking after the 90,000 of our province against a score or less who would like to fleece the taxpayers by securing an exorbitant price for their holdings."[63] The money that the park was bringing to the province helped win Islanders over to it as well. The province was granted $40,000 in 1938 alone for a "Tourist Roads Programme," fulfilling Campbell's wish that the roads leading toward the park would be paved. This fund, in addition to the regular park construction budget and a special work relief grant in aid of park development, meant that in 1938 the Prince Edward Island National Park was worth over $145,000 to the province. Crowing about this to the legislature, Campbell was careful to point out that the Island would lose nothing in return – in fact, it would be prettier than ever. "All those natural beauties and amenities preserved, polished up a little," he stated, "and the traditional associations of that locality which have been made so famous through the books of our beloved Island authoress Lucy Maud Montgomery will be brought into greater prominence by the beauties and amenities of the new golf course which is being established around that focal point."[64] The Conservatives tried half-heartedly to make the expropriation process an issue in the 1939 election, but it had little effect at the polls. The Campbell government recaptured twenty-six seats and retained a healthy majority. Even in districts affected by the park, the returns were practically unchanged from the 1935 election. For every land-owner angry that his or her land was expropriated, there were un-doubtedly more residents looking forward to the financial benefits of increased tourism.

The land expropriation process on Prince Edward Island's north shore in 1936 and 1937 was bullying, inept, maybe even inequitable. But it was over. The provincial government had survived its own handling of the issue, and landowners had been paid. Yet the expro-priation would always play a part in the life of Prince Edward Island National Park. When, in Parliament, Crerar had said that the federal government was not involved in the expropriation process, a Tory member from Ontario, H.A. Stewart, broke in that the Liberals acted "as though this whole matter began and ended as between the gov-ernment of Prince Edward Island on the one hand and the owners on the other. It is not quite that; it is rather a three-sided proposition, a proposition in which the government of the dominion also comes in."[65] Though Stewart was referring to principles of justice, he might as easily have been speaking to the more practical matter of manag-ing the park. The Parks Branch was not only taking possession of land, it was taking possession of any difficulties that had come in ac-quiring it. By the manoeuvre of having the province buy the park land and transfer it to Canada, the federal government avoided the responsibilities but also the rights of expropriation. Because this land shuffle worked faster on paper than in reality, when the Parks Branch was ready to develop the park it found many of the previous owners still there. Branch staff said quite accurately that the land had been given to Canada; landowners said quite accurately that they had not sold it. More troublesome for the park's future, expropriatees did not distinguish who was treating them poorly. Land was taken so that a park could be established, and it was park staff who now wanted it; therefore, the expropriation was the national park's fault. By excusing itself from the expropriation process, the federal government had traded short-term problems for long-term ones.

FENCING NATURE IN

The physical development of Prince Edward Island National Park promised to be a quite straightforward affair. Unlike at Cape Breton Highlands, where the improvement of roads was a major undertaking, and Fundy and Terra Nova, where (as we shall see) roadways and many facilities had to be built from scratch, much of the infrastructure for Prince Edward Island's national park was already in existence. The Dalvay to Brackley part of the park was to be connected by new roadway, and existing roads built up and gravelled. Kitchen shelters and beach houses were to be constructed to draw more tourists to the seashore. Dalvay and Green Gables were to be beautified and shaped up, and other buildings in the park torn down or moved away. But before all this, the Parks Branch built a boundary fence around its new property. That this was given first priority indicates that staff understood that the most difficult development job would be undoing a century of land ownership and making the park seem a legitimate use of the land. At present, owners could look out upon their land, across the property that had been taken from them, to the Gulf in the distance. Especially before the land had time to grow back from its long agricultural use, the park showed off its arbitrariness. To the people of the area this new park was less a coherent piece of preserved nature than a collection of phantom limbs.

The Parks Branch first tried to take an unyielding stance against the previous landowners. In June of 1937, while the Higgs Commission was still assessing property and several months before owners would be offered a price for their land, Commissioner Williamson reproached Cautley upon hearing that he had asked permission to unload a pile of fence posts on Oliver Bernard's property. He explained, "This is, in effect, recognizing that Mr. Bernard has a right to the land, which, of course, he has not. Any grievances which these people may have must be against the Province for their delay in effecting a settlement for the lands which have been taken over. The point which cannot be overlooked is that the area is *now* a National Park and the Dominion hold clear title to the said lands."[66] Cautley, a forty-year veteran of the Departments of the Interior and Mines and Resources, replied frostily that he knew whose land it was, but the province was utterly failing in its responsibility to purchase it. He thought it not prudent to evict former owners before they were paid; in his opinion neither the courts nor the public would stand for it. Anyway, it was not he who had dropped off the fence posts, "although, if I had been there, it is quite likely that I might have been guilty of such a slight civility."[67] Despite Williamson's rhetoric, the Parks Branch did not wish to do anything aggressive that might draw the ire of local residents, the park's permanent neighbours. The Branch could only maintain pressure on the Prince Edward Island government to make speedy settlements. After a summer of working around the dispossessed landowners, Williamson conceded to the acting superintendent, Allan McKay, that inhabitants could stay the winter, as long as they moved out by 15 May of the following year.

The Parks Branch did what it could to give the park credibility as quickly as possible. The new golf course was fundamental to this. It was related to tourism and its construction was labour intensive, so it signified a commitment to local employment. Also, because the sport had high-class connotations, the course would signify that the park was to be a cultured destination. In 1941, James Smart went so far as to say that "we cannot be satisfied with a course that is not at once considered superior to most of the courses on the Island."[68] As at Cape Breton Highlands, Stanley Thompson was brought in to design the eighteen-hole golf course, which would stretch inland from the sand dunes of Cavendish beach to encircle Green Gables. During 1938, Thompson oversaw the work of forty men, more than were working in the rest of the park combined.[69] Whether because of the Branch's eagerness to create a first-rate facility or simply because of the inertia of development, the golf course overwhelmed all preexisting landscape, including Green Gables itself. Though the course was meant to offer homage to *Anne of Green Gables* by giving holes names such as "Ann Shirley" and "Haunted Wood,"[70] it was the farmhouse that became complementary to the course. A little rise was levelled off for a green almost directly in front of the house – and, as a result, for decades the sanctity of many a pilgrimage to Green Gables was interrupted by an errant Titliest. Earth was pushed up around the house to give it more prominence alongside the course. Green Gables became a tea-room for visitors and golfers, and the barn was originally intended to be made a clubhouse, though it proved in too poor shape and was torn down. In the winter of 1939 the Montreal *Star* slammed the golf course development for destroying the beauty of Cavendish. Perhaps the Parks Branch was listening; rather than making Green Gables the golf clubhouse as had been planned, a separate building was constructed nearby.[71]

Development demonstrated the new park's usefulness to the province in tangible and immediate ways, but the Parks Branch made sure that preservation also proved itself an active process. This even influenced the selection of the park's first superintendent in 1938. Rather than settling for just an overseer of the park's nature, the

Branch hired "a capable young forester," Ernest Smith, to take charge of the park's forest development.[72] Smith was to do all the normal jobs of a park superintendent, plus take time to give demonstrations to Islanders of modern woodlot management. He built up a nursery near Dalvay House with Norway pine, white ash, white spruce, and Norway spruce, and thinned existing stands within the park. The Branch hoped that this would justify the park's existence to Islanders; at the same time, such work would start covering up the land's long history in agriculture.

But the land of Prince Edward Island National Park did not so easily shake the traditional associations that locals still gave it. According to Green Gables caretaker Ernest Webb, Cavendish road overseer Austin Laird ran the winter road right over the new golf course on the grounds that this was "the way the road always went for over a hundred years. ... He went on to say that neither *Smith* nor *Crearer* could stop him."[73] Irish moss harvesters, accustomed to travelling through fields to reach the shore, came upon the new park-built fence and simply snipped the wires and carried on.[74] A farmer sold cranberries from what was now park land, and a flurry of telegrams flew between Ottawa and Charlottetown, discussing how to deal with such larceny. After intensive federal-provincial consultation, it was decided that for the next harvest only, licences would be sold to would-be cranberry pickers.[75]

These were annoying skirmishes for the Parks Branch, but ones that they were ultimately sure to win. Canada had legal title to this land, and the longer the park existed, the more accepting locals would be of its presence. One by one, dispossessed owners settled with the provincial government in 1938 and 1939. A few, embittered by the experience and seeing the Liberals retain power, voted with their feet by leaving the Island. Two, though, stayed and voted with their ploughs. Jeremiah Simpson and Roy Toombs continued to work what had been their land. Simpson had from the beginning been an especially dogged opponent of the park; he had a lot to lose. As he stated, "This land has been in my family for one hundred and fifty years. I would not sell it to my best friend."[76] In the 1920s the Simpsons had been the first people in Cavendish to take in tourists, and still did this as a sideline to farming. Jeremiah Simpson had been involved in the Committee of Dispossessed Landowners, and wrote several letters to the *Guardian* and a few to Thomas Crerar. The farmer challenged the politician to a debate on any platform anywhere, and ended his petitions with apt biblical quotes: "Thou shalt not give the inheritance of thy fathers unto strangers" and "Cursed be he that removeth his neighbour's landmark."[77] Ploughing their way onto federal land was merely the latest of tactics employed. Their pleas for justice denied, Simpson and Toombs elected to deny the park's existence.

Superintendent Ernest Smith, unsure how to react, wrote Ottawa for instructions. Williamson asked Premier Campbell for advice, and was cheerfully told that it was no longer a provincial concern. The Parks Branch finally decided to threaten the farmers with legal action, but to let them take in their crop that year. They did, and promptly seeded the land again the following year. Superintendent Smith admitted to his superiors that he could not help but sympathize with Toombs and Simpson. He believed the men had "just cause for complaint," and the whole matter should be investigated further.[78] It dragged on for several years: Smith warning the farmers, the farmers promising to stop but starting up again each spring, the Parks Branch berating Smith for his timidity. In 1943, the new premier, J. Walter Jones, offered to mediate the dispute personally, and succeeded in making a settlement with Toombs. The following spring, Smith came upon Jeremiah Simpson working his land and, after a long talk and further discussion with the Simpson family, convinced the farmer to seed next year's crop into grass. Simpson never did accept the money waiting for him in the Court of Chancery.

From its inception the new Prince Edward Island National Park accomplished its mission of being the tourism focus for the province. Attendance at the park skyrocketed from 2500 in 1937 to 10,000 in 1938, and over 35,000 in 1939. Five years after being established, it was the fifth most visited national park in Canada.[79] It and its sister park Cape Breton Highlands were not thought of by the Parks Branch as traditional parks, but were considered successful for what they were. A 1938 park tourism booklet noted that obtaining land for these two parks had been a new experience for the Branch, but "The public-spirited attitudes of owners, in making their lands available for this purpose, and the co-operation between Dominion and Provincial Governments, have made possible these valuable additions to Canada's system of national parks."[80]

In reality, of course, it was expropriation law rather than public-spiritedness which had allowed the creation of these parks. It is a sad commentary that of the four parks studied here, Prince Edward Island National Park saw the greatest amount of local involvement in its establishment. Only here did the very meaning of the park become a public issue; only here was there active grassroots lobbying on behalf of favoured site locations. Even the expropriation process became a matter of public debate. Yet none of these had any real effect on the final park product. Provincial and federal politicians and the staff of the National Parks Branch decided the new park's location, boundaries, and development with no place for public points of view. In the Canadian national park that shows more clearly than any other the role of humans in shaping the land, the humans who actually lived there had no role in shaping the park.

5 Suburbia Comes to the Forest: Establishing Fundy National Park, ca. 1947

Drive through the village of Alma, New Brunswick, cross the bridge, and climb the short hill into Fundy National Park. The park headquarters is the sandstone cottage on your right at the intersection ahead, but take the road to the left, the one that runs along the coast. Pull over alongside McLaren's Pond, with the natural amphitheatre behind it. There are woods off to the right, but here on the tableland, the grass is lawn-cut for acres around and flowers are everywhere planted in great arrangements. Look back to your left, to more mown grass leading to the trees lining the bank that falls to the long beach and the Bay of Fundy.

I thought of this view when reading biologist E.O. Wilson's theory of biophilia. Wilson believes that people have an innate biological affinity for nature. Following on the work of Gordon Orians, he also suggests that all of us bear a preference for the same sort of landscape. Around the world, people's ideal habitat is one that is "perched atop a prominence, placed close to a lake, ocean, or other body of water, and surrounded by a parklike terrain. The trees they most want to see from their homes have spreading crowns, with numerous branches projecting from the trunk close to and horizontal with the ground, and furnished profusely with small and finely divided leaves."[1] This perfectly describes the headquarters area at Fundy. Perhaps that is why the view, looking south to the Bay of Fundy, seems so picture perfect, a balanced blend of land and sea, slope and flat, grass and forest.

Today, this very perfection makes the headquarters area Fundy's weak point as a national park. With its clamshell amphitheatre, its planted flowers, and its lawnscape, it is too artificial for nature-minded tourists. It seems entirely contrived, suggesting that natural nature is insufficient, and needs to be prettied up. Even the staff will tell you that the real park lies further removed, in the Acadian forests of the park's interior, or on the trails farther along the coast. Wilson notes that a simulacrum of our ideal landscape, but one devoid of life, would be "a department of hell." Imitation of life is never successful, he writes, and "Artifacts are incomparably poorer than the life they are designed to mimic."[2] Fundy National Park is hardly a department of hell, but its entrance area is so obviously unnatural that it seems inappropriate for a national park. The headquarters has become an artifact to the desires of the National Parks Branch, the New Brunswick government, and the landscape architects who crafted this setting (over the farms already crafted there) when the park was established in 1947. Fundy was designed to suggest post–Second World War cultural and economic progress for New Brunswick and for Canada. At the same time, however, the national park mandate demanded natural permanency. Since then, time has made visible the irony of park creation: the inorganic, most clearly cultural components – such as roads and the visitors' centre – are the most easily preserved, while the organic, most natural components – such as trees and wildlife – will not be kept in stasis. The natural has changed at Fundy since 1947, but the cultural has remained frozen, a monument to the late 1940s.

A SURFEIT OF SITES

Fundy National Park was born of the same regional demands and national exigencies that had led to the establishment of Cape Breton Highlands and Prince Edward Island national parks. In fact, New Brunswickers began to press for a park in the late 1920s, well before their neighbours in Nova Scotia or PEI. New Brunswick had long been a retreat for sportsmen from the United States and central Canada, so the province understood the benefit of creating what was thought at this time to be a glorified game sanctuary. The New Brunswick Fish and Game Association petitioned for a park in 1927, to protect "moose, beaver and other animals which are now threatened by extinction."[3] In 1928, the Association's president, Allen McAvity, a Saint John businessman and a former Liberal candidate for Saint John-Albert, helped form a National Park Committee for New Brunswick.[4] The committee, made up of prominent men from across the province, found itself unable to agree on a single suitable site. They narrowed their selection down to six potential sites: Mount Carleton, a hunter's paradise in north central New Brunswick; an

At the Fundy National Park swimming pool. National Archives of Canada, PA205870.
The Bay of Fundy is in the background.

area in Albert County, along the Fundy coast; Point Lepreau, near
Saint John and also along the coast; the Canaan game reserve, near
Moncton; Chiputneticook, on the American border near St Stephen;
and an area at the head of the Miramichi River. Though the committee was unable to choose a single favoured spot, they were ready to
accept unanimously the Mount Carleton or Albert County site. In
early 1929, the committee asked the National Parks Branch to investigate their suggested locations. And soon, too: "As sleighing is good
we respectfully contend this to be the best time to look over sites
quickly and thoroughly."[5]

The National Parks Branch, ambivalent about park proposals in the
Maritimes, stalled. James Harkin told his superior, "I think that it is
not desirable that the Dominion should in any way be mixed up with
this delegation." Still, he felt that the Parks Branch should prepare a
more definite policy.[6] To the New Brunswick national park committee,
Harkin wrote, "Department's attitude is that when it is notified of any
specific areas which Province is prepared to transfer to Dominion for
National Parks purpose it will arrange for me or some one else to inspect such lands as to their suitability for National Parks."[7]

As had happened at the creation of Prince Edward Island National
Park and would happen again at Terra Nova, the province moved forward by bypassing the Parks Branch and seeking satisfaction directly

from federal politicians. Members of the national park committee along with provincial minister of lands and mines C.D. Richards met with federal minister of the interior Charles Stewart, and it was agreed that someone would investigate the sites. But there was still no common understanding of what exactly was being sought. Stewart told the Saint John *Telegraph Journal* that "A large tract of timberland" would be bought by the province for a park, while Saint John Board of Trade president F.M. Sclanders came away believing that a park would go through "providing some suitable site bordering on the salt water could be secured."[8] Nevertheless, the province had succeeded in having the federal government begin the process of park creation.

Making good on Stewart's promise, the National Parks Branch sent R.W. Cautley to investigate sites in the spring of 1930, just as he would in Cape Breton in 1934. Cautley was not very happy with the assignment. The committee, he soon reported, was made up of members from all over the province, all advocating their own district. They were predominantly interested in the park to promote fish and game, and did not have any real knowledge of what a national park was. Perhaps worst of all, they offered no exact boundaries for the sites, so Cautley had to guess what portions of what lands should be considered available to be purchased by the provincial government.[9]

The surveyor concluded that the Lepreau site was by far the most suitable, with the Albert County site the runner-up. The other four sites should not be further considered.[10] It is worth noting that though Cautley did not say he had sought a coastal site, he had narrowed the field down to the only two on the coast. In all four national parks studied here, the Parks Branch inspectors chose sites on the sea, and in Fundy and Cape Breton Highlands they did so over local preferences for inland sites. Those living in the region did not necessarily define themselves in terms of the sea, and looked to their land for a place scenic enough to be a national park. But park staff assumed that in a place called the Maritimes, the sea must be a paramount feature of any national park.

Cautley thought Lepreau particularly beautiful and "truly representative of the best of New Brunswick's coastal scenery." Best of all, it had a long, scenic beach. Behind the coast, however, were four or five miles of ugly barrens that would add nothing to the park. The Albert County site surrounding Alma – the general location of what would become Fundy National Park – was in Cautley's estimation nearly as attractive, containing "a number of scenic features which are all situated on the coast." Unfortunately, as at Lepreau the scenery a mile or so back from the coast was of very little interest – just "an unrelieved density of timber on a high, rolling surface." Cautley could imagine a

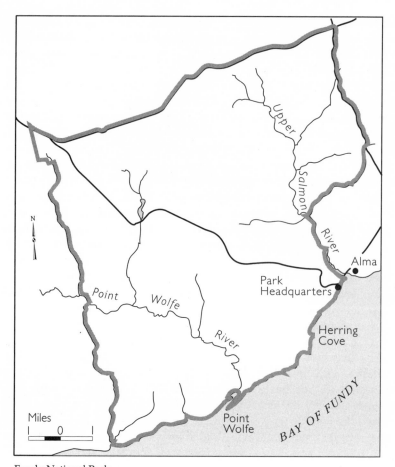

Fundy National Park.

scenic drive leading to the several waterfalls in the interior, but otherwise considered this hinterland useful only as a game preserve, perhaps so good that the area would become overstocked.[11]

Cautley made only cursory reference to the community of Alma, mentioning its single church and its 250 to 350 residents (actually, it had two churches and a slightly larger population).[12] Rather than speak to members of the community, he deduced "From the number of boats on the shore ... that the population includes a good many fishermen." There was fishing at Alma, but the region's economy centred on the timber trade. Most men in the community cut lumber and sold it to the local mills, supplementing this income with farming or fishing. Therefore, though Cautley's suggestion that Alma be excluded from the park ("at least for the present") was intended to allow the community to carry on its traditional existence, the removal of much of the area's timberland for a park would necessarily make this difficult. Cautley was quite taken with the coastline landscape west of Alma, which was to become part of Fundy Park. He spoke of the "lovely, open valley" leading down to Herring Cove, and imagined that on the "very pretty valley" between Alma and Point Wolfe a beautiful golf course could be laid out.[13] Left unsaid was that this land was attractive as well as accessible to Cautley because it had been opened up for farmland. In this area, within a mile of the coastline in such tiny communities as Herring Cove, Hastings, and Alma West, lived about forty-five families whose properties would have to be expropriated. Cautley was satisfied with the Albert County site, but his report in no way matches the enthusiasm he would feel for Cape Breton Highlands when he inspected it four years later.

Cautley saw nothing at the other sites of particular interest. Though prominent New Brunswickers, including provincial historian W.F. Ganong, favoured Mount Carleton, already a centre for hunting and fishing tourism,[14] the park inspector found it thoroughly unacceptable. It was inaccessible, it offered no spectacular views, it had been burned by a major fire in 1923, its remaining forests were too valuable to be preserved uncut, and much of it was under long-term lease to lumber companies. And most important, Cautley "did not see one single feature of which it could be advertised, 'This mountain – (lake or stream, as the case might be) – is one of the finest and most beautiful of its kind in Canada.'"[15] The other sites at Chiputneticook, Canaan River, and the Miramichi River were even less suitable.

That left the Lepreau and Albert sites. To make his evaluation as scientific as possible, Cautley proceeded to rate them out of 100 in eight categories. The Albert County site, future home of Fundy National Park, was given a 60 for scenery – "because of inferior character of coastal scenery, absence of lakes and rather monotonous type of interior" – 50 for accessibility, 70 for recreational possibilities, 50 for fish, 80 for game, 90 for geographical position, 50 for cost and practicability of purchase, and 80 for cost of development. This made a total of 530, leaving the Albert County site a clear second to Lepreau's score of 665.[16]

Harkin accepted Cautley's report and passed the recommendation of the Lepreau site on to his superiors in December of 1930. By then, however, the Conservatives were in power federally, and had little interest in park creation. New Brunswick waited for word from the Canadian government until 1933, when the province was finally told that nothing could be done until it officially proposed a location it-self.[17] New Brunswick straightaway offered the Lepreau site, "subject to the willingness of the Dominion Government to spend a large amount of relief work and to reimburse the Provincial Government the cost value of roads, bridges etc. within the site."[18] But there was another obstacle: the conservative prime minister, R.B. Bennett, was a son of Albert County, and it seemed natural that if there was to be any national park in New Brunswick, it would be there.[19] There is no evidence that the Parks Branch faced any direct pressure to establish a park in Albert County. But on his way back to Ottawa after survey-ing the Cape Breton Highlands in 1934, R.W. Cautley dropped in to re-examine the scenery around Alma and to estimate what expropria-tion would be needed. Notably, he did so at the request, not of Harkin, but of assistant deputy minister R.A. Gibson.[20]

Hopes for the acceptance of a "Bennett national park"[21] were dashed when the federal Liberals defeated the Tories in 1935. The New Brunswick election the same year also brought to power a Liberal administration, one eager for a national park at Mount Carleton. The new government immediately offered the site on the condition once more that Canada start spending money there.[22] The provincial government's enthusiasm for the site was based on a number of factors. The site was largely Crown land, it was relatively free of settlement, it had Ganong and provincial sportsmen as its supporters, and it would soon be accessible thanks to a highway be-ing built from Plaster Rock to Renous which served to bisect the province. The Liberal premier, A.A. Dysart, predicted in the 1936 Throne Speech that his government would soon announce that a na-tional park was to be established.

A week later Dysart backed down, saying, "the Federal Govern-ment might not embark on that project as soon as hoped for."[23] The National Parks Branch had again vetoed the Mount Carleton pro-posal. Cautley called it "foredoomed to failure," a huge black forest in the middle of nowhere.[24] Not only was the proposed area commer-cially valuable for timber, at 765 square miles it was larger than Kootenay, Yoho, or Glacier National Parks – too big for a such a small province.[25] The provincial opposition members jeered the govern-ment's determination to promote a site the Parks Branch had already condemned, saying, "The present Government administration did not care where the park was established provided it was established at Mount Carleton."[26]

Of course the only thing to be done was to have all possible sites re-investigated; Cautley was sent once more. There was supposed to be a "new" Albert County site for him to inspect, but it turned out to be practically the same one as in 1930 with less interior woodland in-volved. Though Cautley agreed that valuable forests should not be permanently incorporated into a park, he noted that this only served to highlight the area's tragic flaw: aside from its coastline, "there is so little Parks value in the interior of Albert County." Cautley once more listed the reasons why Mount Carleton was totally unsuitable, from its inaccessibility to the fact that "It is a bad fly country." Cautley also visited a new site, Mount Champlain along the boundary of Kings and Queens counties, and fell in love with it. He proclaimed it the best of all sites he had seen in 1930 or 1936. It was extremely scenic, he reported, and with the best view in all of New Brunswick. He en-visioned it as an Eastern equivalent of a Western park, with the bene-fit that "it will not only be possible but easy to construct a driveway almost to the very summit. Thus it will be accessible to all kinds of tourists, including invalids, whereas the trouble with most of the splendid mountains in our Parks is that they are only accessible to the small proportion of tourists who are athletic and vigorous." To the engineer Cautley, Mount Champlain's potential even made the fact that the site was inland forgivable. It was close to the Saint John River, and he had explained to his New Brunswick hosts "that the site could not be considered at all unless it included a solid block of river frontage."[27] Cautley's 1936 report concluded that accessible Mount Champlain was the best possible location for a New Brunswick park, followed by Lepreau and Albert County.

To all involved, Cautley's findings were quite frustrating. Nova Scotia and Prince Edward Island were getting their parks and yet New Brunswick, the first Maritime province to ask for one, seemed to be no closer.[28] To the provincial government, Cautley's preferences for sites in the more populous south of the province ensured that a park would be expensive and difficult to expropriate. As the minister of lands and mines, F.W. Pirie, wrote, "there would be a great deal of ill-feeling if some of the old established settlers had to be removed by expropriation proceedings." Pirie glumly stated that there was al-ready general feeling in the province "that everything goes to the south."[29] The government publicized its predicament by tabling in the legislature the mass of park-related correspondence that had been growing since 1928. At the same time, federal politicians were becom-ing increasingly impatient with provincial indecision and threatened to create a park unilaterally by Order in Council.[30] This step was not taken; the park was not to be forced on the province. But even R.W. Cautley was tired of the responsibility of selecting the park's lo-cation. Before heading to New Brunswick in 1936, he had written Harkin that "I should be glad to be relieved of sole responsibility, since the investigation and report on proposed new Park sites is a matter of the first importance, involving, as it does, the future of the Branch, the ultimate expenditure of enormous sums of public money and the fulfillment of the best purposes of the Govt.'s Parks policy."[31]

The New Brunswick government agreed. Surprised that in 1936 it was Cautley who had been sent to investigate the same sites he had investigated in 1930, advocates of Mount Carleton believed he was somehow immune to its beauty. The province asked specifically for someone else to be sent in 1937. The task was given to James Smart, chief inspector in the National Parks Branch. Smart conducted an aerial survey of the sites, and came to much the same conclusions as Cautley had. Champlain was the best site, and the other sites were unsuitable. The Lepreau barrens were just too much of a drawback, and though Smart found Albert "much more scenic," its woods back of the coastline were unusable. Smart began his dismissal of the Mount Carleton site by noting that it "is typical of the original unexploited forest country of New Brunswick"[32] – in this case, typical New Brunswickness was a bad thing. After reading Smart's report, T.A. Crerar officially asked the Dysart administration to purchase the Champlain site for a park.[33]

The New Brunswick government decided not to do anything for the moment. Lands and mines minister Pirie announced that he would be "a good loser … and we will forget about a National Park in the Mount Carleton area." But neither would any other site be selected. Pirie said he was dissuaded by Nova Scotia's recent experience in park building: "I think it cost the province of Nova Scotia something between $2,000,000 and $3,000,000 merely to acquire the park site. They figured at the start that the cost would be around half a million dollars."[34] There is in fact nothing in the Cape Breton Highlands files to suggest that the Nova Scotia government thought it would only cost half a million dollars to establish the park, or that the costs incurred made it regret park creation. Pirie's concern was more likely strategic. Peppered with proposals from all regions of New Brunswick, and discouraged by the Parks Branch from creating a park where it wished, the government chose to postpone resolving this tricky matter. In Parliament, Thomas Crerar spoke sympathetically of the province's decision, or rather indecision: "I am bound to say that I recognize the difficulties which the provincial government had to overcome in getting an area. If they were starting de novo in an area where there was no settlement it would be a comparatively easy matter. That was the situation not only in connection with the Cape Breton Highlands park site in Nova Scotia, but very largely in Prince Edward Island."[35]

The repeated failure to select a single park site for New Brunswick between 1928 and 1938 tells us something of how the politics of park creation worked. In this era the three main parties involved – the federal government, the provincial government, and the National Parks Branch – all had to be in favour of both the park in theory and its proposed location for it to become a reality. Ottawa and Fredericton politicians could unite to compel the Parks Branch to investigate sites, but it was understood that Branch approval was needed for a site to be accepted. J.B. Harkin could endorse the Lepreau site in 1930, but this meant nothing if the governments did not act on it. And in 1937 the Parks Branch and Ottawa could agree to support Mount Champlain, but it was New Brunswick's choice whether to purchase the land. The national park did not become a reality in this period, not because of an intractable position taken by any of the three parties, but because of a lack of sufficient will on all sides. For the politicians, the desire for a park in New Brunswick was never calculated to be greater than the ill feelings it might foster in other parts of the province. And for the National Parks Branch, a New Brunswick park was of uncertain value: around 1930, the Branch was still ambivalent about Eastern national parks in general, and by 1936, no one location had evoked a coherent vision of what a New Brunswick park could be, as had occurred in Nova Scotia and Prince Edward Island. It would be another ten years before politicians and public servants united with a common goal: the creation of Fundy National Park.

INTERMISSION

Just over a year after the New Brunswick government had decided to postpone selecting a park, war broke out in Europe; there would be no thought of park establishment during the next six years. Parks already in existence were called to contribute heavily to the war effort. In the West, parks opened themselves up to war-related resource extraction, and housed Japanese internment and alternative service camps.[36] In the Maritimes, Cape Breton Highlands and Prince Edward Island National Parks tried to maintain services that had just been developed, despite deep budget cuts. The Parks Branch saw its yearly grant drop steadily from 1938 to 1943, falling below $1 million for the first time since the 1920s.[37] No one in the Parks Branch, much less in the rest of the country, considered this the time to expand the park system. New Brunswick member of Parliament A.J. Brooks periodically asked minister of mines and resources Crerar if there was any word for a new park from the provincial front, but he was quick to add, "I would not be absurd enough to ask for it during the war."[38]

Crerar responded that though of course it was quite impossible to create parks at the present, he was confident that New Brunswick would get the park it deserved after the war.[39] That was good enough for everyone. In the Parks Canada files, September 1939 began a long silence on the subject of a New Brunswick park. This was broken in August 1943, when Crerar confidentially wrote his provincial counterpart asking him to think about reconstruction, and stating that a national park was a natural project for New Brunswick.[40] The province took the hint and hurriedly added plans for a national park to the report on reconstruction it was writing.[41]

Once the war was over, politicians returned to the park issue as if they had never left it. Members in the New Brunswick legislature once again praised locations within their respective ridings. As if reading the minutes of a ten-year-old argument, Conservative Fred Squires in 1946 recycled verbatim his 1937 speech criticizing the Liberal handling of the park issue.[42] And in the spring of 1947, James Smart was sent twice to New Brunswick. This time he did not inspect sites; he simply talked with the premier, John McNair, and the minister of lands and mines, R.J. Gill, about the old proposals. Though the politicians temporized, it was clear to Smart by the end of their first conversation that they favoured the Albert site. Mount Champlain may have been Cautley's and Smart's first choice, but the unspoken truth was that it sat in a riding that stubbornly continued to vote Conservative.[43] The Lepreau site was considered too close to the United States border, and so would not lead tourists through much of the province. Albert County was in a better location, it could be an excellent wildlife reserve, and it was Liberal country.[44] Though Smart considered the Albert site "too valuable land from a forestry standpoint to tie up for all time as a National Park," he rejected suggestions that it therefore be pared in size. If this was the province's choice, they would have to transfer it as is. The province accepted.[45] On 25 July 1947 the transfer of seventy-nine and a half square miles in the south of New Brunswick, surrounding the shoreside community of Alma, was made public.[46] The Albert County site had never been the National Parks Branch's first choice, but it would be home to New Brunswick's first national park. After the years of squabbling, this speedy backroom resolution seems decidedly anticlimactic. Rather than praise or condemn the new park's location, New Brunswick politicians and newspaper editors spoke of the park as merely a stepping stone to their next project: a Fundy Trail. Modelled on Cape Breton's Cabot Trail, it would attract tourists to a drive along the southern coast of the province.[47]

And now seemed the right time to ask the federal government for help in such endeavours. Having spent the past six years managing all strands of the national economy, the Canadian government now had the political might and financial resources to pump an unprecedented amount of money into peacetime national development projects. The Ministry of Mines and Resources, working on just $15.1 million in 1945, was granted $25.7 million by 1947, and $47.5 million the following year. The National Parks Branch was just one of many agencies pared during wartime that saw their budgets bloom in these free-spending days. In 1946, the Branch was budgeted over $2 million for the first time. The next year it was given $2.5 million, and in 1948 this more than tripled to $7.7 million – more than it had received during the entire war.[48] In such an atmosphere, that New Brunswick chose a new national park was more important than where and what it would be. Though the Parks Branch had spent considerable energy in the 1930s defining what a New Brunswick national park should be, and communicating this to the politicians involved, the park was nonetheless being established because of a short-term political and economic reality. This reality, more than any pre-existing vision, would shape the development of the park as well.

A WASTELAND

What little has been written about the Fundy National Park region's history has shown it to have been centred on the timber industry throughout the nineteenth century and then caught in the industry's spiralling decline for the first half of the twentieth century. The park's establishment redeemed the area: salvation for the land, deliverance for the community. Historian Gilbert Allardyce writes that creating the park was "completing a process that had begun long before. For Alma Parish was already a region of exhausted resources, shrunken population, and encroaching forests, an area that had been returning to wilderness since the closing decades of the last century." Allardyce does not explain how "encroaching forests" can be a sign of "exhausted resources" in an area dependent on forestry. Elsewhere, he writes that when the nineteenth-century timber and shipping boom was over in the area, "the settlements had nothing left to sell except their scenery. The history of the Fundy Park is therefore an easy one to summarize."[49] In much the same fashion, Mary Majka in *Fundy National Park* explains, "The devastated land had to recover from the exploitation of early history to become once again an area sought after, but for a different reason." There had been a "tragic decline," and she ponders that "No doubt the fact that by this time the area had become impoverished was also a factor" in selecting it for a park site.[50] Leslie Bella writes Fundy off in a sentence, stating, "This wasteland was sufficiently useless to be acquired cheaply for a national park."[51] Only Nancy Colpitts, in a Master's thesis on Alma forestry and in a subsequent article, questions this single-minded narrative.[52] Colpitts points out that sawmill production from 1920 to 1947 was relatively stable in the region, and there was no sign of resource exhaustion. Her work, however, centres on the relationships between government, absentee corporate landlords, sawmill owners, and private contractors in fashioning a lumber industry in the area. She is not directly interested in the state of the forests when Fundy Park was created.

This state is worth examining, however. Without knowing the condition of the region's forests at the creation of the park, we cannot understand the motivations of the National Parks Branch and the two levels of government for establishing a park there, nor can we evaluate what the park has done for (or to) the Alma region. If the woods were seen as exhausted, perhaps the Parks Branch saw an opportunity to help heal a land and a community, making land that had been discarded by foresters useful. Or perhaps the opposite was the case: the Fundy region was seen as seemingly pristine, the woods only lightly and not visibly harvested, and the park staff planned either to keep it from further cutting or to manage closely what cutting there was. My research finds that, though far from pristine, the woods of what would become Fundy National Park were plentiful and productive, and the National Parks Branch and both governments accepted the park with this in mind. From the point of view of locals, there was general acquiescence about the park not because the woods were exhausted but because they provided only a small and uncertain income. When a new industry – tourism – presented itself, the loyalty to forestry was not strong.

Since 1922, when the American firm of Hollingsworth and Whitney bought the lease to much of the timber land in the vicinity of Alma, the town had had little control of its own destiny. Locals hoped the Americans would build a pulp and paper mill, but instead they chose (as the Oxford Paper Company did in Cape Breton) to hold the land as a timber reserve. Through the 1920s, although there were local lumbermen ready and willing to work in the woods, there was little investment able to pay the stumpage fees that Hollingsworth and Whitney demanded. The population of Alma parish dropped 30 per cent in the 1920s to about five hundred, and the mills and dams that had been the life-blood of the town fell into disuse. Such conditions were common throughout New Brunswick, and in the 1930s the Crown encouraged companies like Hollingsworth and Whitney to permit more cutting, in return for the loosening of forestry regulations and the lowering of stumpage fees being paid back to the Crown. The freeing up of more forest land permitted two sets of backers to set up stationary mills near Alma. Judson Cleveland and Hartford Keirstead opened a mill at the mouth of the Upper Salmon River that ran down to Alma, and Jack Strayhorne, Fred Hickey, and a new backer, Fred Colpitts, reopened operations at Point Wolfe.

In terms of production, the years leading up to the park's creation were hardly, as Majka puts it, a period of "tragic decline." The annual cut at Alma jumped from 2.5 million board feet in the early 1930s to as much as 16 million in 1940, levelling off at 8 million in the 1940s. More telling, Alma's production as a percentage of the provincial cut climbed from under 1 per cent in the mid-1920s to as high as 5 per cent in 1937, levelling off at around 2.3 per cent in the 1940s.[53] The population of Alma stabilized, climbing almost one-third in the 1930s, to 650. The lumber industry was working at full gear, and most of the Alma region was directly or indirectly prospering from it.

Still, to those involved in forestry, the industry may have seemed in trouble. The community of Point Wolfe had dried up in the 1920s and was not restored when the Colpitts mill opened in the 1930s. More centrally, the private contractors were in an insecure position, being dependent on the patronage and prices of the sawmill operators, who were in turn dependent on the government's interventions to ensure the good graces of Hollingsworth and Whitney. Just as on the streams that ran to the rivers and the rivers that ran to the Bay of Fundy, there could be log-jams at any point along the route. For these reasons, one can see lumbering in the Fundy Park region in the 1930s and 1940s as both profitable and precarious.

Perhaps the period before park creation is remembered as a time of timber exhaustion because there was a perception that traditional conservationist practices were being abandoned. A number of people spoke to me of their parents' and grandparents' practice of selective cutting, sparing all trees of less than a fourteen-inch diameter. Winnie Smith of Alma remembers proudly that her father was an "early environmentalist."[54] People believe that the government permitted Hollingsworth and Whitney to relax this restriction; as a result, overcutting occurred. County councillor Sam McKinley, the first Alma resident to advocate the park's creation, later remembered that "the trees were so thin it was a sin to cut them."[55] On the other hand, Leo Burns of Alma claims that the traditional practices were followed under Hollingsworth and Whitney.[56] It is hard to know what to make of this meagre evidence on its own. Lumbermen may have feared that the industry was on the wane, and felt that they should use up the finite wood resources before their competitors did. In this case, increased production in the 1930s and 1940s can be seen as a sign of overcutting and coming exhaustion – just as decreased production would also be seen as a sign of exhaustion.

The Parks Branch's records can at least offer an outside opinion of the state of Alma's forests, and as a result demonstrate what the Branch believed it was accomplishing in turning this area into park land. In his 1930 report, R.W. Cautley noted that the site was "densely forested," "with an unrelieved density of timber."[57] He made no other direct reference to the health of the forests around Alma, just as he made no reference to what the creation of a park might do to residents dependent on forestry. Six years later, Cautley was surprised to find that what he had been told was a new site suggestion was in fact roughly the same coastal site with much of the inland forest excluded, since it "contained a great deal of commercially valuable timber." Cautley added, "The Forestry officials informed me that the Albert County coast is the most valuable area in the Province from a tree-reproduction or pulp point of view, and that the timber reproduces itself more quickly than anywhere else."[58] For these reasons he did not approve of any Albert County site which included much woodland, but he continued to believe that in any case a park should exclude Alma itself. In other words, he was concerned that timber would be alienated from use, but not that timber cutters would be alienated from their livelihood.

Parks Branch staffer James Smart was a forester trained at the University of New Brunswick, and he reiterated in his report of the following year that the forests north of Alma were "extremely valuable" and that "From a park standpoint the back area has no attraction except as a contiguous timber area for use as a game preserve." At Point Wolfe, he saw what he estimated to be ten million board feet ready for sawing. He was especially impressed with how resilient the timber was, reaching merchantable size in only twenty-five to thirty years. Interestingly, whereas in the Cape Breton Highlands Cautley had suggested that Pleasant Bay be incorporated into the park and Smart recommended it be excluded, here at Fundy he overturned Cautley's suggestion that the park not include Alma. Smart wrote, "I believe that this would be a disadvantage to Park administration and would also deprive this community of their main source of livelihood, the timber." The thriving timber industry worked efficiently in conjunction with small-scale agriculture and fishing, and "the combination of all the industries accounts for the prosperous appearance of the community."[59]

There is no evidence that Smart saw a different forest when he returned to Alma ten years later, when as controller of the National

Parks Branch he oversaw the creation of Fundy National Park. In fact, the woods he saw were evidently so productive that they would need supervision, and he worked with the provincial government to carry this out. He informed his superior, "I have pointed out to you that the New Brunswick area is a different situation from our other parks and in my opinion certain silvicultural systems of improvement cutting should be carried on from year to year to keep the growth in hand. The growth is very prolific and if no thinning operations are carried on the whole area will become a jungle, interfering with its general use for recreational purposes and also crowd out some forms of wildlife. Furthermore, I think in time it will be considered that sections of the area should be used for forestry experimental purposes as demonstration plots of silvicultural systems and for study."[60] Smart even discussed with the provincial government the possibility of allowing this selective cutting to be done by and for local foresters. The Parks Branch did not make a firm decision on the matter, but the provincial Department of Lands and Mines was so confident of having been granted this concession that it announced in its 1949 *Annual Report*,"Since the area is highly productive from a forestry standpoint it will be developed as a demonstration of good forestry practices as well as a wild life refuge and for recreational purposes."[61]

When the park was announced in the fall of 1947, production at both the Upper Salmon River and Point Wolfe mills immediately plummeted. The Point Wolfe mill prepared to close down, while Judson Cleveland at the Upper Salmon River mill tried to figure out how he could continue, with the timber land he had used having been alienated.[62] The new park would give jobs to the area but it was not filling an empty niche: it moved people out of an industry many of them had known all their lives. W.P. Keirstead of Alma wrote Smart that he had returned from a stint in the RCAF hoping to take up his father's business as a merchant and lumber operator. "The Park project has disturbed all this," he wrote, "cutting off local lumber resources, thus depleting lumber camp provision sales, dispersing families, many far removed from our sphere of business, so that our outlook today is altogether different than in the spring of 1945."[63]

The creation of Fundy National Park meant the permanent protection of a pocket of New Brunswick's forest, and a guaranteed tourist industry of some scale for the people who lived nearby. These may rightly be considered good things in themselves. But it misrepresents the history of the area to suggest that at establishment the land was denuded and the people were without jobs or futures. When Fundy National Park was established, forestry was still alive and well in Alma.

A HAPPY, HAPPY TIME

Winnie Smith of Alma was working in Moncton in the summer of 1947, helping her sister take care of her newborn baby. She walked into Staples' Drugstore in Moncton one day and saw the park announcement on a newspaper headline. "I was jumping up and down," she recalls. "It was a happy, happy time. Joyous."[64] She thought the park would bring prosperity to the area through tourism, and she was proud that her home had been considered beautiful enough to become New Brunswick's first national park. Not everyone to be expropriated was so happy, but there was in the establishment of Fundy a greater acceptance among locals than at either Cape Breton Highlands or Prince Edward Island national parks. Those forced to relocate were in many ways more prepared than those in the two earlier parks had been. They were much more likely to know what the new national park was all about, having heard about or visited the two existing Maritime parks. They knew that the Alma region had been discussed as a possible park site since 1930. As well, there existed in Alma a community that would take in those moving out of the park area. Beyond all this, there was a willingness to give up the forestry industry that had kept Alma, in the words of one resident I spoke with, a "dogpatch."

The New Brunswick government also gave residents less reason or opportunity for protest or complaint than had occurred in Nova Scotia and Prince Edward Island. The government settled with landowners quickly, and relatively generously. Unlike in Prince Edward Island, expropriated landowners were able to appeal their settlements. Unlike in Nova Scotia, landowners were dealt with all at once, with a firm date for their removal. There is no evidence, however, that the state's action was a result of having learned from the experiences at the first two Maritime parks. It more likely demonstrates a changing Canadian sense of how expropriation should be handled, as well as the provincial government's awareness that the park issue was so public that land acquisition had to be done straightforwardly. In any case, the expropriation proceedings to create Fundy National Park lacked the messes and controversies that surrounded the proceedings at the first two Maritime parks.[65]

The park location was made public in late July 1947, and federal and provincial Orders in Council cemented the transfer that fall. The New Brunswick government worked quickly, promising the Parks Branch that all occupants would be removed by 31 October 1948. Surveys were made, valuations determined, and the owners offered a price for their property. If this was refused, the government would invariably make a higher, final offer. If the owner still refused, the case would go to arbitration; only two did.[66] The Parks Branch meanwhile began work in the spring of 1948 to develop the park.

Nearly half of the new park area was Crown land under timber licence to Hollingsworth and Whitney, and the remainder was privately owned in about 130 properties – the largest of which was also Hollingsworth and Whitney's.[67] Not surprisingly, the New Brunswick government's first order of business was achieving a rapid settlement with the forestry company.[68] Unlike Oxford Paper Company in Cape Breton, Hollingsworth and Whitney quickly and amicably accepted a settlement: $325,000 for about twenty-one thousand acres of land it owned and another almost forty square miles it held in lease. In the eyes of the Saint John *Telegraph Journal*, this was a major concession by the company. The paper noted that the land was "one of the best lumber areas in New Brunswick," where "seven or eight million feet" of lumber and a lot of pulp were cut the previous year.[69] But Hollingsworth and Whitney was probably quite satisfied with the price. The Maine company knew that the New Brunswick government was beginning to promote the pulp industry and would in the future favour companies that had mills within the province. Also, it had owned the land for twenty-five years and never put it to much use anyway. The New Brunswick government even offered to buy the company's small parcels of land outside the park boundary, which would be of little future commercial use.[70]

Once Hollingsworth and Whitney had settled, the other property owners had to follow. There was now nowhere in the region to get timber enough to keep the private lumbermen cutting, and the mills supplied. The overnight loss of the region's major wood supply underscored how tenuous the industry had always been, how residents had never had control of their own livelihood. And as Nancy Colpitts writes, "Even if objections from Colpitts and Cleveland had blocked the park, Hollingsworth and Whitney was still free to retaliate by denying the two entrepreneurs access to the land."[71]

The two sawmill owners, Judson Cleveland and Fred Colpitts, had their own reasons for agreeing to close the mills without a fight. Cleveland was over eighty years old and seems to have accepted that the park would mean the end of his career as a lumber operator. Fred Colpitts had been the Liberal MLA for Albert County from 1930 to 1939 and was, though not a vocal proponent of the park, a loyal member of the government that was establishing one. The two men's different situations resulted in quite different settlements. K.B. Brown of the New Brunswick Department of Lands and Mines, in charge of surveying and assessing the lands to be expropriated, went out of his way to ensure a good settlement for Colpitts. He had originally assessed Colpitts's property at $5000, but Colpitts sought $7900 plus $2000 for the loss of the use of his mill. Brown went back and found ways to justify a more favourable valuation. He made a new offer of $9652, "as an illustration of an attempt to meet Mr. Colpitts' claim and to justify the amount on a basis of facts."[72] At some point not noted in the files, the assessment was raised again and Colpitts ended up receiving $16,000 for his property. Judson Cleveland, a Conservative, was less fortunate. His mill was just outside the park boundary at Alma, so the province had no obligation to compensate him for it. He received almost $6000 for his 170 acres within the park, but nothing for the loss of his business. Though he complained to the province, "The park has taken my job – made my mill worthless. ... People say I was struck harder than anyone else," he was awarded no additional money.[73]

The small property owners now had to accept expropriation. Most of the community had relied on Hollingsworth and Whitney's leased or owned woods and on Cleveland's and Colpitts's mills to make a living; with those gone, it was now the park or nothing. The province recognized its advantage,[74] and sought to satisfy the landowners speedily – in particular those forty-five or so families who would have to relocate. Most homeowners received between $4000 and $10,000 for their property, depending more on the size and quality of their land than on the value of their house and barns. Cottagers received about $2000 for their holdings. For most residents this was enough to rebuild in Alma or nearby. Not surprisingly, there is still a wide variance of opinion as to the fairness of the prices given. Bob Keirstead told me that he and his brother obtained quite different settlements on the two thirty-acre woodlots they owned. The government inspector flew over the land, and set a price of $600 for his brother's property and only $97 for Keirstead's own. "It must have been a foggy day," Keirstead said dryly.[75] Generally, though, and in contrast to the expropriations at Cape Breton Highlands and Prince Edward Island national parks, the residents seem to have received a fair market or even above market price for their property.

The provincial government accepted no responsibility for those who did not live on the park land but who depended on it for their living, however. Alma residents who worked in Hollingsworth and Whitney's woods received nothing, and those who owned woodlots which they harvested received a sum for the land, not for its value to them (as park residents did for their home and farmland). Thus while Mary Jonah of Sussex received $2022 for a lot and cottage, Wilfred McKinley of Alma received only $760 for eighty-seven forested acres.

Fred Colpitts, past MLA for Albert, was the only property owner to receive extra money for the loss of potential income.[76] While it is certainly understandable why the province sought to avoid a debated and expensive procedure of compensating for loss of livelihood, the incident demonstrates why establishment of a national park means more than just the preservation of a block of land. It shapes the land use and economy not only of the park itself but also of the surrounding area. An expropriation process which does not speak to this is incomplete.

For a time, the National Parks Branch considered allowing some occupants of the area to stay, not in recognition of their right to the land but because of their potential benefit to the park. Prior to the park's establishment, Claude Bishop kept a fishing business and small house at Herring Cove and asked to be allowed to stay and fish there. James Smart considered the matter and decided that this would be in keeping with the new park. He told K.B. Brown, "we would be prepared to let him continue his fishing operation at Herring Cove and give him a permit of occupation. … I told Mr. Bishop that I thought it would be a great advantage to the park development to have an operation as his in the park when it is established."[77] Smart was also interested in letting the half-dozen families who cottaged each summer at Herring Cove stay on, in government-issue cottages. Smart had spent part of his career at Riding Mountain and Prince Albert national parks, where summer-long visitors were permitted to put up shack tents. He believed that Fundy should permit such occupation because "those people are the best boosters for the Department in its efforts in promotion of National Parks."[78] They would give the park a small permanent clientele, and would be more inclined than ordinary tourists to safeguard the park's interests. Until 1952, the Herring Cove cottagers were allowed to summer there for $10 per year, while the Parks Branch debated whether to build them a subdivision of identical new cottages. Bishop and the cottagers were considered acceptable by Smart because they would not detract from the park's image. That is, they would be either seasonal recreationists or quaint fishers. Farmers or loggers were a different story. The Parks Branch ultimately decided to forbid all residence in Fundy, recognizing that it might be politically inexpedient to expropriate some people while letting others stay, and fearful of creating a permanent administrative concern. Neither Bishop nor the cottagers were invited back in 1953, and their buildings were demolished.[79]

The strange case of the Baizley family wonderfully demonstrates the paternalistic power the forest industry wielded in the Alma area. When park settlements were being negotiated, Judson Cleveland asked K.B. Brown for an additional $1500. He explained that he had allowed the poor Baizleys of Alma West to live on his property for twenty years, and he would like to continue to manage their interests. The New Brunswick government was uncomfortable with the idea of paying Cleveland to keep the Baizleys, rather than reimburse them itself, so it instead granted the municipality of Albert $2000 for the family's care. It is not clear whether the Baizleys saw the money; in 1951, the daughter wrote the province asking why her family was never given anything when the park came in.[80]

The great majority of the expropriated families moved to Alma and built homes there. For those living on the tableland of Alma West, where the park headquarters are today, it was a move of just a half-mile or so to the east. This undoubtedly made things easier, but it also meant less public sympathy for any signs of disaffection. Reporter Ian Sclanders, who had advocated a New Brunswick park since the early 1930s, wrote in the Saint John *Telegraph Journal* in 1948 that the golf course designer, Stanley Thompson, expected great things of Fundy's course, "but he probably wishes that the residents of Alma West, so soon to be evacuated to other parts, were a bit more understanding. The green for the second hole is right in Mrs. Jim Armstrong's backyard, where her garden used to be, and she frequently reminds him that, thanks to him, she won't have any vegetables to can this year." Armstrong complained that "We'll have to live in a little tarpaper shanty" while rebuilding in Alma.[81] (This was an unforeseen part of the expropriation: everyone who could possibly help you build a house in Alma in 1948 was already working in the park.)[82] Some of the landowners were angry at being dispossessed. Pearl Sinclair recalls one family who did not set foot on the park land again, though they would drive through on the way to Sussex.[83] Norval Martin never forgave the park its takeover of the family farm, while his brother Murice became the park staff's main authority on local history.

What most bothered residents, and what has stuck in many of their memories, were the efforts of the incoming park developers to eradicate all evidence of past human presence. Homes, barns, and outbuildings were not removed or torn down; they were simply vacated, bulldozed over, and the remains burned. The Parks Branch did not keep a single pre-existing structure for the park's use. Even a local petition to have the two community churches left standing was unsuccessful.[84] Winnie Smith, who found it a "happy, happy time" when the park announcement was made, is saddened that her family farm is now the cesspool for the park's Chignecto Campground. She notes, "They could have left signs saying, 'This was where there was a house.' It's sickening. I just feel awful about that."[85] Perhaps the most symbolic change involved Point Wolfe. The point had been named for the British general James Wolfe, and there had been a little community there from about the 1830s to the 1920s. The Parks Branch knew its name, and referred to it correctly during the years of park acquisition and development. But the official booklet that went with Fundy National Park's opening in 1950 refers throughout to "Point Wolf," and the name stuck for a number of years in park publications.[86] This may have been an accident, but more likely was an attempt by the Parks Branch to associate the park with natural rather than human history.

THE FACE-LIFT

As in Prince Edward Island and Cape Breton Highlands, at Fundy the Parks Branch did not simply remove all sign of human presence, but rather provided its own vision of what a national park should be. That vision had evolved since the 1930s. The first two Atlantic Canadian parks had been chosen to resemble classic turn-of-the-century resort settings. They had seaside features, rich summer homes turned into hotels, picturesque drives along the water, and attractions for both wealthy and middle-class tourists. Fundy had none of these. Most of it was densely wooded, it had only one noteworthy "view" (from the tableland, looking down on the Bay of Fundy), and whatever cultural heritage it had was considered not worth preserving. But the Parks Branch was not concerned. Approaching the mid-century, it was time for a new national park, one that symbolized the present and the future rather than the past.

The first two national parks in Atlantic Canada had proven successful. In 1948, Prince Edward Island National Park's attendance was the fifth highest in the park system at over eighty-four thousand, up from forty-eight thousand in 1945. Cape Breton Highlands drew twenty-six thousand visitors, thirteenth in the system, and up from nineteen thousand in 1945.[87] The two parks were more accessible to the population core of Canada and the United States than were the Western parks, and they had in no way weakened the park system's reputation or image. But while Eastern parks were more readily accepted by the Parks Branch, their scenery was still considered unremarkable by Branch standards. They were attractive and would attract tourists, and that was sufficient. The main question was not whether to have Eastern parks but rather how to market them.

This was increasingly important in the postwar tourism boom. Park attendance had doubled between 1945 and 1947, and would jump another 50 per cent by 1949.[88] To satisfy the growing class of affluent, mobile middle-class consumers, a park system which had survived on little upkeep and development for most of the decade now began spending heavily to improve facilities. And the government provided the funds to do so. As mentioned earlier, the loosening of the purse-strings began in earnest in 1948, the very year that most of Fundy was developed. Parks Branch could conceive of Fundy in grander terms than had been possible for the first two Atlantic parks. There, limited funds demanded that the Parks Branch accept the restrictions imposed by nature and culture. Cape Breton Highlands and Prince Edward Island national parks relied on existing roadways, elaborated on existing tourist views, and recycled existing cultural allusions; they came to resemble beachside resort settings because that was what they were already trying to resemble. But at Fundy, this limitation did not exist. Alma West and vicinity had never attempted to be a tourist centre, and more importantly the Parks Branch could afford for the first time to start development from scratch. Only $1.1 million had been spent at Cape Breton Highlands in the first four years of development, but at Fundy over $2.2 million was spent between 1948 and 1950.[89]

The archival record unfortunately lacks any Parks Branch member's description of what the Fundy Park development was to be, or what the hopes for the new park were.[90] But this in itself is telling. There is no sign of disagreement either within the Parks Branch or between levels of government about the park plan before development, or much indication of unhappiness about how development progressed. Once the site was selected and the properties expropriated, work went straightforwardly ahead. There is thus no better way to begin describing the development than to show the park as it existed shortly after opening. New Brunswick writer Lilian Maxwell tells of visiting it in her 1951 'Round New Brunswick Roads. She offers a rather perfunctory description of Fundy's natural scenery, but this proves only fitting: it is of secondary interest to the clearly new, precision-built environment. Maxwell writes,

Near the Administration plateau are facilities for people who want to have a good time. There is a 100-foot-long swimming tank full of warmed sea-water, bathing-houses which contain bathing-suits, 1000s of them of every description, a natural arena where band concerts or church services are held, and near at hand some 40 cottages of the Swiss chalet type, each containing all the modern conveniences even to propane gas for cooking. We took the road going to Point Wolfe several miles west, and on the way passed the golf course of nine links, tennis court, and a ball field, then the club-house. The clubhouse is a magnificent building, fitted for the enjoyment of all the sports in- cluding bowling. There are, besides the lockers, a snack bar, sports shop, a handicraft shop, etc. The main room of the club-house is 50 feet by 30 feet in extent and there is a fireplace 13 feet wide.[91]

Fundy National Park was designed to provide all possible amenities for the travelling family that was expected to dominate postwar tour- ism. The swimming pool and golf course would fulfil sporting needs; an amphitheatre and handicraft school would fulfil cultural needs; cottages, hotels, and campgrounds would fulfil accommodation needs; and a "townsite" of gift shops and restaurants would fulfil consumer needs.

While most of these amenities were present at other national parks, what makes Fundy unique is that the developments were to be the fo- cus of the visitor's experience. At Cape Breton Highlands and Prince Edward Island, development was meant to spread tourists around and help them see more of the park's natural area. Because staff saw Fundy as having only one marketable view, development here would need to have a different purpose: to keep tourists busy in this one spot. The Parks Branch had the money, the motivation, and the op- portunity (once the people of Alma West were gone) to develop this national park to an unprecedented degree. The editor of the Saint John *Telegraph Journal* complimented the park designers for their in- novation: "And New Brunswick's site – from the description of its numerous features – promises to be distinctive among Canada's na- tional parks. Not just an attraction to be looked at for a few minutes, like a curiosity of Nature, Fundy National Park will be a place for vis- itors to stay in and spend as long as they wish."[92]

The park headquarters area was developed at the expense of both its cultural past and its environmental present. Once the residents had been moved and their properties bulldozed, a preliminary land- scaping was needed. The Parks Branch contracted this and much of the construction work to Caldwell and Ross of Campbellton, New Brunswick. The new Fundy National Park, populated for almost 150 years, was too wild to be currently acceptable. Roads needed to be straightened, hillocks flattened, ugly and misshapen trees cut down, stones removed, grass planted. It was a big project that in- volved the use of over fifteen thousand cubic metres of topsoil[93] (enough to cover a twelve-acre field a foot deep).

The provincial media reported on the tidying operation in a num- ber of articles, returning again and again to metaphors suggesting beautification. "It requires men and machines to put the finishing touches" on the landscape, one writer suggested, and it would be a long-term job: "It will take years to completely landscape and trim the Albert county countryside."[94] This was to be a "tremendous face-lifting operation," "an improving facial," "an estimated $3,000,000 face-lifting," wrote others.[95] It was "the effect which may be obtained by polishing an already gem-like setting. Improvement touches have been largely in the nature of removing man-made blotches in a beauty-blessed area."[96] No reporter saw this landscaping as out of the ordinary. One even congratulated the Parks Branch for retaining existing scenery: "the camping ground is taking shape in what was once a jungle of cut-over trees, stumps, and stones. Here, too, the accent on natural setting is evident. As many trees are being retained as possible without interfering with camping ground plans. ... Rustic picnic tables and benches, hewn from trees removed in the park make-over job, abound."[97]

The Parks Branch fully supported the scale of this landscaping. Ian Sclanders noted that the person in charge, Superintendent Ernest Saunders, was relaxed about the work, "as though moving scenery around were mere routine. 'We're cutting the top of that hill off,' he said, 'and we'll use it to fill in that bog down there. We did a little job on that stream yesterday – shifted its course a couple of hundred yards with a bulldozer.' " It was all part of "turning rough wilderness into a gigantic playground."[98] James Smart was a little more apologetic about the changes being made, but he too saw them as necessary and the damage as temporary. In a 1950 Christmas card to New Brunswick historian Esther Clark Wright, he wrote, "At the present time our development area looks a little raw in spots due to construction work. ... To some people it may appear that we are not keeping the area inviolate as is our general aim in connection with the administration of National Parks but it has been my experience that nature is a great healer and a lot of the rough spots will be cured and in a few years the whole layout will again take on a more natural appearance."[99] Nor was Smart above helping the great healer along. After a landslide in the administration area, he had park wardens plant trees along the scarred incline – in irregular clumps, the better to imitate nature.[100]

On top of the topsoil, Fundy National Park was built. Stanley Thompson's company designed and set up a nine-hole golf course along the road on the way toward Point Wolfe, just as Cautley had suggested eighteen years earlier. Golf was still considered de rigueur for national parks, and $100,000 was spent on the new course.[101] A swimming pool was built almost right on the beach at Alma West. It was an engineering showcase, with water directed from Spring Brook Hill through a pipe system overhead on the road to Point Wolfe down to the beach where it was warmed. The pool was meant to ensure swimming regardless of the weather, and soon after its opening a large screen was put up to keep the cold Bay of Fundy winds out; of course, it also blocked the view of the Bay.

Potentially the most innovative development was the opening of the New Brunswick School of Arts and Crafts in the park headquarters area. Headed by Dr Ivan Crowell, director of handicrafts for the province, the new school would teach single-day to eight-week classes in weaving, wood turning, candle making, and the like. "Courses in weaving," it was noted, "are taught in English, French, Estonian, Norwegian, and Swedish."[102] Really! (Ausõna!) James Smart hoped the school would give the new park cultural cachet: "we would like to see this activity established on a scale similar to the Banff School of Fine Arts in Banff Park."[103] This once more shows Smart – who had toyed with the idea of keeping the Herring Cove cottages as a townsite – trying to create community within Fundy Park in the image of a Western park with which he was familiar. Smart also established in Fundy a business subdivision which he occasionally called a "townsite": seven building lots where a service station, restaurants, and gift shops were planned to serve the national park's needs.

There was no longer an attempt to segregate tourists by class, as there had been at Prince Edward Island National Park a decade earlier. Fundy would keep everyone in its headquarters area. As a result all park components had to be first-class. Campgrounds were developed for tents and trailers, but there were also twenty-nine cottages built on a row overlooking the golf course. These were fully furnished two-room chalets with modern gas stoves and refrigerators. The Parks Branch expected that the new park would encourage local initiative into tourist-related businesses. They offered business lots and long-term leases to potential investors. But the Branch was disappointed to find no local interest. Of course, the park's own development work made investment in the park unlikely. People in Alma simply could not afford to invest to the high standards that the Parks Branch had set, particularly in what was considered a risky and unfamiliar venture. At a time when some in the village still did not have refrigerators, it was difficult to imagine setting up seasonal accommodations with refrigerators in every unit. Yet Alma citizens were also wary about developing at their own expense in town; they felt they would not be able to compete with the prime facilities and landscape of the park headquarters area.[104]

Also, the residents of Alma, like the New Brunswick government, were quite happy to have the feds foot the entire bill for park development. The short-term employment the park brought overshadowed thoughts about long-term economic effects. Particularly in the peak years of development between 1948 and 1950, everyone in Alma seemed involved, from construction workers to old lumber cooks to groundskeepers. It was steady pay for labourers, at 60 cents per hour if hired by the park, 65 if hired by the contractors. Winter projects also kept over a hundred men working. Smart could announce proudly early in the 1950s that "Almost every able bodied person in Alma available for work has at some time or other during the past three years been employed on the park work."[105] And to the delight of the New Brunswick government, it was entirely a federal expense. This had been an issue whenever the national park had been discussed. As Russell Colpitts bluntly told the legislature in 1946, "It has been said that each county should have its own Provincial park. I have no fault to find with this, but one thing we must keep in mind is that these small parks would have to be developed by monies from the Provincial Treasury, whereas all the expenditures in connection with the development of a national park are paid by the Federal Government."[106] This was still the thinking. Though the province spent over $850,000 for expropriation of the park land, it could demonstrate its economic savvy by pointing out that it had convinced the Canadian government to spend several times that in development.[107]

Only in one memorable case did the Branch's willingness to spend money (and everyone's knowledge of that fact) result in embarrassment to the park system. The superintendent's residence was traditionally seen by the park service as an important building: evidence of permanent watchfulness over the park, and one of very few perks for men who took the job.[108] At Prince Edward Island and Cape Breton Highlands, existing homes had been fixed up to accommodate superintendents. But at Fundy a new home was to be built, and on the most prominent location imaginable, on the bluff at Alma West. It actually cut off some of the view of the Bay. The first designs from Saint John architect H.S. Brenan for what was authorized to be a $12,000 house were, in Smart's eyes, "absolutely unacceptable. ... an old-fashioned style which one would see on almost any street in Saint John."[109] Just as the Parks Branch wanted the scenery in Eastern national parks to be somehow reminiscent of Western landscapes, it expected the buildings to be Western-style structures. Brenan was given designs of existing structures in Canadian and American parks as inspiration, but he slavishly followed them and the cost jumped to $24,000. In charge of construction on the ground was Superintendent Saunders, who would benefit from whatever was built. The Treasury Board soon noticed the cost overruns, demanding to know why there had been so much winter work, why an intricate cobweb brickwork design was needed, why the kitchen had so many cabinets. And still another $4700 was granted in early 1949 to complete the project. By now the Fundy superintendent's residence was an embarrassment to the whole Parks Branch, the story of its construction shared throughout the system as a warning about overspending.[110] Assistant controller J.A. Wood visited the park in the summer of 1949 and reported, "When I arrived I found a number of men laying a flagstone driveway. When I asked Saunders where the money was coming from to pay for this work he told me it was being charged to landscaping. I stopped the work immediately."[111] By the time the house was completed it cost $30,000, the price of three or four farms in Alma West in 1948.

Perhaps the overruns were worth it: fifty years later, the superintendent's residence looks to be in fine shape. Other park developments have not fared so well. The handicraft school is gone, the business subdivision never worked out, the golf course is considered anachronistic by park staff, and in general all development on Alma West seems an unnatural legacy. Yet older residents and expropriatees all remember Ernest Saunders, who oversaw development, as the superintendent who did the most for the area. He didn't let the politicians or anyone else push him around, they say; he was autocratic but fair (that also translated into "He wouldn't last five minutes today, the unions would crucify him").[112] Of course, he also gave people jobs, making locals' hopes for the new park as a place of employment a temporary reality. As Leo Burns told me, "He accomplished more in four or five years than the rest put together."[113] This makes the relative extravagance of the superintendent's residence seem somehow fitting. To the people of Alma, the park superintendent stepped into the shoes left by Colpitts and Cleveland, as a patron who could distribute or refuse community employment. From Alma, the view of the superintendent's residence perched on the bluff above suggested that the park would oversee the community's future. From the road to Point Wolfe, the superintendent's residence blocked the view of the Bay of Fundy, development getting in the way of nature.

6 Sawed-off, Hammered-down, Chopped-up: Establishing Terra Nova National Park, ca. 1957

Created just ten years after Fundy, Newfoundland's Terra Nova National Park seemed to signal a rapidly evolving National Parks Branch. It was a new park for a new province and a new sort of Atlantic park. Park documents show it to have been chosen more for its typical Newfoundlandscape than in honour of any more formal aesthetic standard. Its blending of land and sea was considered fine in itself, and its bogs, its miles of spruce and fir forests, its inaccessible islands and dark shorelines, and its sometimes dismal views never threatened its creation. Plans were in place to do a better job accommodating those who had owned or depended on park land, and maintaining good relations with those who lived near the park. The new park was not to be heavily developed: there was to be no tennis court, no heated pool, and – the sharpest break with tradition – probably not even a golf course. Terra Nova National Park was established without the cultural associations of Cape Breton Highlands, the obvious scenic attractions of Prince Edward Island, or the sheer developmental imperative of Fundy.

Cabin design is as sure a sign of changing park philosophy as any. The cabins at the Newman Sound headquarters area are strikingly different from cabins at the other parks studied here. They are not on flat ground, nor are they in a regimented, suburban configuration, facing one another as if to find community in nature. Instead, the Terra Nova cottages are at odd angles, built into a hill and almost hidden from one another amid the trees. Each offers a view of Newman Sound and the Bread Cove Hills in the distance. A visitor looking out a front window does not see recreation, progress, or development, only nature.

Terra Nova National Park symbolized a burgeoning modern environmentalism and an increasingly "nature-minded" national park system, as will be seen in a later chapter. But this is only half of its story. When the park was established, the federal and provincial governments agreed, and the National Parks Branch accepted, that the new park would be used to supply lumber if Newfoundland could attract investors for a third pulp-and-paper mill. The yearly growth in the park, estimated at twelve to fifteen thousand cords, would be available for cutting. This promise hung over the new Terra Nova National Park like a chainsaw of Damocles. The history of Terra Nova's establishment and design illustrates that the Parks Branch of the 1950s increasingly defined its work in terms of nature preservation, while it experienced decreasing autonomy owing to political interference and ministerial control. This separation of interests foreshadowed the showdown over the park system that would occur between the proponents of use and preservation in the 1960s. For now, the Parks Branch was able to bend awkwardly to accommodate both interests; in the next decade, a firm declaration of policy would be needed.

SAWED-OFF

There was talk of Newfoundland having a national park even before it was part of the nation. The colony's Department of Mines and Resources hired American sportsman Lee Wulff throughout the 1940s to help develop sport tourism on the island, and Wulff encouraged the creation of what he called a national park.[1] When the Newfoundland National Convention met in 1946 to discuss the possibility of entry into Canada, its Transportation and Communications Committee also heard recommendations for a park from their publicity consultant (they were advised to talk to Robert Moses of the New York State Park Commission for assistance).[2] At the time, though, such planning seemed the most wishful of thinking. Tourism was practically nonexistent on the island. There were very few accommodations and only short, bad stretches of roadway, none of which crossed the colony. The Tourist Board resigned in protest in 1946, saying it could not work without government assistance. When the Board formed again in the 1950s, it took the rather extraordinary (and self-abnegating) step of advising tourists away from Newfoundland, saying that the new province was just not ready for them.[3]

Newfoundland's entrance into Confederation made the creation of a national park much more feasible. Canada already had a strong national park system and a Parks Branch able to find, develop, and

Newman Sound. Photo taken by James Smart during 1950 inspection of potential
Newfoundland national park sites. National Archives of Canada, c146803.

maintain a park site. The federal government also had money to
fund the project. When the Newfoundland delegation visited Ottawa in 1947 to discuss the terms of union, they included among
their questions, "Would Newfoundland become entitled to a national park? Does the Department of Mines and Resources establish
as well as maintain National Parks?" and, perhaps most important,
"Approximately how much does the federal government spend in
setting up a national park?"[4] Happy with the answers, the delegation asked for a "definite statement" in regard to a park in the report
on the working basis for union.[5] The federal government assured
Newfoundland informally that after Confederation a national park
would be forthcoming.

Talks between St John's and Ottawa continued after Newfoundland joined Canada in 1949. In December of that year, the province's
minister of natural resources, Edward Russell, asked that a number of
potential sites be inspected. The province particularly favoured a
thirty-five-square-mile section in the Serpentine Lake area near
Corner Brook on the west coast. It was already a well-known salmon
fishing area with beautiful scenery, and the province hoped that as a
national park it would attract more hunters and fishermen from away
(whether the province understood that there would be no hunting in
the park itself was unclear).

It may be seen as evidence both of the perceived importance of satisfying Newfoundland after Confederation and of the National Parks

Branch's unfamiliarity with the newest province that Controller James Smart himself inspected the proposed sites. After a visit to the province in May 1950, he reported on his findings. He began by outlining the factors he had considered, primarily following the realtors' mantra of location, location, location. Any Newfoundland park should be accessible to Newfoundlanders, it should be accessible to off-province tourists visiting by air or travelling along the soon-to-be-completed Trans-Canada Highway, and it should disturb as little settlement as possible. As for its appearance, Smart simply stated, "it should be an area typical of the province, embracing sea-coast country, the habitat of indigenous wildlife, forest and fisheries, and with scenic values."[6]

Smart found most of the proposals unacceptable for one reason or another. Serpentine Lake was not on the sea, was too small to sustain wildlife, and had too much privately owned land. Salmonier was not on the coast, and Placentia was too settled and lacking in wildlife. The Upper Burin peninsula was too remote. Only the suggested site on Bonavista Bay, 250 square miles enclosing all of Newman Sound and running back beyond the Terra Nova River to the west, seemed acceptable. Though portions of it had been burned by fire in 1950, most of it was wooded and healthy. It was dotted with lakes and rivers and would be good moose and caribou habitat, if they were reintroduced to the area. The park site was on the proposed Trans-Canada route and neither too far from St John's (which would make the park inaccessible) nor too close (which would focus Newfoundland tourism in too small an area). The head of Newman Sound would make a fine headquarters area, being well wooded, close to water, and sheltered. And the scenery? "The scenery from the point of view of spectacular and rugged terrain is not so pronounced as other areas, but the combination of land and sheltered waters and many islands shoreward is most attractive and what one would consider typical of Newfoundland. Probably there are more small boats in use in this section of the coast than in any other part of Newfoundland."[7] Smart chose to see the prominence of boats as evidence of the area's Newfoundlandness, rather than evidence that local inhabitants used and depended on the land. In any case, Smart recommended a number of "special conditions" on behalf of those people who would be dispossessed by a park. Fishermen should be allowed to maintain camps on the park seashore. Timber and wood permits should be available to locals. The Terra Nova River should remain open to log runs that begin outside the park area, and the river should be considered for possible hydroelectric development to serve both park and community.

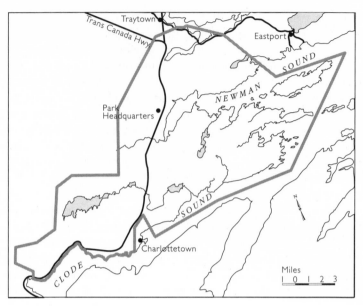

Terra Nova National Park.

These are much greater concessions to past (and future) land use than any that Smart had been willing to make at Fundy National Park. It is not clear what made the Newfoundland case different. Smart likely knew that all Newfoundlanders traditionally enjoyed liberal access rights to land, such as the right to come ashore anywhere and take wood as needed, and that anyone could cut wood up to three miles in from the shore on Crown lands. Perhaps he felt in retrospect that the compensation paid to landowners at Fundy had been insufficient. In any case, Smart concluded, "The above conditions would not be a detriment to the National Park but in most cases a benefit to Park administration through the cultivation of friendly and appreciative co-operation of the neighbouring settlements."[8]

Smart had noted that the park should be set aside even "If it is felt that the time is not opportune due to the international situation to allot funds for the initial development."[9] Just as Fundy National Park had been delayed by the Second World War, the Newfoundland park was delayed in the early 1950s by the Korean War and the arms build-up of the cold war. As more and more money was poured into defence – in 1952, it made up 40 per cent of the federal budget – appropriations for other departments suffered. The Parks Branch, which was granted $9 million in 1950, was allotted only about $6.5 million each year until 1955. With increasing pressure on the park system, owing to a near-doubling of visitors during the same years, there were only enough funds to maintain existing parks and no opportunity for expansion.[10]

Nor was Newfoundland pressing for a park.[11] Premier Joey Smallwood, in his tenure from 1949 to 1972, was fixated on the idea of Newfoundland's economic development, primarily by attracting foreign manufacturers and developers with provincial investment. His administration foolishly sank money time and again into foreign companies, only to have them go under or never even open in Newfoundland. An entire chapter of Smallwood's memoirs, *I Chose Canada*, deals with twenty-three failed business ventures – 1 for each year in office – and concludes with a proud declaration of ineptitude: "But these 23 aren't even 10% of the total; they are only the most serious ones."[12] Smallwood was too mesmerized by the thought of developing every bit of the province's hydro potential and forest resources to favour a national park which would lock up a portion of the woods and water forever.

Yet two factors pulled the Branch's interests closer to Smallwood's. First, the federal Department of Mines and Resources was retooled – into Resources and Development in 1950, and then Northern Affairs and National Resources in 1953. This second shuffle by Prime Minister Louis St Laurent prefigured an increased developmental ethic within the ministry under the new leadership team of the minister, Jean Lesage, and the deputy minister, Gordon Robertson. Since the National Parks Branch was part of this department, it could expect to be caught up in the development drive. Second, Jack Pickersgill was elected MP for Bonavista-Twillingate in the summer of 1953, giving Newfoundland a bridge to the world of federal bureaucracy. Pickersgill had been the most influential civil servant in the country in the Prime Minister's Offices of King and St Laurent and later as clerk of the Privy Council; he had even entered Cabinet as secretary of state in 1953. This last move was clear proof to the Conservative opposition that the supposedly non-partisan civil service was a Liberal stronghold. Pickersgill needed to be elected if attacks on his integrity were to end in Parliament. When Smallwood asked the Manitoba-born Pickersgill to become "the Newfoundland minister in Ottawa," he jumped at the chance. Since the proposed park was not only in the federal district where Pickersgill was elected, but also part of Smallwood's provincial district, chances of its establishment improved considerably.[13]

Joey Smallwood was wary of anything that stood in the way of development, but he knew that with a national park came federal money.

The Trans-Canada Highway was being constructed at the time, paid for on a fifty-fifty federal-provincial basis. As Pickersgill pointed out to the premier,[14] if the highway were to go through a national park, it would become a completely federal responsibility, and save the province twenty to thirty miles of highway costs. Smallwood asked St Laurent if the Canadian government would agree to build the road before the park itself was established, and after some debate this was agreed to in principle – though it admittedly contravened the Trans-Canada Highway Act. The feds would begin road-building as soon as the park location was definite.[15] Smallwood was so pleased that in a rush of good feeling he offered to extend the proposed park to the west and the north, making it 425 square miles in all. The Parks Branch declined the offer, thinking this extension unhelpful.[16]

It would seem that the provincial government was now ready to donate land for a park, but the idea of losing valuable timber land still made it hold back. Smallwood's greatest dream was of a "third mill," a pulp and paper mill to make use of more of the province's forest resources and to bring in more tax revenue. The two existing mills owned by Bowater and the Anglo-Newfoundland Development Company were privileged with the most agreeable long-term leases imaginable: their logging gains were not subject to tax, and they paid only about 5 cents stumpage per cord, as compared to the $3.50 – $4.25 rate in other Eastern provinces.[17] Smallwood vowed that Newfoundland would get better terms in any agreement next time, and he travelled the globe to find a company to start up a new mill. The problem was that no one thought it could make them money – which would have led a less driven premier to suspect that it was a bad idea. But as Smallwood later recalled, "I must confess that as the years passed, my determination to build a third and a fourth mill increased as one concern after another failed to come through with a firm commitment to build."[18] The premier's obsession meant that every tree in the province was seen as a potential offering to the third mill, and every tree preserved in a national park a threat to the province's economic future.

As if to remember why Newfoundland would ever want a national park, in the fall of 1954 the province's deputy minister of mines and resources, P.J. Murray, wrote his federal counterpart, Gordon Robertson, asking him to recap the tentative park negotiations of the past five years. Why, again, was the Newman Sound site chosen? And what would become of park resources? Murray flatly stated that the Terra Nova River would have to be left available for hydro power, and if not it would have to be excluded from the park. He also asked whether "a forest management policy for the park woodlands based on the maximum sustained yield principle" could not be practised in a park, to "lessen to some degree the disadvantages which we would suffer in passing over this particular area as a national park."[19] In other words, could the park not permit forestry which would harvest mature timber at the same rate as natural regeneration? Robertson asked the advice of J.A. Hutchison, director of the Parks Branch, on these matters. Using the National Parks Act as his guide, Hutchison responded that hydro development could not be permitted in a national park, and if this meant the exclusion of the Terra Nova River, so be it; perhaps the province could be convinced to offer an alternative river system in the area. As for forestry, it "cannot be considered" in a national park.[20]

Robertson had not asked just the parks director to comment on Murray's letter. He had also solicited the advice of Jack Pickersgill (who, not surprisingly, favoured development in Newfoundland over park inviolability) and that of the Canadian Forestry Service. This latter move in particular demonstrates that the Parks Branch's theoretically autonomous management of parks was not in fact practised. Because Northern Affairs and National Resources contained both the parks and the forestry agency, the ministry could rely on the advice of both and then make its own policy decisions. Forestry director D.A. MacDonald of course saw the matter as a forestry rather than park issue, and argued that national park policies did not necessarily preclude sustained yield if cutting over-mature timber could be proved beneficial to the forest and the park in general.[21]

Robertson's reply to Murray's letter offered a precedent-setting redefinition of national park management. "Under the [National Parks] Act and Regulations," he wrote, "it would be possible to carry out a fairly extensive program of cutting in the interest of good forest management and protection." While Robertson clarified that this would depend on the forest conditions, that the trees would have to be selected and marked by park staff, and that cutting would be closely controlled to maintain the forest in a "substantially natural state," it nevertheless was a major concession.[22] It blithely offered a redefinition of park policy that could conceivably result in extensive forestry within a park, and did so in a way that suggested that such forestry would benefit some parks. As well, the way this decision had been reached signalled a willingness to remove park policy-making from the Parks Branch and make it the responsibility of all branches of the ministry, with the minister and the deputy minister the ultimate arbiters.

Still, the province balked at approving a national park. The government waited for word from the Royal Commission on Forestry headed by Howard Kennedy, which had been called to explore the feasibility of a third mill. The Kennedy Report was released in the spring of 1955, recommending that a third mill (preferably one co-owned by the two present mill-owners, Bowater and the Anglo-Newfoundland Development Corporation) be built on the eastern end of Newfoundland. The report stressed that practically all of Newfoundland's forests would have to be available to supply such a mill. At the end of their report, the commissioners affirmed that in principle they were agreeable to a national park, but reminded the government that alienation of any land from forest or even hydro use could affect the third mill's chances.[23]

What this meant to the national park plans was unclear. Pickersgill believed it irrelevant, since to his mind the feds had already agreed to allow timber cutting. The *Western Star* reported Pickersgill as saying, "I am anxious to stick the treasury here for the cost of that road all the way through the park, so naturally I am impatient at the delay." The newspaper added, "More recently the Newfoundland government has balked on the question of timber. There is no sound reason for the delay on the latter point, Mr. Pickersgill said, since the timber within the park would be open to commercial use, and would be managed by the Federal government at least as ably as the Provincial government could expect to manage it for itself."[24] But this was not quite what Canada had offered: no one had promised that commercial lumbermen could do their own cutting. Having read a number of misleading press reports, Jean Lesage felt obliged to send Pickersgill a copy of Robertson's 1954 promise to Murray, underlining the important passages: "I should point out that any cutting under such a plan would have to be in accordance with the best forestry practices and *would be done under the immediate control of the officers of the Park. It would be a cutting of trees selected and marked by Park officers.*"[25] The Newfoundland government understood this. When asked at a provincial tourism conference (a softball question from Pickersgill's friend in Gander, Edgar Baird) why the provincial administration had not yet accepted Ottawa's offer of selective cutting in the park, the provincial minister of mines and resources, Frederick Rowe, stated that they were concerned that the Parks Branch might find this selective cutting expensive and pass on the high cost to the mill, making the park forests in practice unavailable.[26] The Kennedy Report validated the provincial government's belief that a national park would be a danger to development. Fearful of losing use of the Terra Nova River and its surrounding forests, the province announced it was excluding this drainage basin from any further discussion on the park.[27]

Having failed to negotiate effectively at long distance, the federal government sent Gordon Scott, chief engineer of the Parks Branch, to St John's to keep talks moving. Scott's involvement at this point was invaluable from the Parks Branch perspective. It had been locked out until now, forced to respond to concessions and compromises made at the departmental level. Scott could ensure that the Branch remained a primary force in park decision-making. And he could turn the focus of the discussion away from deal-making and toward explaining to the Newfoundland government why, when it came to resource preservation, the parks had to be unyielding.

In a letter back to Ottawa and a subsequent report, Scott told of the "confusing and involved" factors that were slowing down talks: the competing interests in hydro power, a third mill, and the Trans-Canada Highway, plus the fact that the proposed park was in the constituencies of both Pickersgill and Smallwood. Above all, Scott explained that when Smallwood and company thought of the Newman Sound area, they thought of the estimated 330,000 cords of merchantable timber. Scott wrote, "If I heard the phrase 'the economic development program' spoken once I have heard it a dozen times since coming here. They want the National Park, please don't misunderstand me, but the conclusions of the Kennedy report & the idea of a 3rd mill is the magic words to these people – much more so than a National Park." To the province, there was no reason why a park and forestry could not coexist. Scott disagreed. He told the Newfoundland government that though the federal ministry had promised some form of experimental forestry, the Parks Branch would not approve an out-and-out pulp operation. Scott was "most emphatic we would not tolerate any idea of controlled tree-cutting for pulp purposes in any Park. The Director of Forestry thought this was a dog-in-the-manger attitude" but Scott rejoined that they "want their cake and eat it too."[28] It is clear that relations between Scott and P.J. Murray, provincial deputy minister of mines and resources and a fierce opponent of the park, were very strained. Scott stated that Murray's attitude "seriously jeopardized" the chance that a park would be created.[29]

Scott still tried to make the most of his Newfoundland visit, however. Against his superiors' instructions, he scouted out other sites and even drew up an alternative proposal that traded the head of Newman Sound and the Terra Nova River watershed for the Northwest Brook watershed to the west and south of the original boundaries. This extension, stretching down twenty miles to the woods behind Clarenville, would give the park salmon streams and caribou habitat missing from the area currently under discussion.[30] Scott pleaded with Ottawa not to accept an "emasculated," second-rate park: "If we were ever foolish or short-sighted enough to do so then the recrimination which would definitely be directed at us by both the Province and the public at large, would certainly be justified."[31]

Chief engineer Scott's report buttressed the Parks Branch's opposition to politically directed planning. Parks director J.A. Hutchison, quiet throughout early negotiations, wrote his deputy minister with full support for Scott's findings and warned that "unattractive, indeed unacceptable, features are present either in the actual lands offered or in the reservation that the Province wishes to couple with the offer."[32] But if the Branch position was now clear, it was also clearly antagonistic to provincial wishes. This left the Branch little opportunity to negotiate further, and allowed the federal ministry to act as mediator between the stubbornly idealistic Parks Branch and a development-minded provincial government.

The Trans-Canada Highway kept the Newfoundland government interested in the national park when nothing else did. By the fall of 1955, all of Newfoundland from Port aux Basques in the west to St John's in the east had been connected by roadway, except for a twenty-two-mile stretch from Alexander Bay to Bunyan's Cove – the proposed national park area. The province had purposely left this, waiting for it to become a federal responsibility once the area was made a park. Smallwood did not want to wait any longer. He wrote St Laurent, going over the many points of contention that were slowing the park establishment and asking that in the mean time Canada reimburse the province for a highway to be built in what would soon be park land.[33] St Laurent agreed, on two conditions: one, that the Parks Branch be allowed to ensure that the road fitted its requirements; and, two, that the park be agreed to within the year. Smallwood accepted these conditions, but recognized this was a gamble. On the one hand, it would save the province a considerable amount of money during a very trying time in the provincial economy. On the other hand, the road committed him to the park, and Smallwood believed that the park would jeopardize the mill. And as he told the legislature, "we regard the third paper mill as incomparably more important than a National Park."[34]

Smallwood felt certain of this when the Jenkins Report, a provincial government- and industry-sponsored assessment of the park's forestry potential, was presented in late 1955. Forester F.T. Jenkins had found that 60 per cent of the proposed park was forested, containing about 384,000 cords of pulpwood. Though locals had cut from it extensively for decades, Jenkins concluded that it could be logged again if cutting was postponed for twenty-five years or so.[35] This clinched things for Smallwood: the wood would be there, the wait did not bother him, so the Newman Sound area could not be a park. The premier wrote Pickersgill, "These facts certainly appear to rule out the possibility of our turning over this area for the purpose of the

National Park, unless something can be done to assure the timber to the third paper mill. ... I think that any other course would be criminal." He left the problem for Pickersgill's "fertile brain."[36] Smallwood's reaction to the Jenkins Report demonstrates a notable longsightedness about the park land that the Parks Branch itself rarely exhibited. Whereas the Parks Branch saw its job as maintenance of a park's existing nature, one day at a time, Smallwood imagined the area as it would exist twenty-five years down the road and planned accordingly. Of course, Smallwood's vision demanded that the area be preserved today, held in trust until its usefulness would be made manifest after a period of natural restoration.

With the federal and provincial governments no longer negotiating for a park, the issue seemed dead. It took Jack Pickersgill's fertile brain to keep it alive. The Ottawan-turned-Newfoundlander still hoped to accommodate the park system's philosophy while satisfying Newfoundland's needs. After failing to interest the Forestry Service in managing the park as a kind of experimental forestry station, Pickersgill met success after bumping into deputy minister R.G. Robertson in an Ottawa hallway. Pickersgill said that he was sure Newfoundland would be won over if Canada confirmed Robertson's 1954 concession to cut in the park, plus extended this to agree that if a third mill was built the province could buy mature and over-mature timber up to the yearly growth of the park's forests.[37] Robertson took these conditions to his minister, Jean Lesage, who tentatively accepted them.[38] On 13 November 1956, Smallwood also agreed to these conditions and the park was all but created.[39]

What made the Department of Northern Affairs and National Resources accede to such heavy cutting in the park, which even Lesage said would "go further than anything we have done anywhere else"?[40] The main reason was that the ministry wanted the deal made, wanted Newfoundland to have the national park it had long been promised, and this seemed the only way to satisfy Smallwood's government. But the agreement also reflected dissatisfaction with present forestry policy in national parks. In a memo to his assistant, Robertson wrote,

As you know, there have been some growing doubts in the Minister's mind, as well as in our own, as to whether our present policy with respect to forest management in the parks is entirely satisfactory. There have been allegations that the spruce budworm is getting an increased hold in Cape Breton Highlands Park and there have also been suggestions of disease in Yoho Park. While the objective may be to retain the parks in their natural state, I do not think this means that we have to refrain from taking measures that will maintain the forests in as healthy a condition as possible. There is no question but that we will be severely blamed if the "natural state" in Cape Breton Highlands, say, after a few years, is one of dead forests.[41]

This is to some degree a rationalization: there is no indication that forest policy for Terra Nova would have been any different from that at other parks if Newfoundland had not demanded it. Nevertheless, it suggests that the department was dubious about a belief in preservation that called for a completely hands-off approach.

The Parks Branch had no opportunity to defend its policy, or to voice its reaction to the policy being thrust upon it. In fact, it was not even the first group consulted. The Forestry Service's D.J. Learmouth was called in to estimate the proposed park area's annual growth and decide how it could be cut in a way that would not detract from the scenery – in fact, so that tourists would not be aware of it. Learmouth found that the yearly growth was an impressive twelve to fifteen thousand cords, making it impossible to limit forestry operations to segments of the park. Nevertheless, he was optimistic that one could maintain a timber operation that was "beautiful in the truest aesthetic sense." He wanted the department to understand, though, that

A project which frankly attempts to reconcile large scale commercial exploitation with the preservation of the forest and other values in a condition suitable for a National Park has not, to my knowledge, been attempted on an area of this size before; certainly not in North America. This experiment – and experiment it will be – will certainly attract the attention of Park authorities, foresters, and others concerned with National Parks and the management of forest lands on a world-wide basis. Its success or failure could have considerable influence on future development policies of National Parks both in Canada and in other countries. This in itself is no reason for not proceeding with such a project. ...[42]

Learmouth's claims to science are somewhat disingenuous. This project was experimental only in the sense that it was untried. It did not arise from a careful, independently made decision to test whether this degree of forestry should be tried in parks. Learmouth and the department ignored forestry director D.A. MacDonald's earlier report, which both rejected the idea of mixing parks and forestry and stated that Terra Nova's small stands and shallow roots would demand clearcutting if forestry did go on.[43]

What the correspondence shows is that no one knew what intensive forestry in this proposed national park would mean. Would it cost the park millions of dollars each year to operate a cutting operation, hurt tourism, weaken the image of parks nationwide, and harm the ecological integrity of the park land? Or would it prove a precedent-setting merger of parks and forestry, help the Parks Branch deal with over-mature trees, beautify the park, and open up a source of revenue to all Canadian parks? No one knew, and the department rationalized its gamble by stating that the problem might be moot: there might still turn out to be so much wood in Newfoundland that the park forests would not be needed, or conversely not enough wood in Newfoundland to justify the building of a third mill. But the federal department was absolutely unwilling to let this policy become a precedent. When Newfoundland deputy minister Murray asked for a formal agreement, he was told there was "no question whatsoever of any contractual relationship in this matter."[44] When Murray pressed, Lesage responded, "I do not think we could cover the forest policy question in any more precise fashion without risking the whole future not only of the Newfoundland Park but of all the forested National Parks across Canada."[45] This is hardly a vote of confidence in the policy.

It is striking how completely the Parks Branch was removed from this vital stage of the negotiations, as the federal and provincial ministries worked out a deal, with Pickersgill greasing the wheels. Though Smart helped locate the park site in 1950 and Scott ensured that worries about the proposed park were expressed in 1955, the Parks Branch had no opportunity to voice its opinion of the conditions that were being placed on this park's creation. The agency was relegated to noting that there would be no clearcutting: the Parks Act and Regulations only permitted selective cutting, even if it was more expensive. This declaration was directed at Lesage as much as to the Forestry Service or the Newfoundland government, in that it reminded the minister that the agreement he had made would potentially involve the Parks Branch (and thus his ministry) in a heavy annual expense.[46]

One might think that Newfoundland would be generous following Canada's accommodation of the third mill. After all, Smallwood more than anyone had needed the park to go through quickly: the federal payment for the section of Trans-Canada Highway depended on his ability to establish the park within the year. St John's *Evening Telegram* columnist Harry Horwood called Pickersgill the province's man of the year for "bailing the Newfoundland Government out of an extremely embarrassing financial situation (The local boys had built a strip of highway in the park area; the contractors were hammering on their door for payment, and there wasn't a solitary quid left in the treasury to pay them. Now this road, and all others in the park, will be financed by Ottawa)."[47] But even if the Smallwood government breathed a sigh of relief, it did not become a more giving negotiator. In drafting the land transfer, the province presented the land in two parcels, with the provision that the smaller one, "Schedule B," be withdrawn by the federal government if the province later needed it for hydro development.[48] And P.J. Murray announced that the Northwest Brook watershed, proposed by G.L. Scott as a replacement for the Terra Nova watershed, was itself needed for water power. The feds conceded to this, Lesage saying that "the area could always be added to the Park at any time the provincial government asked."[49]

Should one read this as naïve or sarcastic? Throughout the establishment proceedings, the Newfoundland government made it clear that their ideal national park would contain no marketable natural resources whatsoever. From Smart's initial estimate of a necessary 500- to 1000-square-mile site, the province whittled the 250-square-mile area he proposed down to a final 150 square miles. The federal government had even conceded to a provincial request to swing the Trans-Canada close to the coast (to accommodate communities like Charlottetown), in violation of the Trans-Canada Highways Act, which demanded that roads go in as straight a line as possible between well-populated areas. Newfoundland then pushed the park's western boundary closer to the coast, knowing that the park would still enclose the highway.[50] The new park designed to be typical of Newfoundland scenery contained no river systems, no noteworthy salmon streams, no caribou country.

The federal government capitulated to provincial wishes for a variety of reasons, but an especially important one was the countdown to a federal election in the spring of 1957. The Liberals in Ottawa and St John's wanted what they had negotiated for several years, and considered for a decade, to bear fruit. As the election drew nearer, Ottawa became more conciliatory, and Smallwood's hedging became an even better strategy for getting what he wanted.[51] In his memoirs, Pickersgill remembers the 12 April 1957 proclamation of Terra Nova National Park because it was "the last day in the life of that Parliament. I was pleased. I was even more pleased after 10 June 1957, because I doubted if the new Conservative government would have established a national park in my constituency."[52]

To the different players in the park creation, the new Terra Nova National Park represented different things.[53] To the Smallwood government, the park was 384,000 cords of merchantable timber held in trust. To Pickersgill, it was security that his new riding would remain faithful to him for elections to come. To the federal deal-makers, it was a national project helping a beleaguered provincial economy, proof of the wisdom of federalism. To the *Evening Telegram*, it was, according to the headline the day after proclamation, "A Disappointment." The editor wrote,

The Newfoundland national park, according to latest Canada Press reports, has been cut in area from the original 400 or more square miles to 150 miles on either side of the proposed transinsular highway. The effect ... will be to destroy a great deal of the park's value as a wildlife refuge, a place for scientific study and for holiday recreation. We can only guess what motives are behind the move. We know it springs from the Provincial, not the Federal Government; and it is most likely connected with the old "third mill" dream that our current crop of dreamers refuses to allow to die. It is a pity that out of Newfoundland's 160,000 square miles a mere 400 or 500 could not be found somewhere for conversion into a national park without putting our whole industrial future in jeopardy. ... Their attitude has been that they'd be delighted to set aside an area for a national park provided the Federal Government would allow them to build a pulp mill right in the middle of it, or open a mine, or start several saw mills, or build dams and canals for a hydro-electric development.[54]

To the National Parks Branch, the new park was one smaller than originally planned, with a timber policy that might soon make it a clearcut with picnic tables. But the Parks Branch was not free to voice its displeasure: as a new member of the park system, Terra Nova had to be idealized in official literature to the same degree as any other park. Years later, when long-time Branch staffer W.F. Lothian wrote of Terra Nova's establishment in the commissioned *A Brief History of Canada's National Parks*, he did so in his usual crisp, non-judgmental fashion, neither defending nor protesting how this park came about. But he ended his piece with the long passage from the *Evening Telegram* editorial quoted above. Lothian let the newspaper editor make his criticism for him.[55] The National Parks Branch may have been disappointed with the latest jewel in its crown, but it had to put on a brave public face and proceed with park development.

HAMMERED-DOWN

Expropriation was an even more straightforward exercise in Terra Nova than in the first three Maritimes parks. Almost all of the park land was either owned by the Crown or part of the Reid Newfoundland Company holdings, given at the turn of the century as partial payment for building the colony's railroad. Even before the park was proclaimed, the Smallwood government was able to convince Reid's to trade park land for equivalent land elsewhere on the island. Other than this, there was very little Terra Nova land actually owned by locals and apparently none that was permanently inhabited.[56]

That is not to say that there was no human presence. Local communities depended a great deal on the land in the new park for their livelihood. As was noted earlier, Newfoundlanders enjoyed a traditional right to cut from Crown land up to three miles from shore, which included just about every square inch of what was to become park land. So families in communities from Port Blandford in the south to Eastport in the north relied on the region's woods for firewood and their other wood needs. The forests also served commercial purposes. Much of the park area had been extensively lumbered for up to seventy-five years.[57] In fact, in the thirty years before park establishment, many people in the area switched from fishing to forestry because it offered greater returns.[58] Wood along the Terra Nova River had been cut for pitprop and shipped to England since early in the century. Commercial operations along the head of Newman Sound began in the same period, with lumber shipments headed to St John's. The King family of Eastport used Minchin's Cove on Newman Sound as a base for cutting operations along the top of the peninsula; from Bread Cove on Clode Sound, the Spracklins of Charlottetown worked north to meet them. Bread Cove had about two dozen year-round residences until the 1940s, actually. And the community of Terra Nova itself, inland and seven miles to the west of the new park, was the eastern point in the Anglo-Newfoundland Development Corporation's supply route; wood from the area went by rail to the company's pulp and paper mill at Grand Falls. At park establishment, the Newfoundland government estimated that there were fifteen sawmills relying on the forests of the park land, producing about $60,000 of total output and employing 40 to 50 men.[59] But ten years later, it was concluded that there had in fact been seventeen sawmills, and nearly 120 men dependent on forestry.[60] Such small-scale seasonal operations were tucked in such out-of-the-way places that they were easy to overlook.

Though no one lived year-round in what would become Terra Nova National Park, for some it was a second home. Owners of small sawmilling operations like the Kings' would move to the woods in the winter with a half-dozen or so hired men to cut and saw timber.[61] Clayton King remembers most of his family moving to Minchin's Cove from November to May while he stayed in school in Eastport; he would only see his parents at Christmas and Easter, when he travelled to the Cove to visit. His future wife, Mildred, spent one summer up Newman Sound herself, babysitting her sister's child.[62] The Clode and Newman Sound winter homes were Spartan affairs, made warm with a single wood stove and sawdust insulation. The little communities were only temporary, so no one took the trouble of getting a school or church going. But for at least part of the year the park area was inhabited; Mark Lane says that at times there seemed to be as many living up in Newman Sound as back home in Eastport.[63]

When the park came in, the lumbering economy was in something of a decline, a victim of its own success. With more sawmills working around Newman and Clode sounds, there was more competition for the same wood and, to some, not enough wood to make lumbering worthwhile. The King mill at Minchin's Cove ceased operation around 1950, as did the mills at Bread Cove. Clayton King told me that his father stopped sawing because the area they traditionally used was worn out and all the woods nearby were spoken for. He planned to wait eight or ten years for some natural regrowth, and begin work again.[64] However, other businesses believed that there was still wood if one was willing to work at getting it. The Lane mill at Big Brook continued cutting, while the Squires began sawing at Piss-a-Mare Brook (at the present-day park wharf on Newman Sound), as did Don Spracklin at Bread Cove. There is nothing to suggest that forestry and the forests would not have returned just as strong.

Though the staff of the Parks Branch were well aware of the area's past forest use, they spoke of Terra Nova at park establishment as if it were virgin territory. There is never a sense in the park files either that the Branch was getting damaged goods, or conversely that it was saving a forest from exploitation. In his 1950 report, James Smart made no mention of the forests except to say that fire had denuded some sections, yet his accompanying photos plainly show lumber stacked on the shore. Five years later, engineer G.L. Scott was told by his guide, a woods foreman with forty years' experience, that "the proposed park area had been logged over by the local people for the past 75 years and he could not understand how anyone could possibly conduct an economic pulp or saw lumber operation in the park area."[65] This did not suggest to either Scott or his superiors, though, that one could not conduct a park there. When forestry director J.D.B. Harrison inspected the newly created park, he found that only 10 per cent was mature growth and 90 per cent was either young or very young growth. This was a young forest, most of the mature and over-mature timber having already been cut down or burned.[66] The Parks Branch was well aware that it was adopting an area that could not possibly be considered wilderness, that was in fact the product of a long interaction with humans. But the prominent natural effects of human industry could be ignored because there was no obvious evidence of human presence. So when the park was just coming in, parks director Hutchison could confidently write that the first work crew would be "starting from scratch in a completely undeveloped area. ..."[67]

The Parks Branch was pleasantly surprised with its new ward's appearance. Two of the Ottawa men greatly involved in Terra Nova's establishment, parks director J.R.B. Coleman and deputy minister R.G. Robertson, visited it for the first time in 1957. Each man independently praised the designers for accomplishing the "Maritime theme." Coleman called the park "a much more attractive area than I had imagined," and Robertson wrote that he was "very agreeably surprised by the area and the possibilities in the national park. For some reason I had expected that it would have a good deal less attractiveness and less possibility for development than it seems to me to have."[68] They apparently had not expected a typical Newfoundland Maritime theme to be very attractive. (Would they have negotiated differently with Smallwood if they had?) Chief engineer Scott thought Terra Nova could be the best and most beautiful park in the east.[69] Branch staff were so content with the new park that they wished there were more of it: what complaints there were concerned its relatively small size, whittled down through political interference.[70]

More so than at the other three parks studied, the Parks Branch was concerned that development would change what was now pleasing. Robertson wrote, "As I know you are well aware, administrative problems, forestry problems and aesthetic problems will all have to be taken into consideration in choosing sites for buildings, in determining road locations, and so on."[71] This was an obvious caution, but one that now needed to be voiced. The great concern was overdevelopment. The Parks Branch was much more involved in overseeing park development at Terra Nova than it had been in the first Eastern parks – a reaction to what had gone on in the building of Fundy, as I will document in chapter 8.

There was purposefully less intrusion on the natural landscape at Terra Nova and simpler construction using cheaper materials. The Parks Branch staff did not simply put contractors to work at Terra Nova; they ensured that it was done without ensuing destruction. As early as August 1957, Scott was already cautioning the contractor about over-zealous tree-cutting.[72] A year later, Coleman had a similar criticism and demanded that the contractor improve the look of an area which had been roughly bulldozed.[73] Park developments were also to be simpler. As was mentioned earlier, the clearest demonstration of this was that the tourist cabins were built not only to be part of their natural surroundings but also to be of simple design. The Parks Branch even chose not to build all of them right away, since tourists were not really expected in great numbers until at least the early 1960s.[74] Simplicity was an economic choice, as well as a philosophical and aesthetic one. In the late 1950s, attendance at national parks across Canada was rising rapidly and park finances were straining to catch up. It made sense to spend cautiously on new developments.[75]

Most significantly, at Terra Nova the Parks Branch refrained from setting up the recreational facilities that had traditionally been the heart of new national parks. It was decided that there would be no immediate development of a golf course, and that a heated pool was unnecessary with the sea so close by. Newfoundlanders familiar with the amenities found at other parks grew upset. The St John's *Evening Telegram* lobbied for a heated pool, and letters to the editor cried for pools, ski runs, stadiums, and so on.[76] The Gander Parks and Play-grounds Association along with the town's Chamber of Commerce petitioned for a park golf course, saying, "there is absolutely nothing to do there."[77] Even some within the Parks Branch and ministry quietly agreed. Recommending that the Parks Branch reconsider its rejection of the horseshoe pitch and shuffleboard that the cabins' concessionaire had proposed, deputy minister Robertson argued, "there is a serious need for some recreational amusement amenities of a modest kind and of an outdoor character in the park."[78] The park's first superintendent, J.H. Atkinson, suggested shortly after opening that "At the moment we are like the artist who concentrated too much on the frame for the picture, and did nothing about the picture." He called on "every device known to our planners" to make Terra Nova extraordinary.[79] But the Parks Branch resisted such demands, arguing that they had a real opportunity to plan this park and wanted development there to proceed slowly. Park staff were beginning to view recreational development as detrimental: it demanded a sizeable initial investment and a perpetual draw on park funds from then on, and it detracted from the park's preservationist aims.

While there was some dissatisfaction with the lack of permanent recreational development, the scale of work at Terra Nova in its first years quieted most complaint. Between 1958 and 1960, the park kept as many as 250 men at work in the summer months. Even in winter, park projects that were little more than relief work were organized to maintain a year-round federal presence, though they were known to be inefficient.[80]

To local communities, however, it was not only a matter of how many were hired, but from where. This was a concern at all four parks studied here, particularly Cape Breton Highlands, Prince

Edward Island, and Terra Nova, where a number of communities bordered the park. The people of Traytown, Charlottetown, Port Blandford, and the other little communities that surrounded Terra Nova, who had depended on the park forests for their livelihood, believed they had sacrificed enough just by accepting the park's existence. They saw the new park jobs as their right, and saw anyone hired from elsewhere as taking a job from a local boy. The Parks Branch, the federal ministry, and the provincial government all wanted locals to be hired: it was simplest, fostered good relations, and acknowledged what all groups saw as an obligation.[81] The difficulty was that, according to federal employment regulations, skilled men had to have at least a high school education, and many who had gone to the woods at an early age did not. Though locals now worked at unskilled jobs, the more permanent jobs would not necessarily be given or even made available to them. Moreover, politicians did not share Newfoundlanders' own definition of the term "local." The editor of the St John's *Evening Telegram* zeroed in on the problem: "Many of those hired during the present building phase are from outside Newfoundland, and recently local men were passed up in favour of a group from Bonavista, which is so far away from the park that the people there regard it as being in a different world."[82]

Even among the surrounding communities there was fierce competition for jobs. Charlottetown was most affected by the park – the park completely surrounded the town, and cut it off from traditional wood sources – and its citizens expected jobs as compensation. During the 1957 election campaign, Jack Pickersgill and Joey Smallwood assured the people of Charlottetown that they would be the first hired. Parks director Coleman understood the special obligation owed Charlottetown, and suggested that the park favour the little village in recruiting seasonal labour.[83] But this did not happen. When the Conservatives won the election, they selected local supporter Carson Stroud of Glovertown to "advise" on hiring; more jobs subsequently went to people in and around Glovertown. Pickersgill complained in the House of Commons of such blatant patronage, suggesting that the clergy be sought for employment advice instead.[84] When the Liberals were returned to power in 1963, Pickersgill forgot his own advice; he submitted a list of names, organized according to village and hiring preference, to the National Employment Office in St John's.[85]

In coming years, relations between park staff and local residents would be strained by politicians' early promises that the park would provide jobs for all, for life. Expropriatees and local residents at all four of the parks studied here spoke of such assurances. Often these reminiscences are hazy and at second hand. But at Terra Nova it is clear that such promises were given. Several residents recalled Smallwood and Pickersgill flying in by helicopter to deliver a speech, and promising jobs. The park files refers to the Branch having to renege on promises made "by individuals, not at present responsible in the Dominion Government, but who were responsible for the formation of the area as a National Park."[86] The Parks Branch in Ottawa became so accustomed to deflecting locals' employment complaints that in 1960 they wrote to a new superintendent, "The problem you are having with men coming to your Park stating that their livelihood has been taken away has been faced by Mr. Atkinson since he took over. For your information I am attaching two copies of letters which are the only commitments that were ever given from this Department to the Government of Newfoundland."[87] Perhaps local residents should have been more savvy, and understood that development would necessarily peak in the first few years and decrease thereafter. But the Parks Branch did little better in considering the park's long-term effects on local employment. In 1957, Robertson cheerily stated that, thanks to the new park, employment was up in the area; as early as 1960, with most development completed and locals still clamouring for jobs, Superintendent Atkinson thought that staff should explain – to the people of Charlottetown in particular – that most of the necessary work was done: "Conditions at this time are very much better than they ever will be, and yet we receive many complaints."[88]

But men continued to arrive looking for park work, just as they had arrived in years past to work in the Terra Nova woods. On the eve of the first winter of park development, thirty to forty men from as far as 150 miles away showed up. The engineer in charge could do nothing for them but take their names and send them home.[89] Hundreds sought Pickersgill out in the late 1950s, and though the Branch tried to keep on as many people as possible, they were expressly told by their department not to let Terra Nova become a relief project. Complaints spilled into the newspapers. One letter writer maintained, "If there is no work this coming year, I fear the United Nations will have another problem to cope with and they might even have to send an emergency force to clear it up."[90] It never came to that. The park was hiring some people, and such complaints were one way of encouraging it to hire more. Even the Conservative-leaning *Evening Telegram*'s Harry Horwood admitted that Terra Nova National Park appeared "one of the best 'new industries' started in Newfoundland since Confederation." (Now there was a competition with few entries.) In fact, it proved "how valuable even a sawed-off, hammered-down, chopped-up national park can be to a province."[91]

But to many local people, this park that had promised so much growth now looked like a lot of unused land and a few hired labourers. Sitting talking with me at his home in Eastport, Clayton King said quietly, "A lot of people thought it was going to be a major industry, but ..." and trailed off. I waited, and finally asked, "Tourism, you mean?" "No no no," he said. "Employment in the park. But that didn't materialize."[92]

CHOPPED-UP?

Terra Nova superintendent J.H. Atkinson reported to Ottawa in early 1960 that he had had visitors. "I was approached by 25 men from Charlottetown with a request for work immediately," he wrote. "When we were unable to give them work they asked for timber permits. Since all the men concerned were on Unemployment Insurance, we did not consider it necessary to issue them permits, and so stated. The men then stated that they would enter our timber lands on the following Thursday in a body to cut timber with or without a permit." Atkinson relented somewhat, giving permits to those who used to work on this land, and who were now cutting for personal use. He then advertised that the park was selling timber permits. "The response to this call was very startling," he told Ottawa, "since we did not receive even one application. The whole point is that the timber angle is used to get work, and very few if any, are really interested in timber permits."[93] This incident nicely reflects the different ways that local residents and the Parks Branch regarded timber cutting at Terra Nova. To the people of the area, the woods were a way of supplementing income, obtaining their family's wood needs, and keeping busy. It seemed logical that if the park was not prepared to offer replacement jobs, men should be free to use the woods in the traditional manner. Atkinson, however, thought that timber was just an "angle," a way for locals to gain leverage with the park. He could not accept that unemployment insurance benefits did not adequately replace traditional forest use, nor could he comprehend that locals might simply find the forest of government bureaucracy something to avoid. Atkinson's reaction above all signals the Branch's larger paranoia about threats to the Terra Nova forest. From 1957 on, the Parks Branch worked to fulfil the needs of locals who had lost their traditional use of the woods, while waiting nervously for word from the Newfoundland government that a third mill was on the way, and that logging should begin in the park.

More so than at either Fundy or Cape Breton Highlands, the Parks Branch was prepared to allow cutting by dispossessed locals at Terra

Nova. This is somewhat ironic, considering that in this case people who had lost access to the woods had not even owned them. But perhaps it was considered reason for more largesse: most of the men who had worked on park land had not been compensated for it. More likely, the Parks Branch was coming to understand not only that past resource users had a right to compensation, but also that happy locals made for a smooth-running park. James Smart had directed the establishment of Fundy National Park and saw at first hand how non-owners dependent on park land were left out of remuneration. As early as 1950, in his first evaluation of Terra Nova, Smart insisted that local residents should be permitted to continue to cut for their own use.

Though there was no discussion of locals' cutting rights during park planning, when the park was first established the Parks Branch allowed local lumbermen to continue their work, under supervision.[94] During development, sawmillers were allowed free salvage from any clearing being done, and for additional supply they were given permits for wood in out-of-the-way areas.[95] Special preference was given to those who had made their living lumbering in the park prior to establishment. Wardens were even told to look the other way when it was discovered that a mill in Minchin's Cove was still in operation.[96] No further action was taken while the Parks Branch waited for the Forestry Service to set up a management plan that would decide which mature, out-of-the-way timber should be taken.

The park staff's concern about the amount of wood local residents would demand turned out to be groundless. In 1959, only sixteen permits were sought and granted, for about 40,000 fbm (board feet) and 17 cords fuelwood. In 1960, this dipped to only five permits for 3550 fbm and 10 cords, and by 1961 only three Charlottetown families were applying for timber permits, totalling 36,000 fbm. In the years to follow there were never more than ten timber permits sought.[97] By 1967, a report on logging in Terra Nova National Park stated that "the situation is well under control and there is no harm done to the park."[98]

Superintendent Atkinson's claim that locals had no real interest in lumbering was not accurate, however. A variety of factors made sawing in the region, and particularly within the park, less and less practical. In the late 1950s, it became increasingly difficult to find a market for lumber, and stockpiles began to grow. Labour difficulties also affected the provincial industry.[99] But closer to home, the park regulations on timber permits hampered local woodsmen and convinced many to give up on the woods. Permit holders were expected to pay $6 per one thousand board feet, and though the Parks Branch apparently lowered this, it was still a new charge for lumbermen.[100]

By the early 1960s, the park had forbidden any cutting along the coast – presumably because it could be visible to tourists – and set aside an isolated timber block. This site offered space for four hundred thousand fbm of saw logs to be clearcut, but it was far from the usual cutting zones, required loggers to dispose of slash, and worst of all was unreachable by water. As the Parks Branch well knew, lumbermen who traditionally floated their logs along rivers would be forced to sell timber along the road or have it trucked to the mill.[101] If this discouraged loggers and sawmill operators, so be it. In establishing such policies, some Parks Branch staff clearly believed that the park would be of great benefit if for no other reason than that it would force the local citizenry to reject their outmoded lifestyles and join the twentieth century.[102] Parks director Coleman noted, "Generally speaking the sawmill industry in this area is made up of a large number of very small, out of date sawmills which produce very low quality lumber. Financially, their operations are marginal at the very best. They are, in fact, prime examples of the uneconomic and wasteful timber-using practices the Government of Newfoundland is quietly attempting to discourage in the province."[103] In 1963, four lumbermen were caught defying park regulations and cutting on their own along the coast of Clode Sound near Charlottetown. Having initially ignored cutting in the new park, the Parks Branch now felt confident enough to press charges. The men were convicted, and a strong statement had been made to defend the park from traditional cutting practices.[104]

Don Spracklin, who had been the main sawmill operator in Charlottetown at Terra Nova's establishment, and his son Larry spoke to me about the problems the park had created for the family business. Don remembers having fifty-three men on the payroll when the park came in, but once it did he could not find people to hire, let alone wood to cut. Don knows irony when he sees it. He points to Spracklin Pond on a map of the park, named for his family but now a place he cannot fish without a permit. He feels that the guarantees of jobs and concessions that came with park establishment were never fulfilled. Smallwood, he remembers, flew in to Charlottetown and guaranteed its people that "the timber's in the bank," still available to people of the area. Now, Don Spracklin laughs, "Well, they lost the combination."[105]

What Spracklin and the people of the park area never realized was that Smallwood's plan for the park never included their cutting of it. If timber was in the bank, it was gaining interest for the third mill. At a 1964 provincial forestry conference, an audience member asked Smallwood why cutting was so much more limited in the park than outside it. Smallwood expressed surprise that any cutting was going on at all. According to Superintendent Doak, who was sitting in the audience,

Mr. Smallwood replied that Terra Nova National Park is the only National Park in Canada where any cutting will be permitted at all. Through negotiations, permission was obtained for the Government of Newfoundland to cut the annual increment in the Park to be used only for a third paper mill. This is the only wood allowed to be cut under the Laws of Canada. Park authorities have not the right to issue permits to cut. … The Premier also stated "I realize that the Park was cutting and thinning in order to beautify the Park," but he did not know that they were issuing permits to cut timber and that he would investigate the matter.[106]

Doak later talked to W.J. Keough, minister of mines, agriculture, and resources for Newfoundland, who confirmed that "We do not want those people in there hacking away at the Park."[107] This came as a surprise to the Parks Branch; it had not considered that the province might see the park timber agreement as superseding National Park Regulations on timber permits. And it had never imagined that the Newfoundland government would take cutting privileges away from its own people. The Parks Branch chose to continue quietly granting timber permits to those locals who met park regulations, particularly those who had depended on those woods previous to park establishment. And why not? The locals' needs were small, and their claim to a small part of the large park seemed reasonable. It was the shadow of the third mill that hung over the future of Terra Nova's forests.

Part Two
A Pious Hope

Cover of *Playgrounds of Eastern Canada*, 1952.

7 Accommodations and Concessions: Use in Four National Parks, 1935–65

When the proposed National Parks Act was under discussion in the House of Commons in 1930, Conservative member George Gibson Coote complained that the bill's wording was unnecessarily grand. Why was the old clause that the parks "shall be maintained and made use of ... for the benefit, advantage and enjoyment of the people of Canada" replaced with the "high sounding" statement, "The parks are hereby dedicated to the people of Canada for their benefit, education, and enjoyment"? And more than this, Coote said, "I do not think it is possible for us to control the actions of future generations. Just how we are going to administer and take care of these parks and leave them unimpaired for the enjoyment of future generations, it is rather difficult to see." Charles Stewart, the minister of the interior, rose in reply: "Does my honourable friend think there is any very great difference between the old section and the new? In both it is a pious hope. I would not care very much if you wiped both of them out. I do not think either section will do much harm or much good to the national parks."[1]

This was a fitting christening for the National Parks Act. For its entire history, the Parks Branch has depended on a public acceptance that it, more than any other governmental body, has a philosophy which transcends short-lived administrations.[2] The parks are always to be parks, and to be unchanged. In 1930 the Branch hoped that a carefully worded statement of purpose, tightly balancing preservation and use, would win over the public to an eternal fulfilment of the parks' twin mandate. But the minister of the interior understood that even an allegedly rigid philosophy would in practice have to bend to changing situational concerns and philosophical interpretations. Stewart was right to call the Parks Branch's plans to seamlessly marry preservation and use "a pious hope," just as the Branch was right to try to marry them.

To this point, I have focused on the acquisition and early development of Atlantic Canada's first four national parks. Each park has been viewed as the Parks Branch saw it at the time: as pure potential, ready to be transformed into whatever vision of a national park the Branch then thought appropriate. The following two chapters will examine the policies designed and implemented day-to-day after this initial period of detached abstraction. This chapter will be about use, defined as any human demand on a park, and will focus particularly on tourism (with special reference to its relationship to class and discrimination), business concessions, and resource extraction. It will describe the complicated ways in which questions of use could compromise principles of preservation, defined as any action designed to maintain a park in as natural a state as possible, leaving to the next chapter a detailed consideration of the formulation of policies on wildlife, fish, and vegetation.

Though "preservation and use" is a handy phrase to express the two-sided mandate of the national parks, it misleadingly suggests that these constitute a dichotomy. After all, the two were never meant to be mutually exclusive; both were to be fully realized. Nor were they necessarily in contradiction; as goals, they may complement or not even affect one another. Reintroducing a species to a park would be an act of preservation which need not affect park use. Adding a storey to a park hotel would be in aid of use, but would not change the park's ability to preserve nature.

So rather than examining preservation versus use, it is much more helpful to look at both in terms of "intervention" – the Parks Branch's active attempts to ensure that both goals were met. It soon becomes apparent in examining the history of these four parks (and, by extension, trends in the Canadian park system generally) that the barometer of intervention moved up and down for both preservation and use simultaneously. When the Branch was proactive in preservation work, it also tended to be proactive in developing use; when more passive, it was passive about both. The next two chapters will show that from the mid-1930s to the mid-1940s the Parks Branch took a relatively "hands-off" position, with little interest in actively managing park preservation and little money to develop park use. After the war, the Parks Branch became much more active in both preservation and use. The ecological science of the time encouraged the government's experts to become more managerial, and a more intervention-ist state offered them unprecedented funds to do so. Greater funding for public works and a North American culture which promoted rec-reational democracy likewise meant increasing development for tour-ist use. By 1960, however, the public was growing concerned about human destruction of the environment – including that found in parks – and the Parks Branch had begun to question the logic of its own past interferences. The Branch as a result pledged a more pas-sive management style for the future. However, with use by visitors climbing precipitously in the 1960s and the nature in national parks under greater threat than ever, a clash over the question of interven-tion in the parks was almost inevitable. The Parks Branch of the 1960s had to decide not only which style better managed preservation and use compatibly, but which better represented the agency's philoso-phy to the public.

Cape Breton Highlands, Prince Edward Island, Fundy, and Terra Nova serve as excellent case studies of Canadian national park poli-cies on preservation and use in this period. These parks maintained the same regulations, depended on the same scientific and planning experts, and adopted the same practices as the rest of the park sys-tem. But more than this, as new parks they were not fettered by the management decisions of past decades: they were given the most up-to-date ideas to adopt. This in turn resulted in constant discussion be-tween Ottawa and its people in the field over how best to interpret and implement these changes in philosophy. The following two chap-ters cannot claim to be an exhaustive examination of preservation and use in the Canadian national park system from the 1930s through the 1960s, because they do not cover all the parks. Nevertheless, they can be said to offer reasonable conclusions on the directions taken by the Parks Branch to make these seemingly incompatible responsibili-ties compatible.[3]

PARKS FOR PROFIT

Park historian Leslie Bella begins her discussion of national parks in Atlantic Canada with the statement, "Every one ... was a park for profit."[4] Each was created first to draw in federal development dol-lars and over the long term to give focus to its province's tourism in-dustry. Having defined the parks in this way, Bella gives the first four Atlantic Canadian parks only a single page in her history of the park system. She misses a significant part of their story in the process. Re-gardless of why these parks were created, they have had histories of preserving nature and restricting traditional resource use like any other parks. And in disdaining parks created solely for use, Bella misses the point that use is not some contamination of national parks' nobler purpose; parks have always been meant to be used. Flora and fauna benefit from the mandate of preservation, but the mandate of use ensures that the human species does not miss out. And because so much of our society's activity is regulated by capitalism, it is not surprising that these parks, like all others, are to some degree parks for profit.

The following discussion examines the ways that human use affected the management of the first four Atlantic Canadian parks. Use, like preservation, was a part of every moment of their existence, making a day-to-day record of it impossible to chart. Instead, I will focus on issues that in particular reflect on the meaning of the parks, and the ability of the Parks Branch to translate that meaning. The chapter begins with a broad discussion of tourism, the most obvious park use. Of special interest here is how the Parks Branch changed the ways it promoted the parks, depending on its understanding of society's wants and the parks' ability to satisfy them. This is followed by a short section on how, having sold the parks as cultured destinations, staff dealt with cases of discrimination. The next two sections deal more specifically with the parks' relationship to the economy. The first discusses the arrangements made to entice concessionaires to do business within the parks. The second discusses another sort of concession: the Parks Branch's inability to protect the parks from private enterprise and provincial governments' infringements on inviolability. The chapter concludes with the collision of use and preservation in the mid-1950s as the result of overcrowding and overdevelopment in the parks.

Canadian national parks have at various times and places been used as timberland, mine sites, laboratories, and work relief camps. But they were not primarily created with these functions in mind; parks were first and foremost to be used as tourist attractions. It was hoped that visitors would come to parks to admire the splendour of nature, and take in the spiritual, educational, and health-giving properties they possessed. In doing so, visitors would also stimulate the Canadian and local economy. This had been the case ever since Banff Hot Springs had been reserved, and Sir John A. Macdonald had suggested that the spa would help "recuperate the patients and recoup the Treasury."[5] But tourism had more than just an economic value: it did not just financially justify parks, it psychologically justified them. Preservation was a gift Canadians gave themselves, proof that they were civilized enough to leave a parcel of nature free from resource extraction, even though so much of their society was founded on seeing nature's value solely in economic terms. But for the country to take credit for this benevolence, visitors had to see the parks. Without the promise of people travelling to the parks to see scenery and wildlife, there would have been insufficient human use of them to justify a national park system.[6]

Canadians' fixation on American tourists is a wonderful example of how the desire for tourism transcended economic matters. True, there were solid, financial reasons for seeking American tourists. The United States provided a huge market that Canadian tourism promoters had yet to exploit fully. The dollars brought by American visitors were injected directly into the Canadian economy, whereas tourism within Canada just saw money change hands. Cape Breton Highlands and Prince Edward Island parks were justified in terms of drawing not just tourists, but American tourists, and Fundy was selected in part because it was far enough from the US border that American visitors would have to travel across much of New Brunswick to reach it. But financial considerations are insufficient to explain the Parks Branch's special eagerness to attract and please Americans. For example, the licence plates in park publicity photos of the 1950s and 1960s were usually of American cars. Even a *Canada–West Indies Magazine* article on Fundy National Park hoped to impress its readers with the observation, "As indicated by the licence plate on their car the campers are from somewhere in Ohio, USA."[7] The Saint John *Telegraph Journal*, reporting on Fundy's grand opening, carried a photo of an average New York family under the headline "Good Neighbours Visit."[8] In reality, though attendance figures were notoriously inconsistent,[9] it is clear that Americans never made up more than a significant minority percentage of total tourists. Early Prince Edward Island park statistics show that 20 per cent of those who signed at Dalvay and Green Gables in 1940 were American, though this dropped down to 3 per cent later in wartime.[10] As Fundy became better known to tourists, it reported a rise in Americans from about 10 per cent to 25 per cent of total attendance in the late 1950s. Still, it is highly unlikely that the *Telegraph Journal* was accurate two years later when it reported that Fundy's visitors were "mostly from the US."[11] Such enthusiasm for Americans is perhaps the one strand that ran through the tourism history of the four parks discussed here. In 1961, Superintendent Doak of Terra Nova reported with evident disappointment that all the new campsites were being used "but to the best of our knowledge no Americans were staying there."[12]

By studying the Parks Branch's approach to tourism through time, then, we can learn about more than just the economics of tourism: we can better understand the agency's sense of itself and the parks under its care. Though theoretically everyone was a potential visitor, Branch staff sought to attract people whom they imagined most likely to visit and enjoy the parks. They also chose sites and buildings they hoped tourists would consider appropriate. (These developments were considered permissible elements of park creation, and once completed the park would be "locked in" as it was. MacEachern's law states that a national park preserves for all time whatever it happens to include during its opening ceremonies.) But such judgments were necessarily rooted in the tourism of the time, and tourist preferences are fickle. By tying park policy to something as ephemeral as public taste, the Parks Branch was guaranteeing that the parks would become dated. When that happened, staff would have to decide to what extent revamping the park facilities to attract tourists was justified, considering that the parks were supposed to be preserved unchanged.

As has been discussed, Cape Breton Highlands and Prince Edward Island national parks were designed in the 1930s to have crossover appeal. They would have the trappings of a seaside resort – development centred on their beaches, a large Victorian hotel – so that well-off tourists would be enticed to come. Wooing the wealthy was how national parks had always been sold. But because these parks would be so much closer to the Eastern seaboard population base, because the automobile was expanding the tourism base for parks, and because nature in these parks was in itself considered insufficient to attract the discriminating tourist, developments were also planned to entice the less well-off. Prince Edward Island's Dalvay was selected, as has been noted, to draw the best people and in doing so draw the rest. Cavendish would prosper when mass followed class. Although the Parks Branch toyed with the idea of building entertainments appealing directly to the middle class, staff chose to rely on traditional, more upper-class park facilities. The most important of these were golf courses, which were considered compulsory for national parks in this period because golf was the most enjoyed outdoor summer sport for white, male, middle-aged, middle-to-upper-class Canadians – including, notably, James Harkin, Thomas Crerar, and James Smart.[13] The first Maritime parks were also given bowling greens and tennis courts, because they too spoke of outdoor, upper-class entertainments. Conversely, the Parks Branch forbade "Coney Island" recreations that spoke too openly of lower-class culture. The only part of the park seen specifically as a middle-class attraction was the beach itself, but it was acceptable because it was natural.

But after the war the Parks Branch began marketing itself directly to the middle class. This change is attributable to the continent-wide rise of what might be called "recreational democracy," a belief that recreation time, including a yearly vacation, was a right that everyone needed and deserved. In the postwar years, recreational democracy became reality for middle-class North Americans, born out of economics and technology. A strong economy and a quickly rising standard of living gave North Americans unprecedented wealth and opportunity for leisure time. Automobiles were within family budgets, and the public road system was being greatly expanded and improved with work under way on the Trans-Canada Highway and the US interstate system. The technologies of tourism – everything from ice coolers to RVs – were improved yet less expensive. There were also simply more potential travellers: the baby boom meant larger families were taking vacations. Tourism grew by about 10 per cent per year across North America during the 1950s.[14]

The National Parks Branch saw itself as ready to meet this demand. The park system, in fact, was understood to have been a pioneer of the notion of recreational democracy, and to be an intrinsic part of its success. In *Playgrounds of Eastern Canada*, a booklet of the early 1950s, it is said, "the National Parks fulfil a fourfold purpose. They are conserving the primitive beauty of the landscape, maintaining the native wildlife of the country under natural conditions, preserving sites memorable in the nation's history, and serving as recreational areas. Their value in the last category becomes more apparent each year, for they provide in ideal surroundings, unequalled opportunities for outdoor life."[15] It made sense that the author would draw special attention to recreation. It was the great social equalizer, not demanding an aesthetic sense, compassion toward animals, or a knowledge of history. And it was becoming increasingly important to the park idea. Even by the time Fundy was created in the late 1940s, it was no longer considered necessary to develop the park with luxuries expressly favouring the wealthy. Fundy did not need the sort of cultural associations the earlier Maritime parks had provided, or a scenic drive or a resort-style hotel. Instead, the New Brunswick park was designed around a single recreation development area. The Parks Branch did not conceive this as a lowering of standards, but as a recognition that more people could now achieve those standards. The dream harking back to James Harkin that the public at large would enjoy the parks was in the process of being realized.

As historian Lary M. Dilsaver says of the American case, "Incessant promotion of parks as valued destinations and places almost holy in America's splendid landscape convinced the public."[16] Evidence suggests that his words apply equally well to Canada. Attendance at the

Canadian parks, which had hovered around one million in the mid-1930s and dipped as low as 415,000 during the war, climbed. By 1950, there were almost 1.8 million visitors and by 1955, 3.3 million. The first Maritime parks stayed quite consistent in terms of their ranking within the park system, but like those at other parks their actual attendance numbers were way up (see appendix 2). The Parks Branch had succeeded in making parks especially attractive because they provided the best of natural scenery alongside affordable, modern amenities. They were still considered high-culture destinations accessible to the masses. To meet increasing tourist demand in the 1950s, the Parks Branch worked to expand facilities and increase accessibility. In other words, it continued to implement the sort of policies which had made parks so popular in the first place.

Recreational democracy seemed such a unifying concept that parks staff assumed once tourists arrived they would want to spend time together. Fundy headquarters was given an assembly hall, which in coming years mostly saw use as home to the New Brunswick School of Arts and Crafts. Recreation halls were also built at Cape Breton Highlands and Prince Edward Island national parks; the first, called a "casino hall" in planning documents, was put up by the Keltic Lodge management, while the latter was designed for a park recreation program. Publicity stills from the 1950s likewise indicate a belief that parks were very much social places. Whereas earlier photographs of the first Maritime parks were predominantly landscape shots, with perhaps a golfer or a fisherman in middle distance, more of the newer photos were close-ups cluttered with people. Families at park campgrounds and bungalow camps were not usually pictured alone, but with another family nearby, separated by just a patch of grass. The playground at Prince Edward Island's Stanhope Beach shows twenty-five children and adults crowded around the slide and the (gratuitous) sandbox. A photograph of Fundy has several hundred people standing fully dressed around the park's saltwater swimming pool, while the shore of the Bay of Fundy a hundred yards away appears deserted.[17] The Parks Branch's use of such pictures for publicity purposes suggests that staff thought them representative as well as attractive to tourists. The bureau understood tourism to be a social activity, and the close-knit accommodations and mass recreation developments of the 1950s responded to this.

The Parks Branch, in equating increased park attendance with visitors' presumed love of crowds, overlooked the initial appeal of parks as natural antidotes to civilization. In postwar North America, attitudes to parks and to nature in general were more complicated than the Branch seemed to acknowledge. The period saw more people living in cities, with less direct physical involvement in obtaining their food, their heat, their water. Historian Alexander Wilson writes that "Nature was newly out of reach of most North Americans." And yet, technology was serving to fill this gap. Nature shows on television reintroduced people to the world around them, if only for an hour at a time. Automobiles brought people back to nature, if only for a weekend at a time. Of course, these were different relationships from those past; in Wilson's words, they emphasized the experience of nature rather than its use.[18] National parks played a significant role in this new understanding of nature. Parks were popular places for family recreation, but this was grounded in the fact that they were places where families could expect to view nature. Thomas Dunlap credits the US National Park Service with having the greatest impact in the popular understanding of wildlife conservation in that country, and Alfred Runte states, "No institution is more symbolic of the conservation movement in the United States than the national parks."[19] The same could be said of the Canadian national parks.

This is what we might call the optimistic side of the fermenting ecology movement. But there was also a pessimistic side. The 1950s saw North Americans develop a sense that the nature they were enjoying was in danger and that humans were to blame. On the planetary scale, the prominence of nuclear fallout in the air, the presence of strontium-90 in milk, plus the ever-looming potential of nuclear war created anxiety for the future safety of both nature and humanity.[20] But people could not see these threats. On a level that people were more likely to experience at first hand, national parks were starting to show signs of crowding. From the Parks Branch's perspective, a rise in attendance from 600,000 in 1945 to 3.3 million in 1955 meant many logistical headaches, but it was nonetheless a sign of accomplishment, an embarrassment of riches. For a visitor to a park, such an increase meant that where there had been one tourist on a trail, there was now a family; where there had been a family, there was now a crowd. Visitors could not help but take from this that humans posed a threat to nature. And, not surprisingly, those who wanted parks to be natural experiences felt that those who did not appreciate the parks in the same way were the greatest threat.

Just as the Parks Branch's appropriations to meet the presumed needs of its expanding clientele started spiralling upwards – from $2 million in 1946 to $9 million in 1950, and on to $22 million by the end of that decade – a significant minority of park visitors began rebelling against the development-intensive sort of park experience. The greatest evidence of this was that camping, which since the Depression had been associated more with Okies than tourists, experienced a great resurgence.[21] There were only 22,000 campers in the entire Canadian park system in 1952, but ten times that just five years later. The Maritime parks experienced the fastest growth, which indicated just how different this new tourist was: prior to the 1950s, the region's weather had been thought too cold and wet to permit camping. Prince Edward Island National Park had only 243 campers and Cape Breton Highlands 183 in 1952; ten years later, they would welcome 39,000 and 30,000 campers respectively.[22] The Parks Branch was caught off guard by this phenomenon, and did not have nearly enough campsites ready. And the campgrounds the parks did have were not what the public wanted. They were, as publicity shots promised, mini-communities set in tight, symmetrical formation. This was the most common written complaint that the Parks Branch fielded about the Atlantic parks. A visitor to Fundy stated, "Frankly, when I go camping I do so because I want to get out in the woods, and more or less on my own."[23] Another noted that this was not just a Canadian park problem: "You are making the same mistake USA has done in their national parks – campsites are too close together and there is no privacy. Camping area looks like a gypsy congregation – not people enjoying nature!"[24] It took some years for the Parks Branch to recognize the significance of such complaints, and still more to act on them. By the end of the 1950s, the system would be under even greater stress from skyrocketing attendance and the Branch's own attempt to match demand with development. Staff would have to re-evaluate just what kind of tourism the parks were intended to supply.

The Atlantic Canadian parks received more than their fair share of attention, because they recorded such phenomenal gains in attendance – significantly above the park system's national average. Fundy's attendance jumped 360 per cent during the 1950s to 227,000; Prince Edward Island's jumped 470 per cent to 412,000; and Cape Breton Highlands' jumped 1120 per cent to 323,000. It was easy to understand these parks' particular popularity. They were within driving distance of the core population of North America's Eastern seaboard, an essential factor in an automobile-inspired tourism. They had only recently been developed, and with middle-class family tourism in mind. Related to this, their picturesque rather than spectacular landscapes, once deemed liabilities, now made them "safe" and accessible to a public that saw camping as the quintessential nature experience. The Maritime national parks came into their own just as staff were coming to question whether the park system could withstand such mass tourism.

RESTRICTED CLIENTELE

The Parks Branch had the unusual task of creating exclusive places that would not exclude anyone. Parks were to be for the use and enjoyment of all Canadians, and there is every evidence that staff took that mandate seriously; conversely, there is no evidence of any policy designed to discriminate. That being said, the Parks Branch most certainly created parks that were sure to be of less interest to some people than others. Golf and tennis were accepted park recreations because they were sports of the relatively well-off; if tourists did not possess the wealth and social standing to play them regularly in their daily lives, they were unlikely to enjoy them on vacation. And if they were Jews or blacks routinely barred from places to play such games, they were even less likely to. Park brochures, filled with the smiling white faces of nuclear families and heterosexual couples, likewise did not tell anyone to stay away, but they did signal to readers who would most likely be found there and thus who would be most comfortable. It is impossible to know how tourists responded to such symbols, or whether such symbols kept the poor and minorities from visiting in the first place. Similarly, it is impossible to know whether over-zealous staff on the ground took the idea of park standards to mean that certain people should be excluded, regardless of the park system's mandate.

Visitors to the parks were not categorized by class or ethnicity, so we do not know the demographics of park attendance. American studies in the 1960s and 1970s found that a disproportionate number of park visitors were white, well educated, middle to upper class.[25] The same was likely true in the Canadian parks,[26] and probably even more evident in earlier decades. Writing in the early 1970s, sociologist Joseph Meeker notes, "Attempts by the National Park Service to attract minorities to the parks assume that these groups will find them pleasant and meaningful in the same way that white middle-class visitors do, but that assumption is most likely false." He concluded, "Poor people, black people, and ethnic minorities generally show little enthusiasm for the park idea."[27] One might add that rural and working-class Canadians and Americans who worked on the land may have felt less need to "experience nature" on vacation – if, indeed, they could afford vacations.

If some visitors were excluded "passively," by being offered amenities that were of little interest to them or to which they felt they would not belong, others were excluded in more overt ways. The most flagrant cases generated complaints to the Parks Branch in Ottawa, resulting in a paper trail detailing the discrimination.

I should immediately note that the cases which follow are the only ones I found in researching the four parks' histories: they are not meant to be representative of how minorities were treated in the Eastern parks. But they do indicate, by way of the Parks Branch's responses to these incidents, staff's quiet awareness of subtle and not-so-subtle forms of discrimination both within the park system and society at large.

In 1942, parks controller James Smart wrote PEI superintendent Ernest Smith concerning a visitor's list of complaints after a stay at Dalvay. Problems included an absence of rugs on the floor and an abundance of over-active children. Number seven on Smart's list was "Jews are being allowed into the hotel."[28] Smith confirmed that this was true: "Under the terms of the lease the hotel is a public hotel. ... It might be added that during the course of the season approximately one dozen people of Hebrew extraction were guests and they were very fine people and well behaved. The question of their admission is one for the Department to decide in the light of the policy of other National Parks hotels. ... I believe the management of Dalvay would not accept them if they could avoid it. Personally I see no objection as long as they are of a suitable class."[29] The matter was considered minor enough that Smart did not even respond.

The question of allowing Jews in Prince Edward Island National Park arose again in 1948. It came to the attention of parks publicity man H.S. Robinson that Dalvay's brochure advertised "Restricted Clientele" – that is, a visitor could trust that no Jews would be there. Robinson called the brochure "objectionable and dangerous" to the Parks Branch, but the assistant controller, Spero, was not so sure. He felt that, since a concessionaire, not the Parks Branch, ran the hotel, the brochure did not reflect poorly on the department. Spero stated, "Other concessions operated in the National Parks have fairly strict rules concerning the admission of certain races and while these are not advertised they are strictly observed. In any event, the operator of a hotel can always refuse admission to a guest under the pretext of lack of accommodation."[30] Deputy minister Keenleyside vehemently disagreed, believing that such advertising did sully the park's name.[31] The concessionaire was told to remove this phrase. The following year, however, it was found that nothing had been done: the concessionaire had thousands of the offending brochures and wanted to use them up. The Parks Branch received complaints from the Canadian Jewish Congress and from the Quebec head of Women's Voluntary Services, who wrote, "I'm afraid I was not merely puzzled that this sort of thing should be associated with a National Park, but rather horrified as well. The recent war does not seem to have taught people much of a lesson if this sort of attitude persists."[32] Ottawa ordered the brochures destroyed. The Parks Branch's conduct in this incident suggests that it was concerned with discrimination, but primarily in terms of how it affected public relations. Staff recognized racial bias to be antithetical to the park system's aims, and took action to snuff it out. Yet Spero clearly knew of a well-developed system of discrimination within the park concessions. Neither he nor those with whom he corresponded did anything to address it. Once the brochures were destroyed, the problem was solved as far as the Branch was concerned; the offence was only offensive if it was public.

The Parks Branch reacted in like manner to another incident of racial discrimination in the summer of 1960. A professor of theology at Boston University, Harold DeWolf, wrote the owner of the Fundy Park Chalets to confirm a reservation for himself and his wife. There was also, he added, a young couple he wished to invite to come with them: "Canada's history being what it is, we feel confident that you would treat them well, but we want to make sure, to avoid any possibility of embarrassment. The friends of whom I speak are a fine Negro minister and his wife. They might want to take also their young children, but would more likely leave them with relatives. The young man is university-trained, with four degrees, an author and in every sense a cultivated gentleman. His wife is also a cultured person of superior character."[33] DeWolf waited for a response throughout May, throughout June, and finally received it just before leaving for New Brunswick at the beginning of July. Robert Friars, who had bought the bungalow court from the Parks Branch in 1957 and now operated it, confirmed the DeWolfs' reservation, but noted, "With regard to your friends whom you mentioned ... I feel that I cannot accept the possibility of embarrassment which may arise from this situation. Each day we have over one hundred guests at our site of which a great many are from the New England States, as well as those farther South. For this reason we feel that it would be better not to accommodate your friends."[34] So the DeWolfs travelled to Canada alone, without their young black friends, Dr and Mrs Martin Luther King, Jr.[35]

Harold DeWolf was King's professor and mentor at the Boston University school of theology, and they had stayed friends after King received his Ph.D. there in 1955.[36] It would seem that King periodically relied on DeWolf to help him get away from the rigours of the civil rights movement. Biographer Stephen Oates recounts that in 1956, King was feeling run down from a long speaking tour and talked to DeWolf "about arranging a retreat for him in Boston, a sanctuary where he could be alone for 'spiritual renewal and writing.'"[37]

In 1960, DeWolf (who eight years later would be the only white person to speak at King's memorial service) again tried to help relieve his old friend of the stresses that came with being Martin Luther King. It was a very hectic spring for the young preacher. He was the inspirational leader of student sit-ins in the South against lunch-counter segregation. He had been arrested on trumped-up charges of income tax fraud in February and acquitted by an all-white jury in late May. In June, he met privately with presidential hopeful John F. Kennedy.

King never learned why the planned summer vacation to the Maritimes fell through. Harold and Madeline DeWolf travelled alone to Fundy, and while there tried to meet with Robert Friars and Superintendent J.D.B. MacFarlane, but without success. Upon his return to the US, DeWolf wrote the Parks Branch in Ottawa and told of the offending correspondence, concluding, "Mr. Friars' letter, when it came, while putting the blame on possible guests who might object – as is customary in discrimination practices – quite flatly declined to accept our friends at the Chalets."[38]

To its credit, the Parks Branch responded quickly and unequivocally. Chief B.I.M. Strong told Superintendent MacFarlane, "The action of Mr. Friars in refusing accommodation on the basis of colour is certainly something we cannot condone. ... I want you to make it clear to Mr. Friars that his arbitrary decision ... was certainly not in keeping with the democracy and freedom of which Canada is so justly proud. He should also be informed that if another incident of this nature comes to our attention corrective attention will be taken by the Department."[39] The superintendent, perhaps feeling that this episode reflected poorly on him, defended Friars by noting that his rejection, though poorly worded, had only been intended to avoid embarrassment. As to Fundy Park Chalets not permitting blacks, MacFarlane stated, "In fact it was just about this time that a Mr. and Mrs. Brown from Philadelphia (negroes) stayed at Mr. Friars' for several days."[40]

Staff in Ottawa were more than happy to have this incident interpreted for them as a misunderstanding. An unsigned note from within the Parks Branch on MacFarlane's letter reads, "This is much 'ado' about nothing. Prepare the usual 'will not happen' reply and thank the Supt. for this."[41] In a mollifying note to DeWolf that both apologized for and rationalized Friars's letter, Strong wrote, "Admittedly, Mr. Friars' statement ... was in bad taste but it was not by any means intended as a direct refusal to make accommodation available. ... His action can only be attributed to overzealousness on his part to avoid possible embarrassment being precipitated by other patrons of his establishment. I am satisfied this will not happen again."[42] Whether DeWolf accepted this explanation or not, he chose to pursue the matter no further.

Can we judge Friars's intentions with complete certainty? Most likely, Friars did tolerate the presence of blacks, but was quite willing to refuse entrance to an individual who sought permission beforehand. His slippery response to DeWolf was worded to let himself off the hook: I don't discriminate, but others might. Less slippery is his conclusion: "it would be better not to accommodate your friends." As for the National Parks staff, they were quick to choose an interpretation of the episode that was easiest to smooth over. What was important was to alienate no one, to make Fundy a park that everyone could enjoy, black and bigot alike. It was not considered necessary to draft a directive to all park staff explicitly stating the Canadian National Parks' racial policy. The unpleasantness at Fundy was interpreted as an isolated incident.

There is no way of knowing how isolated it really was. It would be helpful to know if the exclusivity of the national parks often permitted or even fostered such acts of exclusion. What can be said is that the park staff's response to these episodes of discrimination is analogous to Canada's experience as a whole. As historians Robin Winks and James Walker point out, Canadians have traditionally felt a certain moral superiority to Americans based on allegedly more progressive race relations.[43] A major reason for this, however, was that discrimination was always less visible, more anecdotal, since visible minorities here did not attain the population concentration that blacks in particular did in the United States. The only way the National Parks Branch could have proven its antipathy to prejudice would have been to respond more decisively in the instances of discrimination it faced.

THE HOTEL BUSINESS

Indecisiveness on questions of exclusion was of a piece with the Parks Branch's overall stance on tourist accommodation. Once the Parks Branch had completed the tasks of park selection, expropriation, and development, its official position was that it wanted little to do with the new tourist operations that were the principal justification for the parks in the first place. The feeding and accommodating of visitors would be left to concessionaires who would invest in the right to participate in Branch-approved facilities. As staff were quick to tell provincial politicians and local entrepreneurs, Ottawa had created the park; it was now up to local industry to make it a worthwhile investment. The Parks Branch's thinking on this was understandable. It was a federal agency and not in the business of conducting business. The parks had been established to encourage the local tourism industry, not compete with it.

And yet such claims were somewhat disingenuous. It was the Parks Branch which made sure that each of the first Maritime parks had some impressive form of tourist accommodation within its borders, though in each case there were nearby communities that could have been expected to fulfil visitors' needs. In fact, Cape Breton Highlands and Prince Edward Island parks were selected in part specifically because staff envisioned the houses at Middlehead and Dalvay as reminiscent of grand hotels like those at Banff, Jasper, and Waterton. The Parks Branch believed that national parks must have accommodations within their borders, and ones worthy of the park name.

As a result, in the four parks studied here a similar pattern emerged. The Parks Branch offered land for development of accommodations within its parks. Staff expected investors to provide facilities of the highest aesthetic standards. Investors, though, felt such qualifications impossible to reach, particularly since they would only be making money on their investment a few months each year. Local operators wondered why the federal government did not supply the facilities themselves. Unable to attract any interest when tenders were called, the Parks Branch responded by doing everything possible to accommodate what interest there was. The lucky concessionaire was offered concessions to ensure his continued satisfaction. Staff blamed the situation on a lack of local initiative while, through time, other local tourist operators blamed the federal bureau for creating a subsidized business against which they could not hope to compete.

From the first days of the first two Maritime parks, the Parks Branch understood that accommodating tourists would be a problem. In his 1934 report on the proposed Nova Scotia park, R.W. Cautley noted, "it will probably be found necessary for the Parks Branch to finance the construction of suitable tourist accommodations, and either to put people in charge or rent them to carefully selected contractors."[44] There simply were no hotel facilities in the Cape Breton Highlands area. Indeed, the head of the Cape Breton Tourist Association admitted that visitors thought the situation ghastly: "as modern accommodations go, Cape Breton is at zero."[45] James Smart worked diligently to find investors. He contacted Banff leaseholder Fay Becker, guaranteeing that a bungalow camp near the new park was bound to be a great success since there was really no quality competition.[46] Failing to win Becker over, Smart offered a location in the park's development headquarters at Middlehead to a group of

Sydney businessmen. The Parks Branch's plan for the accommodations may be seen in Smart's suggestion that the investors should "really cater to high-class business and therefore should not stint on providing facilities in the way of comfortable and well furnished cabins and, above all, a first-class central building." Another set of cabins would be built near the golf course for "those with more moderate means."[47] Concerned by the scale of the project, the Sydney group backed out. Knowing that the federal government was having no luck finding sponsors, the Nova Scotia government asked what was to become a common question: since this was a federal project, why did the federal government not supply suitable lodgings? Smart replied, "You seem to believe that in other parks this accommodation has been provided by the Government. Such is not the case. Hotels, bungalow courts, and other tourist accommodation is provided by private enterprise."[48] The provincial government finally agreed to tackle the problem itself, building Keltic Lodge in 1940 and immediately regretting its decision. Though it tried to convince the Parks Branch to take over control of the lodge, the federal agency demurred, saying it was not in the hotel business.[49]

The Parks Branch always declared its willingness to work with would-be concessionaires, but federal wishes tended to surpass local capabilities. Katherine Wyand had opened Avonlea Cottages at Cavendish Beach to tourists in the 1920s, and when word came that her land would be expropriated for Prince Edward Island National Park, she asked to stay. The Parks Branch was interested in having the services of an experienced tourist operator and so tentatively agreed that she could have a cottage concession. But Cautley reported from PEI, "The Wyand cottages are an eyesore. Badly built in every variety of slipshod construction. Located without the slightest sense of order. Painted different colours. ... I recommend that the Provincial Government be required to buy her out lock stock & barrel – NOW."[50] The Parks Branch agreed, and rescinded its invitation. The following summer, the Wyands were hard at work on their cottages, refusing to move. Cautley recommended they be told "there is no question involved of their being given preference as possible concessionaires. On their own showing, they have not sufficient means to build first-class bungalow cottages, and what they have built show bad judgment in construction, location and style."[51] Wyand, though, said that thanks to the park, local land prices had jumped so much she could not afford to relocate. She became a vocal opponent of the park, writing letters to the editor detailing the bureaucratic run-around she had faced, as she travelled from Parks "Controller" to "foreman" to "Director" to "Commissioner" to "Inspector" in hopes of keeping her business.[52]

Surprisingly, in 1938 the Parks Branch offered Wyand the cottage concession. No one else had put in a tender for it, and the new national park badly needed accommodations. The main stipulation was that she fix up the cottages (with the Branch's help) and let them be moved off the beach to a less central location. Wyand happily agreed. The cottages were tidied up and painted white with green trim. A park staffer reported, "While they do not come up to the standard required in other parks they are clean presentable buildings."[53] Visiting the park in 1940, James Smart was less satisfied: "I doubt whether we can expect to ever have this camp present the appearance desired. The buildings are all of such design as to preclude any pleasing or harmonizing finish. The best we can expect is neatness. The final outcome of this camp will be its abandonment and complete removal through lack of patronage."[54] The Parks Branch had every right, of course, to retain only park buildings that matched its standards. But Cautley's and Smart's disgust over Wyand's cottages suggests how deep was the divide between the two sides' aesthetics. Wyand needed the cottages to be simple and serviceable, the kind that could withstand families tracking in sand. Besides, they would only be used a few months a year anyway. The Parks Branch wanted the cottages to possess the order and symmetry that said they were the finest that Canada had to offer. Local tourist operators probably did not possess the funds, never mind the sensibility, to fulfil such demands.

By the late 1940s, the Parks Branch's thinking had changed in a number of ways. Staff noticed a dipping interest in the park system's resort-style hotels. Visitors were instead turning to the parks' bungalow camps for reasonably priced yet comfortable accommodation: the expanding middle class was seeking middle-class facilities. This suited the Parks Branch fine. Projects like Keltic Lodge and Dalvay, which had been thought necessary totems of park status only a decade earlier, were losing considerable money. Dalvay was called a "white elephant" and it was planned that if more accommodations were to be built at Prince Edward Island National Park, they would be ones for "people of moderate means."[55] This was the general plan for all park facility development at the time. "The whole trouble," according to one staffer, "is that the present rates are too high for the average family man."[56]

One might think, on this evidence, that cottages like those Wyand offered would have found their niche. But the Parks Branch still wanted park buildings to possess the same high standards as always. Therefore, when the new Fundy National Park was being designed in the late 1940s, though it was given a bungalow camp rather than a hotel, the camp was of the Parks Branch's own construction and design. Twenty-nine cottages were built along a single line on a bluff overlooking the golf course. All were an identical twenty feet by twenty feet with a high French-style roof and fetching wooden shutters with hearts carved in them. All came with the most modern of conveniences, including a refrigerator, gas stove, and bathroom. A New Brunswick newspaper article of the time suggested that such accommodations were "almost sure to have a family-man appeal. ... [I]t would seem that a low-budget visitor could enjoy a peak holiday at bargain basement figures."[57] The Parks Branch justified such development as necessary to provide Fundy with accommodation when it opened. But more important, staff meant to set "a standard for the type of layout we would like to see undertaken by private enterprise."[58] The Parks Branch built similar bungalow camps at Cape Breton Highlands and Prince Edward Island parks in 1949 and 1950. Again, it was made clear that such developments only provided a "model establishment which might encourage others to invest capital in a similar scheme. It was never intended, however, that the National Park Service should embark in the hotel or bungalow camp business on a wide scale."[59]

The Parks Branch put out a call for tenders on the three new cabin developments, but received only a few expressions of interest for each. In one sense this seemed to justify the building of the cottages in the first place, since local people obviously did not want to invest in tourism. But there were reasons for locals to act cautiously. Tourism was still a rather new enterprise in areas surrounding the parks (particularly at Fundy and Cape Breton Highlands), and people did not trust that they could make money on it in just a few months a year. If new to the industry, they may have doubted their ability to run these first-class park facilities. Many people also simply did not understand that such facilities were intended for local investment; others who did know this felt that, considering how long the government had funded Western parks, it should invest the money in these new parks itself.[60]

From small lists of applicants, the Parks Branch chose Robert Friars to lease its bungalow camp at Fundy, R.S. Humphrey at PEI, and Charles Fownes at Cape Breton Highlands. Each would be charged 20 per cent of the gross rental revenue for the year, a sum that suggested a reasonable return for the government without being onerous to the concessionaire.[61] In retrospect, it is little wonder that these bungalow camps became successes, booked solid by the beginning of each season, and quite profitable for the leaseholders. The cottages were attractive, maintained by the Parks Branch, well advertised in park publications, and possessing a monopoly within the park. And they were priced affordably for visitors, because the Parks Branch wanted to attract the middle class. The agency acknowledged that in setting this low price it was limiting the concessionaire's potential profit, so it offered in return a low rental rate.

The Parks Branch did not find the developments so profitable, having built the fine structures and leased them cheaply. Staff soon calculated that each Maritime bungalow camp was losing the government over $2000 each year. When the concessionaires expressed interest in purchasing the bungalow camps, the agency took them up on it. Though the PEI camp had an assessed value of $67,000, the New Brunswick camp $73,000, and the Nova Scotia camp $60,000, the Parks Branch accepted offers of $55,000, $55,000, and $52,000 respectively.[62] The government had given the concessionaires virtual monopolies at considerable cost to itself, and at the expense of other potential local investment in tourism accommodation.

From Ottawa's point of view, the entire episode could be blamed on a lack of local initiative. The minister, Jean Lesage, defended to a PEI politician the decision to sell the tourist cabins without public offering by stating, "These courts were built by the department at a time when it was quite impossible to interest private persons in such construction."[63] The same was said of conditions at the other Maritime parks. James Smart wrote that the community of Alma just outside Fundy "has not yet come to the idea that they are overlooking the opportunity passing by their doors of bettering conditions in their town and developing it as one of the most attractive tourist towns in the Maritimes."[64] But the situation was not as simple as park staff might pretend. As Fundy superintendent Saunders noted, no one in Alma had the money to invest in tourism on any scale, and if they did, Fundy's facilities were hardly ones they could aspire to. Locals felt that the Parks Branch itself should provide facilities, since it had the location offered by the park itself and obviously had the funds to do the job right. In this way, the facilities which were intended to encourage local investment more likely served to discourage it.

Considering the money the government had lost in the bungalow camps of the first Maritime parks, it is surprising that exactly the same method was followed in the development of Terra Nova National Park. Nineteen cabins (of simpler design) were built at Newman Sound by the department, and tenders were called to manage them. Only two submissions were received, and the Parks Branch chose St John's businessman Norval Blair to take the concession. The cabins were an immediate hit, especially with vacationing Newfoundlanders. Blair soon asked if he could provide some services himself, and the Branch permitted or refused on the basis of applicability to park taste:

a laundry room (yes), a playground (yes), a passenger boat service (yes), a hot-dog stand (no), a shuffleboard court and horse-shoe pitch (no, no), and a miniature roller coaster (no).[65] Blair also asked that the Branch itself inject more funds in the site by building tennis courts, a swimming pool, more cabins. Meanwhile, Superintendent Doak was receiving complaints from locals who "did not know what the Government was doing in the accommodation business and taking trade away from them."[66] The department decided that the best course of action would be for Blair to buy the cabins, as the concessionaires at the other Maritime parks had. After all, noted the superintendent, "we should not be in the 'hotel business.' "[67] Blair expressed tentative interest in the idea but, thinking like a good businessman, said he would prefer to wait until the government had built the twenty or thirty more cabins he was asking for.[68]

No clear lessons emerged from the Parks Branch's foray into the business end of tourism. Perhaps it was worth it to have accommodations in parks, even those with communities nearby, because they improved the visitor's experience of the park. Perhaps the accommodations built within the Maritime parks really did serve as models for local tourism operators and thus ultimately helped local industry become stronger. Such results are hard to determine. More interesting is the manner in which the Parks Branch justified its actions by claiming a lack of local initiative, while ignoring both its own motivations for park development and the knotty reasons why locals found investing so difficult to imagine. And then, after investing considerably in park accommodations, the Parks Branch felt obliged to show its uninvolvement in capitalism by selling them at a loss to a businessman who would enjoy a valuable monopoly. The Parks Branch *was* in the hotel business, and it would have cost taxpayers and local business less if it had admitted so.

INVIOLABILITY

Throughout the national park system's history, parks were created because the benefits they provided their provinces and Canada – nationalistic pride in the place's beauty, an influx of federal development money, and, above all, tourism – were thought more valuable for that specific land than the small-scale economies that they replaced. But these benefits were not believed to measure up against those of large-scale economic investment. If alternative uses for park land were discovered to hold out the prospect of more money for a province, politicians lobbied for them. In many instances, the national parks were threatened by provincial governments and private interests which thought of parks as little more than land held in trust, ready to be reclaimed when needed. And the federal government quite often accommodated. Coal was taken from Jasper and Banff; lead, silver, and zinc from Yoho; and timber from a number of parks, including Banff, Riding Mountain, and Wood Buffalo.[69]

For the first decades of the park system, staff seem to have shared with the rest of Canadians the belief that preservation should not be allowed to interfere with economic development. But in the 1920s, the Parks Branch began to declare in its annual reports that what made national parks unique was the principle of inviolability, the notion that parks should be forever safe from the incursions of either private development or other government departments. National parks were distinct because they were timeless, representing not only a distant past without human interference in nature but a refuge from any future meddling as well. Canadian park historian C.J. Taylor sees the first great test of inviolability to be the Calgary Power Corporation's 1923 proposal to dam Banff's Spray Lakes for hydroelectric purposes. Commissioner James Harkin argued vehemently against the damming on the basis that parks had no intrinsic meaning if they were not free from the economic concerns that affected other land. When it became clear that development at Spray Lakes would go through, Harkin's Parks Branch asked that this and other areas with obvious resource value be removed from the park system. Taylor writes of Harkin, "Principle rather than size had now become his priority."[70] Though Spray Lakes was taken from the park, the incident was in one sense a victory for inviolability, demonstrating that when threatened by economic exploitation the Parks Branch would chop off its hand rather than have its whole body infected. Inviolability has been an intrinsic part of the national park ideal ever since.

The Spray Lakes case demonstrates inviolability's intrinsic weaknesses, however. First, it is ultimately unenforceable, since any statute that attempts completely to fetter the power of future executives is not legally binding. Harkin could speak of parks' singular status, but there was no way for him to make this law; nothing could keep future governments from expanding, contracting, or indeed terminating the parks as they saw fit.[71] Second, the Parks Branch's own place within the federal system made maintaining inviolability very difficult. At various times during the years under study here, parks were part of the ministries of Interior, Mines and Resources, Resources and Development, and Northern Affairs and National Resources, all of which were primarily concerned with land use and resource extraction. As in the Spray Lakes case, the ministry was as likely to side with the proponents of economic development as with its own Parks

Branch. Finally, inviolability could always be sidestepped as it had been at Spray Lakes, by removing the contested land from the park system and from the restraint of inviolability. National parks were inviolable until they were needed.

When Maritime provinces began to express interest in having their own national parks, the Parks Branch firmly and clearly explained that the right to all timber, minerals, and other resources would transfer to Canada with park establishment. It would seem that the provinces understood this well enough. In 1929, Nova Scotia premier G.F. Harrington chose not to push further for a park at that time because "it would be improper for me to recommend to my colleagues that the people of Nova Scotia abandoned all interest in the possible mineral deposits in the northern part of Cape Breton."[72] Seven years later, the next Nova Scotia government made a different decision, and agreed to the Cape Breton Highlands park. The federal government again impressed on its provincial counterpart that this would mean a loss of resources. Federal minister of mines and resources T.A. Crerar, answering a provincial question about mineral rights, was careful to suggest that "If any portion more valuable for mineral development than for National Park purposes such portions should be eliminated now."[73] Yet inviolability was such a new concept that even with such blunt declarations, the federal department fretted that it might still be somehow undermined. When the Nova Scotia premier, Angus L. Macdonald, agreed "that the Dominion Government will exercise full control of the park area in accordance with the authority vested in it by the National Parks Act and Regulations thereunder," deputy minister Wardle's handwritten response was: "Not enough."[74] The department wanted a more specific declaration of jurisdiction. The federal deputy minister of justice, W. Stuart Edwards, assured the department, though, that a general statement was far superior: an attempt to define all that Canada owned within the park would be bound to leave something out.[75] The Nova Scotia act ultimately transferred to Canada all "estate, right, title and interest in the land so purchased or acquired or leased" for the new park.[76]

Almost immediately, Cape Breton Highlands' inviolability was tested. A prospector from Toronto, F.M. Connell, asked the Parks Branch if he could reopen a gold mine at Clyburn Brook, near the centre of park development at Middlehead. He had tried mining in the area when gold was at $20 per ounce but found it unprofitable; now that it was $35 per ounce, it might be worth while.[77] Both the provincial and federal governments were troubled by this. As parks director Gibson noted, "The point of the whole thing is that this Middle Head area is the most desirable feature in the Park. We have known this from the beginning."[78] And there was concern that allowing mining in an Eastern park would set "a very dangerous precedent." Yet neither government wanted to turn Connell down. Neither wished to discourage investment in Cape Breton, particularly in an industry so central to the island's existence. After considerable soul-searching, the federal Department of Mines and Resources refused Connell's request.[79] The Parks Branch periodically received similar petitions to prospect in Cape Breton Highlands Park during the Second World War, but turned each one down.

The Connell incident was a minor one, but it demonstrates how the supposedly uncompromising concept of inviolability could be shaken by the prosaic needs of capitalism. The park system also had to deal with governments wanting to use park land. The Parks Branch was generally receptive, wanting to help the local economy and prove itself a good neighbour. After creating Fundy National Park, the Parks Branch was asked by the New Brunswick government if the provincial Potato Research Station could be allowed to continue operation within the park's borders. It took up just fourteen acres, after all, was tucked in an out-of-the-way place, and would only be needed for a decade. The Parks Branch felt that this would do no harm, and even gave the station a three-acre extension and some recently expropriated farm buildings. But this guest became a nuisance. In 1951, the Potato Station asked for an additional eight acres (and the right to cut the trees growing on them), and in 1954 park staff discovered that it had absorbed eleven more. Though the Parks Branch bristled at letting the research station stay longer, it succumbed to provincial and federal pressure to do so, and even ended up letting it have yet another eight acres for crop rotation.[80] The Parks Branch could not openly complain, because to do so was to admit that it did not have the control on park land that the claim of inviolability suggested. Besides, both the provincial and federal governments – who owed no allegiance to the notion of inviolability – were happy to see the land in use. In his authorized history of the Canadian national parks, W.F. Lothian calls Fundy's Potato Research Station just "an unusual feature," condoned because its research provided a service to Canadians and was relevant to New Brunswick agriculture. In noting that the site was finally moved out of Fundy in 1974, he does not mention that the Parks Branch had been trying to accomplish that objective on and off for the previous twenty years.[81]

The Parks Branch had a difficult time defending the principle of inviolability when either private business or government contested it. When the two acted in concert, the Branch's position was all but hopeless. At Cape Breton Highlands in 1953, staff were surprised to stumble across survey parties at work within the park. Toronto's Mineral Exploration Company soon reported finding high-grade metals in the area, including inside park boundaries.[82] With Cape Breton's coal industry in tough times, the Nova Scotia government did not even bother debating the province's right to the park minerals. It simply claimed them as its own. The provincial minister of mines bluntly told his federal counterpart, "The mineral right within the National Park belongs to the Province and if any mineral occurrence of apparent economic value is found within its boundaries the province would definitely like to have arrangements made for the surface necessary to carry out its development and exploitation to give employment to persons displaced by the coal industry."[83] Northern Affairs and National Resources disagreed with this interpretation, of course, pointing out that the province had lost the right to any minerals when it handed over the park land.[84] But the department was not actually opposed to park mineral extraction, since development would assist the provincial economy and show the federal government to be a gracious partner. The Canadian government allowed a Nova Scotia crew to survey throughout the park, the only conditions being that the work go unpublicized and that the superintendent be kept informed; the Parks Branch had to appear to be in complete control of the land.[85] The survey results for the park's southwest corner showed promise of diamonds and other minerals, making even the pretence of inviolability unsustainable. Nova Scotia's minister of mines stated, "Although the mineral rights belong to the Crown in the Right of the Province of Nova Scotia, yet we realize that National Park regulations prohibit the prospecting or development of minerals within its boundary. ... [W]e feel that it would be a reasonable request to change the Park boundary to the position indicated," and with a red pencil, he cropped a small section off a map of the park.[86] The Parks Branch offered surprisingly tame resistance when discussing this proposal within its ministry. The southern part of the Highlands was too secluded to have any real value for park aesthetics or development, so it was deemed relatively expendable. The federal government agreed in the spring of 1956 to remove 13.3 square miles from the Cape Breton Highlands park. This was the same sort of compromise as had occurred at Spray Lakes in the 1920s: the province got the land it wanted for development, the federal ministry looked pragmatic and benevolent,[87] and the Parks Branch could say that its principle of keeping parks free from resource extraction had held firm. And yet Cape Breton Highlands National Park, created less than twenty years earlier to be maintained forever, had been chipped away as a result of the first significant interest in its resources' traditional uses.

Having accepted one infringement on its Nova Scotia park, the Parks Branch immediately faced another. In October of 1956, representatives of the Nova Scotia Power Commission asked permission to land a seaplane on Cheticamp Lake in the southeastern part of the park's interior, and set up camp on its shore. This puzzled park staff, since the crew would then have to travel some distance if they were doing work outside the park.[88] Only then did the Parks Branch learn that the Power Commission had for months been considering a major hydroelectric project at Wreck Cove, a few miles outside the park's southern boundary. The provincial government of Liberal premier Henry Hicks had set aside $30 million for the utility – which was chaired by Hicks himself – to increase generating capacity and improve services.[89] The Power Commission now hoped to divert the Cheticamp River headwaters drainage basin centred at Cheticamp Lake away from its traditional course down the western side of the island, and instead send it off to the east to supply a hydro plant at Wreck Cove. In asking by telegram for permission to survey the area, the Power Commission explained that it considered "Wreck Cove of deep significance to industrial future Cape Breton."[90]

Park staff were unanimous in their opposition. The proposed project would dry up the Cheticamp River, lessening the aesthetic and recreational attraction of the western side of the park, which was coincidentally undergoing extensive tourism development.[91] It would, according to the Canadian Wildlife Service, most certainly ruin the fishing that the park was trying to encourage, and draw the ire of the federal Department of Fisheries.[92] It would unfairly target a national park which had already lost land that very year, and it would weaken the park system's more general claim to inviolability. Parks chief Coleman compared this situation to the hydro project at Banff's Spray Lakes, citing both the original damage it had caused the park system and the subsequent damage caused by further hydro development there in the early and late 1940s.[93] For all these reasons, the Parks Branch pleaded with its ministry to turn down the province's request. Director James Hutchison told his deputy minister, "The Cape Breton Highlands National Park is one of the outstanding parks in the system. We are all aware of the favourable comment from visitors and of the pride that the people of Nova Scotia have taken in the park. ... If the park is eventually to fulfill its proper purpose the area now enclosed within the boundaries should be defended to the uttermost and the natural features of streams, lakes, mountains, forest cover, should

not be sacrificed for the need of immediate commercial requirements."[94] Impressed, the department supported the Parks Branch. The
assistant deputy minister wrote the Power Commission in early
November that its request to survey the Cheticamp Lake area and ultimately to divert the Cheticamp watershed was denied.[95]

Within two weeks the federal position was reversed. The Power
Commission could have its survey, though there was still no promise
of development.[96] The turn-around was the result of heavy lobbying
by the provincial government, especially by Premier Hicks himself.[97]
The only power that the Parks Branch could exercise in the matter was
to act as if it was in complete control of the situation. It was agreed that
the survey by Power Commission staff would be paid for by the Parks
Branch, so it could "study" what a hydroelectric diversion would
mean to the Cheticamp River salmon population. This may have been
an effort to keep the survey from seeming a precedent for development, but it also prevented the Branch from voicing further opposition. When the Cheticamp Board of Trade heard about the survey for
the proposed Wreck Cove diversion project, it informed the federal
ministry that it would "enter a strong protest on the grounds that it
would be detrimental to the tourist industry, which has a great bearing
on the economy of this area."[98] Alvin Hamilton, the minister of northern affairs and national resources, replied that there was apparently
some misunderstanding: his own department was handling the survey, and "They are engaged only in determining the quantity of water
flowing in the river and are not reporting in any manner upon the possibility of diversion from Cheticamp Lake."[99] By maintaining a pretence of complete control over what happened to the park, the Parks
Branch further limited its opportunity to exercise real control.

The survey reported its findings in the fall of 1957, and the Nova
Scotia Power Commission interpreted them to mean that the
Cheticamp watershed would be valuable to the Wreck Cove project.
Of course, the Commission then asked for the watershed. Parks director J.R.B. Coleman was incensed: "I wish to go on record as being absolutely opposed to either the diversion of water from the Cheticamp
River or any further reduction of the Park area."[100] Coleman was especially angered by the Power Commission's refusal to say whether
the diversion was necessary for the project's success. Deputy minister
Robertson agreed, noting, "It is not at all apparent to me, however,
that the advantages of the Wreck Cove project with the diversion (as
distinct from Wreck Cove without the diversion) are so outstanding
as to justify an encroachment on one of our best national parks."[101]

Though the federal ministry probably would have permitted the
removal of park land for the hydro project anyway, the Parks Branch

was mollified somewhat when the newly elected Nova Scotia premier, Robert Stanfield, offered to give it land in return for the Cheticamp Lake watershed. His original suggestion was the entire northern tip of Inverness county (the northwest corner of Cape Breton Island), which had been originally intended as part of the park. Upon consideration, though, the Parks Branch decided that the area consisted of barrens not conducive to tourism and not sufficiently attractive. Stanfield then offered five acres of provincial land bordering Canada's historic site at Port Royal. The Parks Branch was much happier to get these few accessible acres than a sizeable out-of-the-way corner of the province. "Wonderful!" Coleman proclaimed. A more restrained R.G. Robertson said, "It is better than nothing – which is what we got when the last chunk was cut off."[102]

Having been promised land in return and assured that a token flow of water would continue down the Cheticamp River, the Branch no longer contested the Power Commission's threat to park inviolability. In early 1958, about ten square miles were removed from Cape Breton Highlands National Park's Cheticamp Lake area. The Canadian government held onto the land, waiting for word from Nova Scotia that it wanted the parcel transferred. That word did not come. The Power Commission decided that the Wreck Cove development was not viable after all, for the time being at least.

Fifteen years later, the Power Commission felt ready to move forward. Its plans were largely unchanged: build about twenty dams and dikes to control the water of over eighty square miles of the Cape Breton Highlands, directing this water to an underground powerhouse at Wreck Cove. Environmentalism had grown considerably in the interim, however, and the Commission now faced considerable opposition from those who felt the plant would destroy fish habitat, flood some of the best moose range in the province, and taint some of the last remaining Nova Scotia wilderness.[103] Though the project was sure to affect Cape Breton Highlands National Park, the Parks Branch kept quiet during this debate; it had fought and lost its battle years earlier. Wreck Cove, when completed in 1978, became an integral part of the Nova Scotia's energy system, immediately doubling the power produced in the province.[104] The Cheticamp River is a shadow of its former self, most of its water diverted to the east side of Cape Breton. The large stones which once held the current as it flowed down out of the Highlands now merely hide the trickle as it runs underneath.

The Parks Branch never found improved ways to ensure that the principle of inviolability would be practised. It faced much the same difficulties in keeping park resources sacred in the 1950s as it had in the 1920s. At the establishment of Terra Nova National Park, the Newfoundland government fought even more aggressively for its pro-development position than had provincial governments in the past, demanding and winning the right to pre-empt inviolability by leaving room in the land transfer for the future removal of timber. Newfoundland even forced Canada to take the park in two lots so that, as their agreement read, "If at any future date the Province should require any of the lands described in Schedule 'B' for the purpose of hydro-electric, or other commercial development, Canada will introduce into the Parliament of Canada such legislation as may be necessary to exclude from the Park all or any portion of the said lands described in Schedule 'B.'"[105] Legal scholar N.D. Bankes calls this "a remarkable provision from a legal perspective" because it both tries to tell a future Parliament how it must act – which cannot be legally binding – and then only forces it to "introduce" legislation, not demanding that such legislation ever be passed.[106] Just as the Parks Branch had difficulty codifying inviolability, Newfoundland had difficulty codifying violability.

The concessions that the Parks Branch made to public and private economic interests both willingly and unwillingly in the first Atlantic Canadian parks demonstrate how tenuous its hold on the park system really was. Parks existed because of the value parks gave their land; if a greater value was found, the park's existence was in doubt. The dogma of park inviolability was perhaps needed to convince Canadians of the special sort of value that parks provided, but it was ultimately untruthful. The Parks Branch could no more honestly claim that it was protecting a plot of land forever than it could claim that this land, expropriated within the previous half-century, was pristine wilderness. Just as a statute cannot bind future legislators to a certain act, a concept cannot bind future Canadians to an appreciation for parks. Inviolability will not be achieved through the shortcut of declaring it a timeless park principle; it demands that park staff work at it every day and in every park under their jurisdiction.

EROSION

Throughout its history, the Parks Branch's primary concern was in fulfilling what the minister of interior, Charles Stewart, in 1930 had called "a pious hope": the perfect marriage of use and preservation. Park staff had faith that a balance between the two could always be found, that they could, even as they sought to increase public patronage, also keep the park system intact and inviolate for future generations. But in the late 1950s, the Parks Branch for the first time began to doubt its ability to achieve this balance. The parks were becoming perhaps too popular.

The park system's initial response was its traditional one, to build facilities that would satisfy tourists. Between 1953 and 1959, the Parks Branch's government appropriation quadrupled to $24 million as the federal cabinet approved a major highway construction project throughout the parks, and the construction of better accommodations, particularly campgrounds.[107] The Canadian program mirrored the United States' Mission 66 program, initiated in 1956 to help the Park Service meet growing needs.[108] But the Parks Branch discovered, as did its American cousin, that this time development did not resolve the problem of overuse. Great increases in the number of park visitors meant that the new facilities were insufficient as soon as they were completed. Moreover, as Alfred Runte writes of the American case, "By enabling more tourists to visit the parks, they inevitably came."[109] Better facilities for the masses led to more masses. The Parks Branch's 1959 annual report speaks of the need to know "how many can use the area without gradual deterioration," suggesting for the first time a niggling doubt whether preservation and use could be effectively balanced.[110]

Symbolic of the Branch's coming to terms with park use in the 1950s was the case of Robinsons Island in Prince Edward Island National Park. Robinsons Island was a three-mile-long green-bean shaped island in the middle of the park, between the Cavendish-Rustico section to the west and the Brackley-Dalvay section to the east. Its west end was one long beach, home in the late nineteenth century to the Seaside Hotel and a number of fishing stages. The hotel burned down at the turn of the century, and the island was little used in the years to follow. When Williamson and Cromarty examined the proposed park land in 1936, they do not seem to have noticed the island's existence. The Higgs Commission in charge of calculating prices for the park expropriation recommended that Robinsons Island be excluded, since it would be expensive to buy, but it became part of the park nonetheless.[111] For fifteen years it sat unused, allowed to stay the most wild and inaccessible part of Prince Edward Island National Park. The Parks Branch took to calling it "Rustico Island" – freeing it of any association to past land ownership – and in 1960 the name was officially changed.[112]

Following the war, the Parks Branch considered how to best use all parts of the PEI park. The park's prime feature was obviously its shoreline, so improvements of the roads along the shore were given top priority. Grading, gravelling, and hard-surfacing the roads along the western and eastern ends of the park got under way. Staff began to toy with a more ambitious project: linking the two ends of Robinsons Island with the mainland to the west and east. This would fulfil a number of worthwhile objectives. First, it would make the park a single coherent unit; at present, visitors travelling from one side of the park to another had to exit it at Rustico, drive a dozen miles, and re-enter at Brackley.[113] Second, it would create a single scenic drive, showing the park to best advantage. Third, it would open up Robinsons Island to development. Prince Edward Island supported the project for all these reasons – plus, of course, it would bring federal funds and construction jobs to the province.[114]

There was another reason, a curious one, why linking Robinsons Island seemed a good idea. The harbour between the island and the community of Rustico just outside the park to the west – Big Harbour, as it was called – was constantly silting up. The resulting shallowness was a danger to fishermen working out of Rustico. Engineers deduced that if a causeway was built between the island and the eastern mainland – that is, across Little Harbour – there would be increased tidal flow through Big Harbour, scouring it much more efficiently. Local fishermen would have a deeper, safer harbour.[115] That was the theory, but no one knew for sure. Fishermen themselves were divided. The province's Fisheries Development Committee endorsed the plan, but some Rustico fishermen reckoned that more tidal movement at Big Harbour could just as easily mean more silting.[116] The Parks Branch began work on the Little Harbour causeway in 1953 regardless, while continuing to figure out what would happen. As the project neared completion two years later, parks director Hutchison was still describing it as work "which I understand is of an experimental nature."[117]

The Little Harbour causeway was completed in 1955. Sand quickly accumulated along its length, and Robinsons Island became, in essence, a peninsula. The same year, Parliament approved a major highway program for Canadian national parks to better serve the rising wave of tourists. As part of this program Prince Edward Island National Park received funds for a Gulf Shore Parkway, a reconstruction of existing shore roads to Trans-Canada Highway conditions. The $2.8 million project was to be capped off with a $500,000 high-trestle bridge across Big Harbour to Robinsons Island. This would be in addition to previously planned developments in the park, such as the completion of the new Cavendish campground, improvements to Brackley Beach, and the eventual construction of facilities on Robinsons Island. Despite the speed and scope of these planned changes, the Branch showed no sign of trepidation. After a visit to the park, deputy minister Robertson justified the development by saying, "While it is only a long, thin strip of land, the area that is available for actual use by people is, I think, as great as we have in almost any other park. In the case of Cape Breton Highlands Park, it is really only the coast portions of the Cabot Trail that are available for 'enjoyment' by the people of Canada. In the case of Fundy Park, it is only the headquarters area, Bennett Lake, and for scenic purposes the twelve-miles of road. Prince Edward Island Park has, I should think, more enjoyable and useful space than either of them."[118]

But in the late 1950s, park staff were given an omen that their plans for seemingly unlimited use might be unattainable. With the causeway at Little Harbour having closed off the east end of Robinsons Island, stronger tide movement began to eat away not at the Big Harbour sands but at the sandy west end of the island itself. Great chunks of the island eroded away. Public Works engineers conceded that more study was needed before a bridge to Rustico could be considered.[119] Development on the Gulf Shore Highway and other facilities still went ahead, accommodating and encouraging the mounting visitor traffic, which doubled at Prince Edward Island National Park between 1957 and 1960.

W.F. Lothian writes in the authorized history of the Canadian park system, "During the years following 1955, operations of the National Parks Branch had expanded to an extent that the existing organization was experiencing difficulty in carrying out its functions and responsibilities at a level of efficiency expected by the Department."[120] Among other things, it was felt that too many decisions were being made on a short-sighted, day-to-day basis. A Planning Section was created to look at park policies in long-range terms. The chief of this section, Lloyd Brooks, visited Prince Edward Island park in 1960, and was appalled by what he saw there. "Unfortunately," he wrote, "this Trans Canada type highway is now nearly completed in a manner and to a standard which is unrelated to the whole concept of national park use." He listed its faults: it disregarded natural topography, it encouraged speeding, it permitted shoulder parking, and worst of all it forced a major development straight through the middle of the entire park. As for Robinsons Island, development there had resulted "in the destruction of the single geographic feature which had a chance of surviving in a relatively natural state in spite of continued heavy use in the rest of the park." He hoped that rather than choosing to go ahead with plans to connect it with Rustico, the Parks Branch would help fund a road around Rustico Bay, outside the park. This would be "more consistent with park ideals and much less costly."[121]

The federal ministry could not so easily admit failure. After a 1960 trip to the park, deputy minister Robertson explained that the Parks Branch faced other factors: "The change in the Park since 1957 – and especially since 1955 – is quite remarkable. We have made very great progress indeed in improving the facilities, but the increased use is so tremendous that it is doubtful whether we have really gained much ground. ... Mr. Brooks has expressed the view that it was an error to build the road onto Rustico Island and thus eliminate the chance of preserving it. In principle, this may be right, but I think in fact it is and would be quite impossible. The pressure of use is too great to allow that relatively large area to be untouched." At a time when facility expansion plans set for 1962 were already proving insufficient for tourist needs in 1960, the Parks Branch could not possibly limit development, Robertson concluded. However, having seen the destruction of the west end of Robinsons Island, he acknowledged that it was probably hopeless to build a bridge there. The deputy minister instead suggested a bridge heading southwest off the island, cutting down the amount of travel outside the park. This was preferable to stopping the Robinsons Island road altogether because "It would make the construction of the road (at least at its high standard) along Rustico Island largely a waste. It would also, I am sure, bring sharp criticism from the people of the province."[122] After reading Robertson's report, the park superintendent put it more bluntly: "he feels it will emphasize a 'boob' on the Department's part."[123]

But the deputy minister could not win nature over to the wisdom of his plan. Because of the causeway built on the eastern end of Robinsons Island, the western end continued to erode away and the harbour entrance at Rustico silted up more than ever. Attempts to reinforce the island's western wall proved useless, as did repeated dredging of the harbour. In 1954, the park had stopped construction of roadway 1400 yards from the western tip of Robinsons Island. By 1968, the tide had taken so much of the island that the new tip was now only about 600 yards away. After continuing, constant erosion – including the loss of 100 yards of land in a single October night in 1975 – the widened Big Harbour reached within 125 yards of the Gulf Shore Highway.[124] Rustico Island was little more than half the length it had been twenty years earlier. The Parks Branch had by that time officially announced it was abandoning plans to link the island with Rustico.[125] In Parliament, Island MP Heath Macquarrie called the whole thing baffling: "It is hard to explain why the feasibility studies on the bridge were not done before the expensive, so-called connecting highway was laid. How difficult it is to explain to the people of Canada, the people of PEI, that while the US can get a man on the moon, and the USSR and the USA can place equipment on or around Mars and Venus with great precision, all the forces of the government of Canada cannot cross Rustico Harbour."[126]

The erosion of Robinsons Island was admittedly an unusual case: usually the strain of rapidly increased use caused a figurative rather than literal wearing away of the national parks. But the lesson to the Parks Branch was the same. Visitors' interest in and demands on parks were proving so great that staff could not hope to keep up, and if they did it would mean the destruction of the very elements they were mandated to protect. In the 1960s the Parks Branch would decide that preserving parks for future generations meant limiting to a degree even those uses for which the parks had been created. The alternative, like the Gulf Shore Highway that now ran right off the end of Robinsons Island, was a road to nowhere.

8 Changing Ecologies: Preservation in Four National Parks, 1935–65

The period 1935 to 1965 is an interesting one for studying Canadian national parks, if only because so little is supposed to have happened then. Most park histories offer only a pause between the 1930 Parks Act and the rise of environmental interest in parks in the 1960s. Kevin McNamee's twenty-five-page "From Wild Places to Endangered Spaces: A History of Canada's National Parks," for example, gives these years two small paragraphs.[1] A central reason for this neglect is that the system did not grow much in these decades: the only new parks were the four Atlantic Canada ones. There was a sense within the Parks Branch itself that the system was now complete, demanding only eternal vigilance to maintain.[2] Much the same attitude was present in the United States national park system, which saw the addition of only five new natural areas between 1940 and 1959. An American park historian has aptly titled this period the "we think we've done it all" era – and, also aptly, she then spends little time discussing it.[3]

The general neglect of this period by historians is surprising when one considers that in both Canada and the United States the 1960s is seen as one of the most important decades in park history. In that decade in both countries, the size of the park system mushroomed, there came to be greater public respect for the parks' preservationist philosophy, and skyrocketing park attendance fostered debate on how to curb development. The continent-wide environmental movement is seen as so central to all of this that the park systems' own histories are presumed to be irrelevant. I would argue that in fact the decades leading up to the 1960s were instrumental to the shaping of park preservationist policies. In its treatment of fish, wildlife, and vegetation, the Canadian National Parks Branch took an increasingly hands-on approach during the 1940s and 1950s, until – recognizing the ecological, philosophical, and political damage of its actions – it sought to become less interventionist in the 1960s. At both stages, it justified its approach by invoking the name of science.[4]

This chapter is a response to the interpretations on the role of science in parks offered by American historians Thomas Dunlap[5] and Richard West Sellars,[6] both writing on the interwar years. Dunlap, looking at the American and the Canadian park systems, sees park biologists as essential in transmitting to the public a rational justification for valuing wildlife and wilderness. Biologists could successfully defend such ideas as predator protection and biodiversity because they wore the badge of authority granted by science. Their work encouraged the slow growth of a public environmental sensibility which fully blossomed in the 1960s. Sellars, in contrast, sees the park biologists' influence in the 1930s to be ephemeral, even within the park system itself. As early as the end of that decade, foresters, developers, and landscape architects were gaining control of the parks, and they would enjoy that control unopposed until the 1960s. The environmental movement that then developed, Sellars argues, was not the result of the continuing influence of park scientists; in fact, it was sharply critical of current park management practices.

There is a middle path between these two arguments. As Dunlap contends, science gave preservation issues prominence from the 1930s on, and was always essential in formulating what park policies would be. This does not mean, however, that science advocated a "leave nature alone" policy that we now would consider more environmentalist. As Sellars contends, the spirit of the times called for intervention; in the late 1940s and the 1950s in particular, preservation was an active process. Science was used to justify all manner of intervention, from the stocking of park lakes to the killing of "surplus" wildlife. Whereas Sellars and Dunlap see scientists as antithetical to the landscape architects, planners, and foresters that prospered in the American park system after the war, I see them in the Canadian case as all sharing a managerial ethos. The perceived rationality of science rationalized the Parks Branch's attempt to improve on nature.

WILDLIFE

It may seem surprising that a book on national parks could travel so far with virtually no mention of wildlife. But this reflects the Parks Branch's general lack of interest in the wildlife that the new Atlantic

Bogged down on the Cabot Trail, Cape Breton Highlands National Park, June 1948.
National Archives of Canada, PA205796.

parks contained, a lack evident in the reports on the planned parks in
the 1930s. In passing, R.W. Cautley expressed satisfaction with Cape
Breton Highlands' wildlife potential solely because deer were
present, and because moose and caribou, though absent, could be re-
stocked.[7] No other wildlife was mentioned. After inspecting the proposed Prince Edward Island National Park, Williamson and
Cromarty stated only that "the same yardstick" could not apply to
this park as others, because "It is not essentially a wilderness sanctuary for animal life." They recommended that to compensate for this
the North Shore waters be set off as a fish reserve, "in the same way
as the Western National Parks act as reservoirs for game."[8] Of the first
three parks, the future Fundy had the most wildlife to its credit. In
1930, Cautley called the Albert County site "first class moose and
deer country" and gave it a rating of 80 out of 100.[9] Still, in Cautley's
reports on Fundy in 1930 and 1936, and in Smart's in 1937, deer and
moose were the only two animals referred to. If Fundy had any other
living creatures, they did not merit mention.

   Why was there so little interest in the wildlife of Maritime parks?
The Branch believed that no place in the region had wildlife that satisfied the national park ideal. To an agency used to relying on bear,

moose, elk, deer, buffalo, and even large predators like wolves and mountain lions to attract tourists and to prove that it hosted real nature, the animals in the Atlantic provinces hardly compared. Small animals such as foxes, skunks, and porcupines (not to mention insects, birds, and amphibians) might be abundant in these parks, but they would hardly draw tourists. There were just not enough different kinds of big mammals, and too few of the kinds that there were.[10]

There were other reasons wildlife played a minor role in the creation of the first Maritime parks. For one thing, wildlife management held a weak administrative position for much of the Branch's history. The Dominion Wildlife Division had been established in 1918 as a small agency within the Parks Branch, but its staff was very small and preoccupied with enforcement of the Migratory Bird Regulations Act; indeed, staff had been hired for their expertise in ornithology.[11] Although the Wildlife Division could and did communicate the latest scientific information to the Parks Branch, it was administrators within the Branch itself who then formulated policy. An example of this on the national level was the Parks Branch's management of predators such as wolves and mountain lions in the 1930s. Reading animal population studies from the United States, the Wildlife Division's supervisor of wildlife protection, Hoyes Lloyd, grew convinced that predators offered a natural and necessary check on prey populations.[12] He sought to convince Commissioner Harkin that the practice of killing predators in Canadian parks should be discontinued, and wrote lengthy reports to that effect in the early 1930s. Harkin used Lloyd's reports to fight proponents of increased kills, though he did not take the advice to stop killing predators altogether.[13] The Wildlife Division had little direct power, but because it existed there was no need to hire biologists or other natural scientists as senior administrators. As a result the Parks Branch was headed by foresters, engineers, and career bureaucrats who lacked training in wildlife matters. The men who initially inspected the Maritime parks and assessed the variety and range of wildlife there were the engineers Cautley and Williamson, the architect Cromarty, and the forester Smart.

The Wildlife Division itself leaned toward non-intervention in the 1930s, as a result of simultaneously following the most established and the most innovative strands of biology. Following the tradition of natural history, the Division saw its primary goals to be classifying all park species, and obtaining information about them. Following the latest tenets of ecology – the twentieth-century science that studies relationships between organisms – the wildlife officers believed that all species act together in concert, and that actions affecting one population might seriously affect others.[14] This was why, for example, Hoyes Lloyd opposed predator extermination policies: the results of intervention were unforeseen and so by their nature unwanted.[15] Lloyd won an ally on this issue with the promotion of F.H.H. Williamson as James Harkin's replacement in 1936. More so than Harkin, Williamson was aware of and responsive to the latest findings in ecological research, and used cases of predator eradication leading to prey irruptions as cautionary tales against tinkering with nature. This is not to say that the park system under Williamson practised complete non-intervention: some predators were still killed, animals were collected for museums and zoos, and parks were stocked with "useful" wildlife. But it may be said that when the Cape Breton Highlands and Prince Edward Island national parks were established, North American ecologists were wary of human attempts to regulate wildlife numbers, and had convinced the National Parks Branch to think likewise.[16]

Administrators were at the same time changing their idea of wildlife's role in the parks. In the first decades of the century, Canadian national parks had often been justified as sanctuaries, where game species would have the freedom to grow, prosper, and then wander outside to be killed by hunters.[17] Fish and game organizations came to accept the argument that sanctuaries of all sorts ultimately helped to maintain healthy wildlife breeding stock.[18] In the 1920s and 1930s, fish and game groups in New Brunswick, Nova Scotia, and Prince Edward Island were important lobbyists for parks in their provinces.

But by this time parks were promoted more in terms of tourism, and the Parks Branch was less interested in selling them as sanctuaries. If sportsmen wanted a reserve, they should ask their province to set one up. The presumed absence of wildlife in the proposed Maritime parks, while unfortunate, was thus not catastrophic. A park had to have remarkable scenery; it could, if necessary, import wildlife. Of course, the Branch was not likely to tell fish and game organizations doing much of the local legwork in promoting parks that the park-as-sanctuary idea did not hold the importance it once had. Staff instead complained to Ottawa that the locals did not understand what national parks were all about. For instance, R.W. Cautley complained in his report on his 1930 New Brunswick visit that the National Park Committee was composed almost exclusively of men from fish and game associations, and that the provincial government supported the Committee's recommendation of the Mount Carleton site on the grounds that it was popular with hunters.[19] As each of the first three Maritime parks was established, local fish and game clubs reacted identically: first with delight that their work had borne fruit, then with surprise that the Branch was acting so slowly to increase game populations, and finally with anger when it was clear that the park was uninterested in improving hunting outside its boundary.

In sum, scientific, administrative, political, and aesthetic positions might reasonably have been expected to keep the Parks Branch from an interventionist wildlife policy in Cape Breton Highlands and PEI national parks in the years following their creation. The point is moot, however. The arrival of the Second World War in 1939 allowed little thought or opportunity for wildlife management during most of the next decade. Since wildlife in the first two Maritime parks was deemed relatively insignificant and not even threatened by large predators, for the time being the act of setting up a park boundary and hiring wardens was considered management enough.

Ottawa did receive monthly wildlife reports from the superintendents of the two parks during the war. Having no training in animal biology either before or after taking their jobs, superintendents just listed creatures they saw during hikes. The reports not surprisingly make reference only to large, "significant" animals. The November 1939 report from Cape Breton Highlands, for example, mentions sightings of 70 deer, 104 grouse, 127 rabbits, 6 red foxes, 17 bald eagles, 2 blue herons, 4 muskrats, and 7 Canada geese.[20] Staff were happy to see that wildlife populations were generally on the rise, thanks to the sanctuary offered by the new parks.[21] In 1941, in keeping with the spirit of Cape Breton Highlands, its superintendent sent Ottawa a complete list of the park's animal, bird, and fish species in both English and Gaelic.[22]

The only interventionist policy practised in these parks during the war was the attempted reintroduction of moose and beaver to Cape Breton Highlands. At the Park Branch's request, two colonies of beaver were captured, crated, moved, and released by the Nova Scotia Department of Lands and Forests in 1938. One of the colonies survived the move, though the other one, like a group of moose transferred the same year, was not so fortunate.[23] The justification for these reintroductions was that these animals had been native to the park area, but had been extirpated by the humans there in the past century. They should be returned, then, to restore the park to its pristine condition. This might seem to be an endorsement by the Parks Branch of what was still a radical idea in park wildlife management. In the United States, biologist Joseph Grinnell in the 1920s and the authors of the 1933 *Fauna of the National Parks of the United States* had met resistance for proposing that parks be returned to their original state, with native species reintroduced and exotic ones removed.[24] But the Cape Breton Highlands case was not revolutionary. Though it was true that beaver and moose were native but absent, they were chosen for reintroduction because they were attractive to tourists and to the Branch's idea of what wildlife a park should have.[25] Bringing them back certainly intruded on existing biological conditions, but this was more of a blip in the wildlife policy of the time than a sign of change. These were one-time-only interventions, and once released the new park residents were left on their own. It was taken for granted that nature would help its own, after this initial push from park staff.

Biologist C.H.D. Clarke's arrival at Cape Breton Highlands in 1942 to report on its wildlife was the first real sign of Branch interest in the animals of the Maritime parks. Yet his report also shows the conservative nature of wildlife research at the time. Clarke, hired as the Wildlife Division's mammalogist in 1938 (and its first staffer not trained in birds), offered an essentially hands-off plan in keeping with the time. He wrote that the population fluctuations resulting from the creation of the new park were of great scientific interest but they should only be monitored and not directed. "They are absolutely natural phenomena and we have no concern in trying to interfere with them."[26] Instead, the Branch's management of wildlife should be limited to protecting all species (predators included), obtaining all possible information on their populations, and restoring vanished species – though only moose and caribou were mentioned. Clarke was careful to note that only the native variety of woodland caribou still present in Gaspé, not the Newfoundland caribou, should be considered for introduction: "For one thing, survival would be doubtful; for another, it is not desirable to introduce exotic species into the parks."[27]

The Parks Branch demonstrated a desire to protect native species even in the most prosaic management decisions of this period. From Prince Edward Island National Park, Superintendent Smith complained in 1942 that skunks were rooting up lawns and might soon move to the golf course. The Branch's new controller, James Smart, refused to let them be killed, "As a National Park should contain normal populations of all species of wildlife native to the region," even the lowly skunk.[28] Smart sought C.H.D. Clarke out for confirmation of his views, and was able to write the superintendent in a follow-up note, "It should be borne in mind that the excavations made by skunks in golf courses are actually in search of insects which are themselves quite destructive to turf."[29] The Parks Branch felt sure enough of its position that it stood firm against provincial government complaints about park skunks. Smart wrote, "The claim that the park is harbouring the skunks and actually is a breeding ground for an overflow into the surrounding country I think is rather a far-fetched contention."[30] This is an amazing assertion, since the Parks Branch had long maintained that game animals did just that.

The National Parks Branch began in the late 1940s to take a much more activist role in preservation issues regarding wildlife. Postwar prosperity allowed for an exponential increase in funds for wildlife management, as it did for federal projects generally. The budget for wildlife matters jumped from $200,000 in 1948 to $400,000 in 1954, to $700,000 by 1960, and to $3.9 million by 1969.[31] To handle this increased responsibility, the Canadian Wildlife Division was replaced in 1947 by a new Dominion Wildlife Service, renamed the Canadian Wildlife Service in 1950. The idea was that as a separate agency the Wildlife Service would have greater autonomy and a more objective, credible role in managing the nation's wildlife. Thanks to enlarged responsibilities and loosened purse-strings, the Wildlife Service grew from a professional staff of seven in 1947 to ninety in 1969. In turn, the availability of scientists and funds encouraged more proactive projects in wildlife management.[32]

The Canadian Wildlife Service did not take over the management of park wildlife. Parks Branch controller James Smart wrote all the superintendents early in 1948, "You are aware that during the recent reorganization of the Department and this Branch a Wildlife Service was set up which will act as our technical division to advise us on wildlife management. This does not mean that we give up any responsibility for looking after wildlife in parks."[33] But it did mean that the Parks Branch would be drawing on more and better-funded scientists for future policy advice.

The creation of the Canadian Wildlife Service signalled not only a new scale for wildlife science in Canada, but a new direction as well. Hoyes Lloyd, Harrison Lewis, and others who were the backbone of the old Division retired around the time that the Service was created. A 1971 official account, *Scientific Activities in Fisheries and Wildlife Resources*, hints at the perceived difference between the old breed of wildlife scientists and their successors. The former "were a group of keen, hard-working individuals who contributed a great deal to our knowledge of the occurrence and distribution of the wildlife of Canada. After their retirement, the Wildlife Service replaced them with men who were trained along more formal lines."[34] The difference was as clear as that between natural history and science. The Wildlife Service was now staffed with young men schooled in the United States in the latest precepts of biological science. Ecology in particular was drawing attention in this era because it theoretically studied a whole system at once rather than a single species, and so was seen as a science of unlimited promise. Hugh Keenleyside, deputy minister of the Department of Mines and Resources, proudly announced plans in 1949 for a system-wide inventory of national park wildlife: "It will be no mere cataloguing of plants and animals, a great deal of which has already been done, but will be concerned with the community of living things and with the manner in which the various forms of life affect one another. This will be what scientists term an ecological survey."[35]

Ecology was not a value-free science – or, more accurately, like any science it was not practised in a value-free way. It accommodated different interpretations. This may be seen in the application of its best-known discovery of the 1920s: that predators were essential to habitat health. Knowing that deer and wolf populations were dependent on each other for healthy populations could have taught ecologists that both species had to be left alone – and, indeed, some ecologists did take this lesson. But more interpreted it to mean that to manage deer effectively, both wolf and deer populations had to be actively managed. Likewise, ecology of the 1940s and 1950s tended to be taken to justify interventionism. Historian Gail Lotenburg writes of the period, "most federal wildlife administrators in Canada interpreted ecological concepts as a means towards securing traditional management goals."[36] It is worth stressing that this did not demand a contrived misapplication of ecological theory: it was (in most cases) the product of honest interpretation. Ecology had since the 1930s been moving away from an organic model to a mechanistic one, from a study of individuals and their places in a community to movements of physico-chemical properties within a system; it was natural that this change in metaphor distanced ecologists further from the subject of their inquiry.[37]

The direction taken by ecology would greatly affect preservationist policies of the Canadian national park system. Parks Branch staff relied on the Canadian Wildlife Service to explain ecology and to recommend how it should be applied within the parks (though they made the final decisions – unless upper levels of the department overruled them, of course).[38] As a result, wildlife policies in the national parks would become considerably more interventionist in the 1950s. Rather than merely fencing off an area and letting wildlife thrive, staff now considered it their responsibility to manage wildlife numbers. As early as 1949, the new philosophy – outlined, notably, by the head of the Canadian Wildlife Service, Harrison Lewis – was as follows: "It should, perhaps, be emphasized that it is not the established policy to administer the wildlife of the National Parks by simply letting it alone, letting nature take its course, and trusting in the idea that is commonly referred to as the 'balance of nature'. That would not be practicable, because the park areas are already more or less disturbed by human activities and they are surrounded by areas that are even more altered by man."[39] By the mid-1950s, this idea went still further: animal populations were bound to erupt in the unnaturally natural conditions of a national park.[40] According to a 1957 policy statement, "This brings in its wake such evils as starvation, disease, range destruction, damage to forest regeneration, displacement of desirable plants and soil erosion. Unless these surpluses migrate naturally out of the Parks, they must be removed without hesitation either by careful killing or live-trapping. ... All of the National Parks of Canada are potential danger areas for the development of excessive populations of game, predators and fur-bearers."[41] The Parks Branch had created the environment in which animals could overpopulate; to leave them alone to do so would therefore in itself be a form of management, and an immoral one. There was a real arrogance in taking this stand. Animals were living and dying because of past human choices, so humans must take responsibility by continuing to make choices to ensure their general survival. This reflected the Parks Branch's opinion not only of animals, but of human society as well; just as in past decades the agency had proven it was more progressive than the general public by not interfering in animals' lives, it would now demonstrate its relative intelligence by interfering efficiently.

The swing toward interventionism is evident in the wildlife policies chosen for the new Fundy National Park in the early 1950s. Unlike that at Cape Breton Highlands and Prince Edward Island, Fundy's wildlife received attention immediately upon establishment. By 1951, there were already four reports discussing Fundy's animal population. The early consensus was that the area's potential for wildlife was excellent. Wildlife Service staffer John Kelsall reported that in places where lumbering had been practised, there was unlimited young growth for moose and deer to feed on. He did warn, however, that moose were dying in winter, probably from a combination of moose ticks and difficult travelling in deep snow.[42] Three years later, the chief mammalogist of the Wildlife Service, A.W.F. Banfield, reiterated the hopes that moose, which currently numbered about 120, would increase naturally to the area's "carrying capacity." (This wildlife management term refers to the maximum population of a species that a given area can sustain. Perhaps inevitably, the idea of carrying capacity came to suggest that ecology could determine the right number of *all* species in *all* places, so that managers could maintain populations at just those numbers.)[43] Banfield noted that "Without a moose reduction programme, we can expect continuing heavy moose tick infestations and winter mortality."[44] In other words, Banfield believed the moose population was below its natural maximum, and yet still needed pruning. This is not really surprising, because winter deaths always bothered park staff. Although such mortality in the animal world is as natural as death by predation or old age, it always seemed needless and avoidable.[45] It was not difficult for wildlife officers at Fundy to convince themselves that the ungulate population needed their help.

By 1953, there were reports of "serious overbrowsing" at Fundy.[46] Apparently, the park's character as a sanctuary had permitted moose numbers to balloon just in the park's first four years. At wildlife officer J.S. Tener's recommendation, the Wildlife Service approved the killing of twenty moose "for the purpose of game management."[47] It is not clear from the files whether this reduction actually took place; before it could, Superintendent Saunders warned his superiors that the province felt it owned the moose by virtue of the New Brunswick Game Act. On the advice of the attorney-general's office, the Parks Branch explained that the federal Parks Act overrode the provincial act, thanks to the clause that gave Canada the right to "all profits, commodities, hereditaments and appurtenances whatsoever thereunto belonging or in anywise appertaining" to the park – including moose.[48] In 1955, with the park moose population around 150 and the carrying capacity now estimated to be only 80, another culling program was set in motion. This moose kill demonstrates just how dubious the Branch's claims to scientific objectivity really were. Though the reduction was supposedly intended to lower an overextended and unhealthy population, it was done in such a way as to make such results impossible. Old bull moose – the least likely animals in the herd to affect long-term population – were targeted. The nine that were killed were found to be "in good condition" and almost all had fat on their quarters, though they had been killed because they were supposedly running out of food.[49]

The same management style was visible at Cape Breton Highlands National Park. Deer populations there were reported to be manageable and increasing naturally during the 1940s. But in the early 1950s, it was believed that deer numbers were climbing fast – almost tripling from 1951 to 1952, with sightings up from 140 to 360.[50] Seeking to make the most of this, the Nova Scotia Fish and Game Association called for an open season on deer in the park.[51] Wildlife officer Tener found that "The degree of overbrowsing throughout the Park, with the exception of the deer yarding areas, has not yet reached serious proportions, but can readily do so if the deer population increases further or if unusually severe winter forces the animals to restrict their range more than at present."[52] If so, reduction would be called for. The Parks Branch and the Wildlife Service followed the Nova Scotia case closely. In late 1954, Banfield noted that there were only a few areas of heavy browsing, and the deer generally still had plenty to eat. Nonetheless, he decided that deer were "on the verge of requiring control" and therefore that a small control program targeted at perhaps twenty-five deer should be implemented. "This project," Banfield wrote, "will provide the Park Service with slaughter experience in case a larger reduction programme is required in the future."[53] That is, control was recommended solely as practice for control that might be needed later. It is difficult when reading the notes of the 1955 deer slaughter to see it as anything but a strange form of scientifically ordained ritual sacrifice, a way for Parks Branch staff and Wildlife Service officers to allow themselves the pleasure of killing animals while denying that pleasure to others. Superintendent Doak reported that "hunting conditions were ideal" when the group went out. Twenty-five deer were killed. All were found to have been in very good condition, and during the entire shoot not a single carcass was found to suggest starvation.[54] The superintendent wrote Ottawa offering to harvest more, but the Parks Branch elected to wait and see.

The reductions implemented in Cape Breton Highlands and Fundy were both too much and too little: purposeful turns away from hands-off preservation as a result of questionable scientific reasoning, yet involving such small numbers that they were unlikely to have any impact on the stated desire to regulate populations. In fact, Fundy would experience a moose overpopulation and crash within five years of its 1955 reduction program. Other choices could have been made. "Surplus" animals could have been removed to other places in the province that would be glad to have them (although, it is true, they would then be targets for hunters rather than wardens). Strips of forest could have been opened up for browse (which, admittedly, would have meant destroying vegetation instead of wildlife). The other option was to do nothing, and see if the anticipated overpopulations actually occurred.[55]

Though the Parks Branch tried to maintain a fixed philosophy on park wildlife, its policy decisions were often shaped by the political and pragmatic needs of the moment, and by the perceived value of the wildlife in question. Such was the case with the hunting of wildfowl in Prince Edward Island National Park. Hunting has traditionally been the most forbidden activity in parks. Whereas parks throughout their history accepted the removal of trees, minerals, and fish from their borders, the removal of animals was always taboo.

When the PEI National Park was established, local politicians complained on behalf of area duck hunters over the loss of a favourite shooting spot.[56] But by the mid-1940s, hunters were resigned to the park's existence and lobbied instead that it do everything to build up bird populations so that the birds would spill outside the park to be shot. However, the discovery of a mistake in R.W. Cautley's original survey of the park forced the Parks Branch to re-evaluate how it dealt with local hunters. Cautley had mistakenly set a park boundary at Stanhope to "the line of mean high tide" on swampland not influenced by tide; as a result, the park did not have title to two pieces of land that hunters coveted. The park chose for a time to condone hunting in the contested areas, and even an acting superintendent hunted there.[57] The Parks Branch sought to rectify this situation permanently in 1957 by arranging a land swap with Gordon Shaw, owner of the local Shaw's Hotel. The park received a parcel it had long thought it owned, and Shaw received a piece of land that he could offer his visitors as a private shooting ground. Provincial hunters, however, felt they had lost out on the deal and again had their politicians complain. In researching their claim, locals discovered that Cautley had defined part of the park's shore boundary as the "high water mark of the Gulf of St Lawrence": hunters could presumably shoot on the beach at low tide. Trying to make the best of the situation, Superintendent Browning suggested that though it would seem that the PEI National Park did not own the beaches, "Perhaps we could claim squatters rights."[58] The Parks Branch believed that its position was stronger than this, but to avoid disagreements with locals, staff were told by Ottawa to let duck hunting continue in the disputed areas.[59] In this case, the prime directive of national park wildlife policy was as flexible as the park's boundary.

The Parks Branch was not above advocating the killing of wildlife thought to be pests. The most prolonged example of this was the Branch's battle against insects, which will be discussed later. But larger pests were also targeted. Porcupines were causing damage at Fundy headquarters, so staff requisitioned "Good-Rite zip," a chemical deer repellent, and were later given a more direct deterrent, a shotgun.[60] Five hundred muskrat were authorized to be killed in PEI National Park since they had reached "nuisance-value proportions"; nonetheless, only sixty-four could be found to be killed.[61] Reductions were not always even aimed at any perceived need, but simply so the Parks Branch could demonstrate it was being a good neighbour. In 1950, Wildlife Service biologist John Kelsall discussed killing foxes at Prince Edward Island Park: "The shooting would not be done with a view to eliminating or controlling the fox population but rather to assure any possible sources of complaint that the Park authorities are cognizant of the high fox population and are doing something about it."[62] In this instance the Parks Branch chose not to initiate a reduction program, not because it violated park philosophy, but because it was decided there were not enough complaints to take this step. The most perverse manifestation of the zeal for control arose when the Prince Edward Island National Park staff dealt in 1959 with a very large deceased pest: a beached whale. They blew it up. This may have been the best decision under the circumstances, but Superintendent Kipping's description of the event suggested a certain perverse enjoyment of the spectacle. He corresponded with CBC entertainers "Gentleman and Olga," who had reported on the incident, telling them that staff had to blow the whale "to heaven with 600 pounds of dynamite."[63] And when Ottawa asked Kipping for details, he sent pictures of what he called "the explosion which transported the whale out of the realm of our jurisdiction."[64]

That was exactly the sort of image the head office wanted to avoid. When the Parks Branch felt obliged to kill and remove some park wildlife, it sought to do so with discretion and decorum. On the subject of controlling porcupines at Fundy, for example, Ottawa ordered that "Any control activity should, of course, be carried out as inconspicuously as possible in the interest of good public relations" and "These operations must be carried on with discretion, in order to avoid offense to the public."[65] Keeping the reduction programs quiet became such a standard policy that staff even convinced themselves that such discretion had a managerial purpose. The minister in charge of national parks in 1962, Walter Dinsdale, told the president of the New Brunswick Fish and Game Association that park hunting was not permitted because "the game species involved would become much more retiring, thus reducing the opportunity of worthwhile recreation for a large segment of the public." For the same reason, he said, when staff had to kill certain animals they were careful not to harass or frighten others.[66]

In reality, it was the human animals that the Parks Branch did not want to disturb. The Branch knew that the public was unlikely to understand why different rules applied for park staff than for park users. Why was hunting forbidden in the park, yet proper game management might demand the shooting of some animals? If management was needed, why were locals not allowed a limited hunt? Why, for that matter, were farmers' cattle impounded if they strayed into Cape Breton Highlands, while Highland cattle were kept by Keltic Lodge into the 1950s?[67] Such awkward questions, typical of those prompted by park development throughout the world, were raised repeatedly about these four parks in this period. There were some relatively valid answers to some of these contradictions. National parks are intended to be places free of human effect; however, the parks need humans to help enforce this ideal (as well as to make the visitor's communion with the park as pleasurable, yet as passive, as possible). Moreover, the Parks Branch argued that nature had been affected by humans before the park's existence. Humans were therefore needed to re-establish the natural state of the park, which demanded management. All this is sensible if we accept the original idea that parks have an inherent logic, and if we accept that staff's decisions for the park are free of human interest.

FISH

Writing of the American case, Richard West Sellars states, "In its management of fish, more than of any other natural resource, the Park Service violated known ecological principles."[68] The same could be said of the Canadian National Parks Branch. As with wildlife, fish in Canadian national parks were deemed deserving of preservation, and as with wildlife, some fish deserved more preservation than others. But it was accepted that tinkering with the population and distribution of fish species could be tolerated to a degree that would not be tolerated for air breathers. Of course, the ultimate tinkering was that fishing was not merely tolerated but encouraged. National parks were promoted as having some of the best sport fishing lakes and rivers in Canada. The Parks Branch never tried to justify or sugar-coat fishing, which in itself suggests how ingrained the logic of allowing it in parks was. As with wildlife, the amount of intervention by park staff on fishing matters increased during the 1950s, before abating somewhat in the 1960s.

Fishing was supposed to have an especially important role in the first Maritimes national parks for a number of reasons. First, the Parks Branch believed the very name "Maritimes" promised tourists opportunities for fishing (that it suggested salt-water fishing rather than parks' traditional fresh-water fishing was not insurmountable). Second, it was hoped that an abundance of fish in Eastern parks would compensate for their insufficient wildlife. Finally, inland fisheries were still under federal control in the Maritimes, having been delegated to the provinces elsewhere. The Parks Branch therefore would benefit from particularly close affiliation with the federal Department of Fisheries in managing park fish populations.[69]

Nova Scotia was renowned for its fishing, and the Cape Breton site's potential for fishing was an important factor at park establishment.

Cape Breton's Margaree River was already famous among North American anglers, and the Parks Branch hoped that it could likewise put the Cheticamp River on the map. Even before Cautley's inspection in 1934, Commissioner James Harkin stated that one of his conditions for a northern Cape Breton park would be "to co-operate with the Dominion Government in doing away with all net fishing of salmon off the mouth of Cheticamp River. I am informed that at the present time the salmon fishing in both Margaree and Cheticamp rivers has been practically ruined by the mouths of both rivers being netted by shore fishermen."[70] Harkin was not concerned about depletion of the fishing stock per se, but felt that net fishing was fundamentally un-sportsmanlike and kept anglers from catching the same fish. Cautley's 1936 report came to the same conclusion, and suggested that if the park was to go through there, fish ladders could be installed to make access up the Cheticamp easier for salmon. He even hoped that, if legally practicable, the Parks Branch could restrict net fishing outside the park boundary.[71]

Fish were also important in the planning of Prince Edward Island National Park. Williamson and Cromarty recommended that the water off the north shore of PEI be made a fish preserve, "in the same way as the Western National Parks act as reservoirs for game."[72] However attractive this idea, and however seemingly responsive to contemporary concerns for the viability of lobster stocks, it had more to do with creating a credible park than with preserving an endangered species. The north shore waters were of interest as a reserve because of their location off a possible park site; Williamson and Cromarty had no idea whether or not this was an especially good place to find and protect fish. And it would be a fish sanctuary "except for taking of fish by hook and line as under Park regulations." When the Department of Fisheries concluded that the fish preserve would serve no purpose, and Island politicians complained about how it would affect North Shore fishing, the idea was scrapped.[73]

Once the two parks were established, the Parks Branch investigated the fish stock they had and tried to determine how they could generate more. The findings were not encouraging. Prince Edward Island National Park lakes did not contain trout, as was hoped, but "a serious enemy," white perch.[74] As well, other lakes outside the park supplied much better angling. Cape Breton Highlands was inconvenienced by both white perch in its lakes and netters outside its main river, the Cheticamp. The natural solution was to stock the park rivers and lakes with fish. The Cheticamp had been stocked with salmon from the Department of Fisheries' Margaree River hatchery since 1916; the Parks Branch simply dumped in more fish.

Cape Breton Highlands was stocked with 170,000 fingerlings in the first year of its existence, 180,000 in 1938, up to 250,000 by 1941. After a late wartime lull, by the late 1940s the park was again being "planted" at rates up to 100,000 per year.[75] Around the same time, Prince Edward Island National Park also began to be stocked with trout, even though its lakes were not considered likely to provide good fishing.[76]

The continuing popularity of stocking is a surprise, since limnologists of the day, and biologists in general, were moving away from population control and showing greater interest in habitat improvement. They were coming to understand the complexity of natural systems, and realizing that a wildlife program that focused on one or two species was insufficient.[77] When Parks Branch limnologist V.E.F. Solman reported on Cape Breton Highlands in 1948, he noted that four million fish had been dumped into the Cheticamp since 1916, and still there were few fish. Yet Solman did not blame the netting at the river's mouth, as Harkin had done and locals still did. He believed that salmon were able to get by the nets in sufficient numbers. More troublesome to fish populations were habitat factors such as water temperature and competition from other species.[78]

The Parks Branch relied on stocking because it was straightforward and seemed sensible. Staff needed to be doing something, since the Maritime parks' appeal to tourists depended so much on fishing.[79] Though R.W. Cautley gave the eventual New Brunswick park low marks for fishing potential in 1930 – a 50 out of 100, with the comment "There are practically no lakes"[80] – by 1950 the opening ceremonies brochure for Fundy National Park showed two photos of fishermen hard at play. James Smart directed limnologist Solman to sell American fishermen on Prince Edward Island's perch by using the more impressive name "silver bass."[81] The Parks Branch then paid C.H.P. Rodman of the American magazine *Hunting and Fishing* to vacation in the Prince Edward Island park; he did not catch a single silver bass, but he wrote a nice article anyway.[82] Cape Breton Highlands aggressively advertised its offshore swordfishing, though in 1948 Solman reported, "In spite of all this publicity not one tourist has arrived and demanded to be shown such fishing."[83]

It seemed that locals enjoyed the fishing in the new Eastern parks more than anyone. This hardly consoled the Parks Branch. In 1939, Commissioner F.H.H. Williamson said bluntly that fishing regulations did not "necessarily correspond with the Provincial dates as the parks policy is to cater more to the tourist trade than for the benefit of the local residents."[84] Tourists meant money, meant a park was succeeding; locals spent far less and then had the gall to catch the fish meant for tourists. The Parks Branch therefore adjusted the parks' fishing seasons to approximate the tourist season more closely. For example, the opening of the salmon season at Cape Breton Highlands was moved from the province-wide 1 June to 15 June, since "practically all of the salmon have been caught by local residents before many of the tourists start arriving in the Park, which is usually from early June on."[85] This more than anything else demonstrates that the Parks Branch saw tourists and locals as fundamentally different users of park resources. Though ostensibly for the benefit, advantage, and enjoyment of all Canadians, parks were meant to be especially for tourists, Canadian or otherwise.

As with wildlife, fish management in the national parks became decidedly more interventionist in the early 1950s. This was due to a number of factors: the creation of the Canadian Wildlife Service with a resultant increase in scientific personnel, a move within ecology toward more extensive management projects, increased funding for such projects, and the Parks Branch's own dissatisfaction with fish stocking.[86] The most fundamental change was technical: the discovery and application of rotenone. When Prince Edward Island National Park had been established, foreman A.L. MacKay had noted that, instead of stocking the park's lakes, they could contaminate it, kill all its fish, and restock. He supposed, however, that such a task would likely leave the lakes "tainted for an indefinite period."[87] He was right for his time, but he did not know that the Fisheries Research Board of Canada was currently testing the purification of lakes using rotenone, a natural poison derived from the powdered root of a number of plants (most commonly, derris). In 1936 the fisheries staff succeeded in killing all fish along twenty-five miles of stream and in a six-acre lake near the Cobequid Hatchery in Nova Scotia. Rotenone was found to wipe out all insect and fish life within half an hour, allowing for the introduction of preferred species, and there was no long-term effect to the water.[88] Rotenone poisoning became relatively commonplace in North American fish management in the 1940s, but it was not until the 1950s that the Canadian national parks began to use it.

The Parks Branch was quite proud of its new fish management tool, and publicized its ability to improve on what nature provided. One press release stated,

Eels have a place – but not on the lines of anglers who visit a National Park looking for good trout fishing. So when they multiplied to the point where they played havoc with the trout in Freshwater Lake, in Cape Breton Highlands National Park, Nova Scotia, Canadian biologists of the Wildlife Service took drastic action. With the help of the Park Engineer they dumped a lethal dose of 4,500 pounds of rotenone into the lake to clean out the waters and rid them permanently of these and other hungry predators. The operation has been successful, and a specially-designed barrier built at the outlet of the lake to prevent the eels from re-entering the waters, as elvers in spring or adults in the autumn. Re-stocking with trout has been postponed until next year to allow for the re-establishment of the swarms of water insects that form the main diet of lake trout.[89]

Eels were in fact a secondary consideration, and only a follow-up press release admitted that white perch were the main target. A speech two years later by the minister in charge of national parks, Jean Lesage, referred to the killing of fish matter-of-factly, as if this sort of interventionism was now normal. Lesage stated, "Many people think that fish are where you find them. That was true once, but nowadays, with fishing becoming one of the most popular outdoor sports, it might be more correct to say that fish are where somebody makes sure that you find them. In the National Parks, that 'somebody' is the National Parks Service, acting on the advice of the limnologists of the Canadian Wildlife Service."[90] Prince Edward Island National Park's little lakes got the same "drastic measures"[91] as had those of Cape Breton Highlands, since planted Eastern brook trout had not been hardy enough to depose perch (sorry, silver bass). Once the lakes were poisoned, rainbow trout were introduced precisely because they were not local fish – few Island lakes had them. It was felt that bringing in this exotic species would make the park's lakes interesting and different.[92]

Rotenone was a great success in park lakes. The Branch's chief limnologist in 1960, Jean-Paul Cuerrier, noted that at Cape Breton Highlands, "The absence of competition has favoured the survival of hatchery trout which have provided satisfactory angling returns to Park visitors."[93] Nonetheless, there began to be rumblings of discontent about the drastic measure of poisoning lakes. After an inspection of Fundy in 1956, F.A.G. Carter, the secretary to the minister, questioned why the Wildlife Service planned to drop rotenone in the unfortunately named Lake View Lake. "Apparently by the locals here," he wrote, "there is reasonable fishing (speckled trout) and no one understands why poisoning is necessary. It may be necessary; I do not know. I do know, however, that it would be most helpful if the Wildlife Service could explain the 'whys' to the wardens, etc., and to all concerned within the park – to try to win them over. This may have been tried and may be impossible. There is a considerable feeling throughout the Park that the Wildlife people are long haired experimenters. I have no doubt this is grossly unfair."[94] This nicely shows that to "locals" – a designation that apparently united ground-level park staff with people of the community – the methods and plans of biologists were a mystery. Carter, a federal bureaucrat, clearly felt the same way. However, it would be a mistake to infer from this that the Canadian Wildlife Service pushed the Parks Branch in managerial directions it did not wish to go. Two years later, it was the superintendent of Fundy who asked his superiors whether Wolfe Lake could be poisoned; his superiors considered the request, but opted for stocking it with fingerlings instead.[95]

The use of rotenone peaked and then disappeared in the mid-1950s; it would appear from available archival records that no lake or river in an Atlantic Canadian national park was again poisoned pure.[96] Perhaps this was because of complaints from tourists and locals or from staff within the Parks Branch over this ultimate form of intervention. Perhaps it was part of a general move away from active management of park resources around 1960. Or perhaps, though less likely, the lakes continued to be poisoned for some time, only without the fanfare and press releases. But the Parks Branch did continue to manage the fish populations under its domain, to a degree that would have been considered unacceptable for any other living things in the parks. Populations were introduced or killed off, with little attention to whether they were native or exotic. Staff knew and admitted this. As parks chief B.I.M. Strong explained to a superintendent in 1959, "As you know, fishing seasons in the National Parks are set not too much according to biological consideration of the game fish involved but rather according to the aims and objectives of the National Parks."[97]

VEGETATION

Neither fish nor wildlife preservation was ever completely practised within the national parks; it was always, which fish? how much wildlife? But the underlying philosophy, that national parks were places where living creatures should exist unbothered by humans, continued to hold even when the resulting policies changed. The theory of preservation was more directly challenged by vegetation. It was obvious that untouched vegetation was not necessarily "better" for itself or for the park; if left alone, it could choke out desired species, increase fire hazard, block views, attract disease, and be aesthetically unattractive. As well, thick forests, if cut, were financially lucrative, and not cutting them was wasteful. Finally, humans did not have the sort of relationship with vegetation that they had with animals. There was not, to the same extent, a feeling that plants had a natural right to existence.[98] For all these reasons, the National Parks Branch forever questioned the validity of a hands-off policy with regard to vegetation, particularly forests. In the four parks studied here, the same trend is visible in forest management as in fish and wildlife – increased interventionism in the 1940s and 1950s, followed by a reversal in the 1960s – but more interesting is the volume of discussion about the underlying meaning of preservation.

The Parks Branch depended on the Canadian Forestry Service for expertise on forestry matters, just as it depended on the Canadian Wildlife Service for fish and wildlife matters, but its own knowledge of forestry was greater. Foresters were trusted with positions of general responsibility in the Parks Branch to a degree that wildlife biologists would never have been. James Smart, Parks Branch controller from 1941 to 1950 and director from 1950 to 1953, began his career with the Forestry Service after earning his forestry degree at the University of New Brunswick.[99] Prince Edward Island National Park's first superintendent, Ernest Smith, was chosen expressly because of his forestry experience. Fundy National Park's first superintendent, Ernest Saunders, had previously been in charge of the Acadia Forest Experimental Station, and was paid extra by the Parks Branch because it was understood he would have extensive silviculture work to perform.[100]

Forestry preservation was an issue from the very establishment of Cape Breton Highlands National Park. The Parks Branch had to decide whether people living near a new park should be allowed to continue their traditional use of its forests. James Smart, respecting foresters and their needs, pushed for small woodlots to be set up at out-of-the-way places throughout the park. Controller F.H.H. Williamson was less enthusiastic: "Settlers adjacent to a Park boundary, who have been dependent on timber in the Park area, form a problem, since satisfying their demands is usually opposed to Park regulations. The man who draws out such a plan as Mr. Smart proposes should be sufficiently experienced in National Parks' operations as to know what cuttings would constitute improvement from a Park standpoint, the only legitimate excuse we have for allowing timber to be taken out for settlers' needs."[101] In other words, the Parks Branch needed an excuse to show that helping fulfil local wood needs would be for the good of the park forests. Smart found one by having the proposed woodlots made "demonstration plots" to be cut on a sustained yield basis, with park staff choosing the trees to be targeted.[102] The Forestry Branch approved this idea, so the project went forward.

Allowing ex-residents to cut wood in the park was a noble idea, and a far cry from the usual treatment of their traditional rights once the parks were created. But it caused difficulties when put in practice, while reinforcing the Parks Branch's (and particularly James Smart's) tendency to see park resources as objects to be used. Staff found sustained yield difficult to manage, so they turned to exhausting one stand of trees and moving on – hardly an innovative forestry practice.[103] The Parks Branch and locals also disagreed about the types of trees to be cut. For building small boats, fishermen needed pine, which the Parks Branch specifically wished to preserve, since, in Smart's words, "this is a desirable species from a scenic point of view and there is very little of this species in the Park." Locals wanted hardwood for fuel, because it burned longer in woodstoves, but the Parks Branch was trying to preserve its hardwoods. (Staff even specifically chose softwoods when they needed to build guard rails, cabins, and such.) As Smart told the Cape Breton Highlands superintendent, "it is the opinion of the Bureau and also of our foresters that the main species of wood to encourage on these areas are the hardwoods and any softwoods should be cut if they come within the size specified – in fact, it might be necessary to treat such as balsam as one which should be eliminated entirely from an area, as it is a very poor type of wood for any purpose and, being a very prolific seeder, it is inclined to take over the area at the expense of the hardwood species."[104] This is so interesting because it is so meaningless. The small sections of forest that the woodlots demanded would have had no appreciable effect on species composition for the park as a whole. More important, there should have been no reason to discuss a tree's "purpose," because it did not need one in the park. And if by purpose Smart was referring to softwoods' scenic value, he was admitting to a regional prejudice against the scrubby, coniferous look of Maritime forests; in any case, he was responding to a blatantly unnatural aesthetic the park could not possibly hope to satisfy. His description of succession – that softwoods tend to block out hardwoods – was just plain wrong. The opposite is in fact the case, and for a forester to say otherwise shows an irrational aversion to softwoods.[105] Smart's quotation demonstrates his uncertainty about how to manage the contradictory task of safeguarding and stabilizing a vulnerable and ever-changing resource. In trying to find a consistent position, the Parks Branch would fall back on questionable science, vague aesthetics, and the familiar values of capitalism.

This is evident in the creation of Fundy National Park. James Smart, parks controller by this point, saw the park's future beauty in doubt specifically because of the forest's fecundity. "The growth is very prolific," he wrote, "and if no thinning operations are carried on the whole area will become a jungle, interfering with its general use for recreational purposes and also crowd out some forms of wildlife."

Underlying this comment is a recognition that in terms of park forests, time worked at cross purposes with preservation. A park is chosen because of its scenery as it exists at the park's establishment; consequently, it might be justifiable for the park to be preserved, not in its natural state, but in a way that maintains that scenery as it had been at establishment. Smart did not tackle this paradox head-on, but resorted to a scientific justification for intervention: for "forestry experimental purposes as demonstration plots of silvicultural systems and for study," a perennial program of "improvement cutting" should be implemented.[106] He called in the Forestry Service to develop a forest inventory for the harvesting of Fundy timber.

It is perhaps not surprising that it would be someone from outside the Parks Branch, H.L. Holman of the Forestry Branch, who would make the most intelligent critique of park preservation philosophy in this period. Holman visited Fundy in the fall of 1949 and reported that he was very impressed with it. Of course, "no small part of the natural beauty of this park, in common with most others, is derived from the forest and … without the forest, it would become a commonplace bit of country with little or nothing to recommend it." He was amazed at the forest's rate of grow-back, calling it the fastest he had ever seen. He agreed with Smart that this was a management concern: "An impenetrable jungle or thicket growth, of small spruce and balsam fir is not, in my opinion, a desirable feature of a much-frequented recreational area. At close range such stands are ugly and, in addition, constitute a fire and insect hazard of no mean proportions." Fundy's forest therefore demanded supervision. Openings and viewpoints should be kept clear, deciduous trees should be planted in public areas, conifers should be thinned. Holman recognized that this degree of management flew in the face of traditional park policy. He wrote,

It might be argued that such a program would not be in accordance with accepted Park policy inasmuch as it contemplates disturbing the natural condition of the forest in such a way as to give it the groomed appearance of a city park. If it is true that that has been the accepted policy of the National Parks, then I think it is time that it was modified to some extent at least. Where people congregate in large numbers, they are not usually the sort of people who are likely to appreciate the unspoiled beauty, if indeed it is such, of the undisturbed natural forest and are more likely to be the sort of people who will thoughtlessly toss a cigarette into the brush on a hot, windy day.[107]

And not only did Holman doubt the park visitors' desire to see real nature, he challenged the park's ability to provide it. The undisturbed, primeval Fundy forest had disappeared in the last two centuries, he wrote, the victim of lumbering as well as insect depredation. "In short, the present forest is anything but natural and is, rather, the direct result of man's mismanagement in the past."[108] The park therefore had licence to preserve what it wished. Smart wholeheartedly agreed with Holman's report and moved to adopt its recommendations. To the observer, this episode is a welcome dose of reality for a park system that so often treated the nature it oversaw as its personal discovery. But by accepting the park as a product of history, the Parks Branch once again permitted itself a more active role in shaping the park's future. Underlying its actions was the belief that land touched by people was somehow profane, and did not deserve the hands-off sort of preservation that virginal land did.

Holman and Smart were not backing away from the aesthetic of a seemingly pristine forest, they were championing it. They just felt that such a landscape could be created. Such thinking is evident in Holman's 1949 report on Prince Edward Island National Park, done at the same time as his Fundy inspection. Whereas Ernest Smith had been hired as the PEI park's first superintendent specifically to show Islanders what a managed forest looked like, Holman saw this project as hopelessly misdirected. "Whole fields have been planted with nice, orderly rows of trees," he noted, "as though the sole object were to grow timber for commercial purposes. There are no gaps or openings to serve as picnic places out of the wind and glare of the beach, no mixture of species to relieve the monotony of pure spruce, or pine, planted four feet apart each way and no lanes left unplanted for accessibility. Further, none of the planting has been done on that part of the area that is most in need of it – the open windswept fields and the dunes near Cavendish."[109] In other words, the Parks Branch had not hidden the artificiality of either past or present occupation. Smart echoed Holman's concerns, and in a letter to his superior argued that a policy of intervention could exist in harmony with the appearance of non-intervention. Smart recommended that at Prince Edward Island National Park "we should not rigidly keep to the regular plantation methods but that the planting should be done with a view to keeping to a more natural condition." In the same letter he stated his hope that the park thinning operations would permit the Parks Branch to open a Christmas tree trade.[110]

The 1950s was a decade in which the Parks Branch would continually find need for active forest management in the parks, and each time find justification. At PEI National Park, staff took to thinning and cropping trees that were obstructing tourist lodges' views of the North Shore. This was done to pacify park neighbours, but was not believed to contradict park policy, since, in the words of deputy minister R.G. Robertson, "trees are developing in a very overcrowded and unpleasant fashion. Keeping hands off these areas do *not* preserve them in their natural state because they are all second growth areas in any event."[111] At Fundy National Park, a sawmill (out of tourists' eyesight and knowledge) was maintained to meet the lumber needs of the Maritime parks. The mill on the Upper Salmon River owned by Judson Cleveland before expropriation was used by the park until it burned down in 1952. The Parks Branch considered implementing an intensive cutting plan for the entire park,[112] but in the end decided just to set up another small mill at Bennett Brook. Throughout the mid-1950s, staff yearly logged between sixty and a hundred thousand board feet (about one one-hundredth the amount taken before the park's establishment), and in the 1960s supplied the new historic park at Louisbourg with spruce logs.[113] When asked by a New Brunswick government member about the park system's timber policy, assistant deputy minister E.A. Côté noted, "I indicated to him that I thought our policy was in the state of evolution and that I could not give him an immediate answer."[114]

Spruce budworm infestation in New Brunswick, Nova Scotia, and, to a lesser extent, Prince Edward Island in the early 1950s cemented the belief that hands-off preservation was dangerous to the forests being preserved. This infestation was quite natural, as budworm populations cycle every twenty-five to seventy years.[115] Cape Breton Highlands reported the existence of budworm in 1951, and by 1955 they had spread widely, most noticeably around the park headquarters.[116] The insects feasted on balsam fir and spruce, and, with the aid of the balsam woolly aphid, left most of the park softwood forests, especially in the interior, dead or dying. The result was a park which, in the words of forest engineer D.J. Learmouth, had "a very ragged appearance ... Fortunately, much of the Park stands are mixed-wood stands and ... serve to camouflage much of the dead and deformed softwoods and maintain the aesthetic appearance of many of the fine views along the Cabot Trail."[117] But in the interior, "its present monetary, aesthetic or recreational value is practically nil."[118] The infestation affected so many trees that any sort of clean-up was out of the question. Fundy was similarly threatened, though the budworm that moved throughout New Brunswick in this period had not yet made it to the park. Nothing in the Parks Branch's theory and practice of park management prepared it for the spruce budworm outbreak. Deputy minister Robertson therefore argued that forest preservation in national parks needed to be redefined:

All this leads up to the question whether we should not give serious consideration to the establishment of a more positive forest policy in connection with administration of the forests in the parks.

I want to say immediately that by "positive" I do *not* mean a policy of exploitation. We are all firmly convinced of the rightness of the Park philosophy *against* exploitation. The question is how best to conserve. It could be argued that maintenance of the Parks "so as to leave them unimpaired for the enjoyment of future generations" means that the forests must be left in an utter state of nature, and with nature left to do with them just what she will – destroy them by disease, by fire, or anything else. We do *not* take that position as to destruction by fire, and rightly so. Equally, I think we cannot take that position where a risk of destruction by disease occurs. In other words, we do not hold to the view that the right course is "hands off". What we should aim at, it seems to me, in the case of fire, disease and all other aspects of forest concern, is an intelligent and scientific policy of management and protection, the object of which is to see, insofar as we can, that the areas now covered by healthy forest remain covered by healthy forest, year in year out, in perpetuity.[119]

Robertson's argument hinges on his implicit definition of a healthy forest: one in which there are mostly living, growing trees, so that overall the forest has more growth than death. But a forest cannot expand infinitely over time, so constant growth is necessarily unsustainable. Forests need trees to die. (Luckily, trees continue to look and be healthy long after they consist mostly of dead cells.) Insect depredation, fire,[120] or sheer age kills off part of the forest, and in doing so opens it up for new vegetation and wildlife. This was well understood by scientists of the time and by the Parks Branch, too, but staff could not help but see dead trees as aesthetically unappealing and – reverting to a logic applicable outside the parks – as economically wasteful. Common sense told them that letting insects kill trees was bad forestry management.[121]

The obvious solution was insecticide. Spraying insect pests was an accepted part of Canadian national park management. There was absolutely no consideration that insects deserved any of the protection provided to other living things in the parks; they were simply killed. For example, from the very first years of Prince Edward Island National Park, its stagnant ponds were annually sprayed with hundreds of gallons of furnace oil – the recipe for killing mosquito larvae.[122] And with the availability of war-surplus planes and the massive expansion of the pesticide industry in postwar North America, aerial spraying grew much more accepted and much more common.[123] It was in this climate that New Brunswick began a full-scale attack from the air on spruce budworm in the 1950s, in a program that would continue for decades.

In a rare act of restraint, though, the Parks Branch opted not to take on the Maritime parks' spruce budworm with aerial sprayers. The Branch was likely wary about what such a massive interventionist move would mean to its public preservationist image. And it could hardly spray Cape Breton Highlands when the Nova Scotia government itself had chosen not to treat infested parts of the province, despite considerable pressure to do so. By the time the Parks Branch even considered spraying, it was advised by federal foresters that the infestation was too far developed to bother.[124]

There was another reason to avoid aerial spraying: in the words of Chief J.R.B. Coleman, "In a National Park, the effect of wide-spread spraying with DDT on the fish and wildlife populations must also be considered and may be reason enough to avoid such procedures on Park lands until more information on this aspect of insect control is available."[125] The chemical DDT – dichloro-diphenyl-trichloroethane – had been found by the American military during the Second World War to be a wonderful agent for insect control: it was inexpensive, very toxic, persistent, useful as either a contact or stomach poison, and yet of low acute toxicity to mammals, including humans. When it was made available for civilian use in 1945, it was heavily marketed as a sort of super-chemical that could cleanse the planet of pests.[126] Biologists were more cautious, and the Canadian Parks Branch took note of their concerns. In 1946, James Smart distributed to superintendents in all the Canadian national parks a US Fish and Wildlife circular, "DDT: Its effects on Fish and Wildlife," which discussed twelve studies showing the dangers of the chemical.[127] It was already proven that DDT tended to kill non-targeted populations such as birds and fish, though some of this seemed to be due to overspraying. Smart demonstrated both the national park experience with using DDT and awareness of its dangers when he wrote in 1948, "In using any chemicals we have, of course, to take into consideration the effect on our wildlife population and in particular on the fish as we have found the latter are very susceptible to DDT even in comparatively small quantities."[128] Insect control was nonetheless periodically carried out in Maritime national parks in the late 1940s and early 1950s – including in the first year of Fundy's existence – but it is not clear whether DDT was the pesticide of choice.[129]

When spruce budworm erupted in the 1950s, the park system knew enough about DDT to be distrustful of spraying it by airplane. And yet, action seemed necessary. Cape Breton Highlands superintendent Doak, who happened to be a great advocate of DDT, informed Ottawa that the public was constantly criticizing the park about its trees, and he pushed for a spray program.[130] More important, senior politicians became involved in the matter. The assistant deputy minister of northern affairs and national resources, C.W. Jackson, told the director of the Parks Branch, James Hutchison, that the ministry was under pressure – presumably from the provincial government or federal Nova Scotia politicians – to act. Though the Forest Branch had suggested that the infestation would have to run its course, this was now deemed unacceptable. As Hutchison relayed to Chief Coleman, "I can tell you now the Minister is not prepared to accept such a stand unless it is a matter of last resort." A more activist response was needed.[131]

The Parks Branch decided that a localized DDT spraying program would take place around public parts of the Cape Breton Highlands Park. Vegetation along roadsides, campground and picnic areas, and park headquarters would be targeted. Of course, by spraying only public places, the Parks Branch was trying to keep its budworm problem out of sight. Moreover, it was giving the impression that trees close to tourists were of greater value to the park than distant ones, and that their appearance was worth an increased risk to local fish, wildlife, and humans.[132] Learmouth, a forester, assured the Parks Branch that he had discussed this "informally with mammalogists, ornithologists, and limnologists of the Canadian Wildlife Service" who did not think it would be a hazard, since it would only be a light spray on a small area.[133]

The spraying of spruce budworm at Cape Breton Highlands took place in the summer of 1957. It was a largely pointless exercise, since the bulk of infestation had already swept through the area in past years and was gone. Three thousand gallons of 25 per cent DDT emulsion was mixed on site with twelve thousand gallons of water to form fifteen thousand gallons of 5 per cent DDT, which was then distributed through a chemical sprayer. Superintendent Doak, though warned that people tended to overdose DDT, complained to Ottawa that the small spray coming out of the machine could not possibly be doing the job.[134] The following spring, it was reported that the control program had preserved the appearance and limited the mortality of the sprayed trees, but not to the degree expected. This was not considered a failure, though, since the exercise was constantly referred to as having "limited control objectives" and as being "somewhat experimental in nature."[135]

The simplest lesson to be learned from this episode is that, regardless of references to an unchanging philosophy of preservation, the Parks Branch still had to answer to political pressures. Its policies were as a result not always ones it would have chosen. Another lesson is that once policy moved in a certain direction, it gained its own momentum. In the years that followed the spraying at Cape Breton Highlands, the Parks Branch organized increasingly extensive insect control programs without any external pressure to do so. Perhaps this was because staff found the experiments with DDT so encouraging, or because they simply enjoyed the concept of cleansing nature in such a modern and scientific way. Or perhaps in justifying an unwanted decision, staff convinced themselves of the rightness of their actions. In any case, insect control picked up. At Cape Breton Highlands, DDT was used in 1958 against spruce budworm, and the following year against the birch case bearer. As staff anticipated, spruce mites filled the vacant budworm niche and subsequently had to be killed with Ovotran acaricide: spraying begat spraying.[136] Fundy staff used 150 "Skeeter Bombs" (which the superintendent believed contained DDT) on the pond near headquarters, 2–4-D on weeds, and Black Leaf Forty for insects on ornamental shrubs.[137] Prince Edward Island National Park controlled its "ugly nest caterpillar" with either Malathion or what its park naturalist called the "well-known 5% DDT spray."[138]

Insect control soon became normalized. What had originally been used for epidemic infestation became the weapon against everyday black fly and mosquito populations. Ex-staff told me of walking around the campgrounds a half-hour before spraying, warning campers to cover their food. They then hooked up the fogger to a forty-five-gallon drum filled with a diesel oil and DDT mix and tied it in the back of a truck. Staff worked without masks, and when they drove around, the fogging operation drew crowds of children who would run along behind. In the words of one staff member, the spray would "kill all the birds and squirrels, there'd be nothing left."[139] The Parks Branch's proactive insect control program became a point of pride. In a St John's *Evening Telegram* story on the new Terra Nova National Park, the community of tourist cabins was portrayed as offering Newfoundlanders unprecedented freedom from insects: "Among the appealing features of this little snug town nestled in deep woods is that there are no flies to bother the inhabitants. They'd be there in hordes but for the 'treatment' given the area three times weekly by Ben Roper. Warden Roper sprays the region with DDT. Dense clouds of it bellow from a compressed air gun and floats over the cabin area eliminating any flies which might be found."[140] The park was sufficiently proud of its new machine that Warden Roper posed for a picture of the fogger in action.

In the same period, however, pesticide use in general and DDT use in particular were beginning to draw widespread opposition. Wildlife biologists were finding evidence that chemicals of low toxicity may become concentrated and cause indirect poisoning further up the food chain; how often this reached humans was unclear. It became a matter of public debate in 1962 with the publication of a series of *New Yorker* articles by Rachel Carson, and the subsequent release of her book *Silent Spring*. Carson's message was that today's pesticidal contamination of the environment might be having permanent, irreparable effects on the health of the planet and its residents. Carson even specifically discussed how the aerial spraying of DDT to combat spruce budworm had killed salmon stock on New Brunswick's Miramichi River.[141] Just as Carson's work was becoming a public issue, the Parks Branch received its first written complaint about its spraying program. An American chemist who had visited Cape Breton Highlands wrote,

There was the indiscriminate fogging of the areas around the camp ground with a DDT-oil fog. We were informed by the park warden doing the fogging that a stronger fog was used last year and resulted in the death of many of the birds in the area. Even though he was using a fog reduced in strength by some 37.5% this year I feel the practice is both foolish and dangerous. It is foolish because the fog had only momentary effect on the black flies which were causing all of the campers much discomfort. The inexpensive insect repellents now on the market proved to be more effective, and without any cost to the National Park. It is dangerous for reasons which I am sure you are familiar. ... As a chemist and as a citizen I condemn the widespread and indiscriminate use of any insecticide and urge that your department discontinue the practice we witnessed on Cape Breton.[142]

The superintendent disagreed with this assessment, saying that staff did not feel there was significant bird mortality – especially now that they had reduced the DDT formula to the recommended 5 per cent level.[143] But there were more complaints, and the Parks Branch began to experiment with different types of chemicals. As a narrator of a 1965 CBC television show on the PEI park explained, the program to spray mosquitoes was ultimately discontinued when people noticed that "the song birds disappeared with the mosquitoes ... Now mosquitoes feed on people, and the songbirds feed on mosquitoes: Nature's perfect balance restored."[144] A happy ending.

But nearly a decade of insect spraying in the parks would serve as a ready precedent when future infestations occurred. New Brunswick, rather than letting the spruce budworm outbreak rise and subside naturally in the early 1950s, had chosen yearly aerial spraying and been rewarded with fifteen years of semi-outbreak levels. Frustrated, the province took to blaming Fundy for hosting the budworm, and urged the Parks Branch to spray. In 1968 the Branch relented and Fundy was involved in spruce budworm spraying until 1975. Nonetheless, spruce budworm destroyed much of the park's fir and spruce between 1974 and 1978.[145]

Just as the Parks Branch had, for a variety of reasons, originally chosen certain locations for its national parks over others that would probably have been equally suitable, it chose some parts of the parks' constituent living things over others in managing them. Effect followed effect, often beyond the Branch's understanding. The preference for attractive forests filled with healthy trees led park staff to cut the forests in a way reminiscent of traditional use. It also led to the killing of insects, and secondarily to the killing of birds, small mammals, and fish. Even the most innocent gestures produced unforeseen results: when staff selectively cut down and removed trees killed by spruce budworm, they felt they were doing the forest a favour; instead, they were removing the homes of insects that were budworm predators, and so helping the depredation continue. As the next section of this chapter will show, park staff slowly grew aware of the changes they were bringing to the parks, and grew more assiduous in at least considering the potential consequences of their actions.

But the Canadian National Parks Branch as an agency has never really acknowledged the extent to which it was responsible for decades of rampant intervention, and how it served as an example to the Canadian public. Today, a sign at Fundy National Park concerning its peregrine falcon population reads, "Toward the 1950s peregrine numbers declined drastically. The widespread use of DDT, an insecticide which accumulated in the food chain, was responsible for the decrease." Also, "A pair of peregrines nested on a seaside cliff a few kilometres east of Point Wolfe in 1948, the year that Fundy National Park was established. They were not seen again and decades would pass before others of their species would replace them."[146] The sign never quite mentions that at one time pesticides were used so widely in North America, and thought so safe, that they were dispersed abundantly in this very national park.

A NEW ECOLOGY

The Parks Branch's interventionist policies on preservation matters in the 1940s and 1950s owed much to the science of ecology, even though administrators rarely used the word. If asked, staff explained management decisions in terms of carrying capacity, climax communities, population cycles, and other ecological concepts. But explanation was rarely needed. During this period of its history, the Parks Branch's presumed right to manage the nature within its border and its ability to do so competently were largely taken for granted by the public. Thus staff could distribute press releases describing in detail the poisoning of lakes or the spraying of campgrounds without feeling a need for discretion. Science maintained enough cultural authority in these decades that the Parks Branch, an agency presumably grounded in science, had freedom to make decisions regarding the parks' natural environment as it saw fit.[147]

But in the late 1950s, the meaning of ecology began to be reinterpreted by North Americans. People were developing a new-found attachment to nature, and were growing anxious about its fate under human "management." Though the public deferred to ecology for its ability to help explain the natural world, they began to learn different lessons from this science. Just as ecology could teach that nature was but a complicated machine in need of tinkering, it could also teach that one should be respectful of nature's complexity and should fiddle with its processes as little as possible. This may seem to have been an opportunistic reading of the science, advocating a non-interventionism that people wanted in any case, but it was no less valid than that made by game managers who had favoured interventionism. Nor was such an interpretation new: it had been in gestation for decades, including among ecologists themselves. The 1960s saw ecology the science – reductive, abstract, with a mechanistic view of nature – adapted by North American society as ecology the movement – holistic, value-laden, with an organic view of nature.[148]

The National Parks Branch, as the foremost Canadian agency dealing with humans' relationship to nature, served as the harbinger of this new thinking. Within the space of just a few years in the late 1950s and the early 1960s, the national park system underwent a swift ideological conversion. From actively promoting recreational facilities for the masses, the Parks Branch turned to advocating a far less interventionist philosophy. In doing so, however, it did not give up the science of ecology for the ecology movement; it continued to defend its policies in scientific terms. But it could now also use ecology to justify its decisions ethically. The scientific authority it had once derived from ecology was now supplemented by a no less potent cultural authority, that of ruling on the "natural values" which it was mandated to preserve – even by curtailing the recreational activities of the masses if need be.

There is no better evidence of this ideological shift than the Branch's changing opinion of Fundy National Park, which it had just recently established. With over a hundred acres of manicured lawn, the most modern recreation facilities, a business subdivision, and a community hall, Fundy offered the most ostentatious display of man-made intervention to be seen in the entire Canadian park system. Though the superintendent's residence was subject to internal criticism for its opulence even while it was being built, Branch staff gave no evidence in the park's early years of development of being dissatisfied with the general results. It had been developed as it had been designed. Yet when Terra Nova, the next national park to be established, was underway in the late 1950s, staff used Fundy as a model for how *not* to design the new park. Parks director Coleman specifically stated that the new park's cabins should not be as elaborate as those at Fundy.[149] More generally, facilities should not overshadow nature: the buildings should fit into the landscape, and development should be on a much smaller scale. When work at Terra Nova was complete, deputy minister Robertson commended the builders, noting that the "headquarters seems to have been well laid out – an infinite improvement on the last park we established at Fundy!"[150]

In the coming years, criticism of Fundy grew. A staff member stated in 1960 that the New Brunswick park had had to rely on recreational development because it comprised such average nature. Even so, the park was not appropriately representative: "The beauty of the Saint John Valley, the Atlantic Salmon streams and the great tidal bore are all outstanding attractions of this region not found in the park."[151] Director Coleman agreed that Fundy was not as attractive as it could be, but shrugged off any blame on the Parks Branch's part by claiming, "Fundy was not the Department's first choice for a national park area at the time the matter was being considered."[152] He ignored the fact that though Cautley and Smart had never ranked the Albert County site first in preference, the Parks Branch certainly had not opposed it. Fundy increasingly became the target of open condemnation. Deputy minister Robertson felt that the subdivision was a travesty, writing, "I was even more struck than previously with the lamentable planning that had led to the location of the park compound where it is. … Even worse than the location of the compound is the complete outrage in having the Imperial Oil service station where it is. Whoever permitted this should have been shot, but I suppose it is too late to take action now."[153] The most scathing and most searching reaction to Fundy came from parks naturalist R.D. Muir in 1964. He had visited all the parks the previous summer, and though he had concerns over the scale of development in some, he saw Fundy's problem as much more serious. The park facilities did not seem to be in harmony with their natural surrounding. Reflecting on a previous visit, Muir wrote, "From last year's Fundy experience came a slowly developing uneasiness, a poorly defined feeling that something had gone wrong in this park. Because of the nebulous and psychic aspect of this reaction, it was not included in last year's report. It was easy enough to see that moose had gone wrong [there had been a population explosion and crash], but there was more than that, something very basic had happened, and the basic concept of a National park had foundered. ... The crux of the matter was that the natural features for which the park was set up were being destroyed, obscured, covered up and cheapened."[154] Muir's rereading of Fundy occurred within just fifteen years of the park's creation. Nothing had physically changed: the park had preserved intact the landscape it had created. But what had seemed in 1950 to epitomize a modern, enlightened relationship with nature was by 1963 proof of humans' inability to leave nature alone.

What makes this transition especially remarkable is that it occurred during a period of the park system's history in which intervention of the sort that Muir found so distasteful was practised by the Parks Branch to a degree never seen before. For the sake of park use, a system gorged with tourists was receiving unprecedented levels of funding. For the sake of preservation, the Parks Branch was implementing some of its most egregiously interventionist policies on wildlife, fish, and vegetation.

This seeming contradiction can be explained, at least to a degree, as a reaction against the physical results of this "hands-on" approach. As discussed earlier, attendance in the parks mushroomed during the 1950s, from about 1.8 million at the beginning of the decade to 4.9 million at the end. The Parks Branch tried to match this increased attendance with increased facilities. It became immediately clear to staff that the park system could not keep up to such growth, and that trying to do so would compromise nature in the parks. Discussion took on an apocalyptic tone. The Branch's annual report in 1959 for the first time voiced a concern about the possible deterioration of the park system because of development. By 1964, this concern was much more obvious, so that the report began, "A race against time has developed in the management of the National Park."[155] The ideological shift, then, can be understood as a response to the manmade change caused by rapidly increasing use of the parks. And there was a new ingredient in this interpretation: the Parks Branch was for the first time including itself in consideration of manmade change. The park system had always looked at the effects of tourists, past residents, and locals as potentially harmful to the parks' wilderness and consequently sought to minimize these impacts. But when staff began to see themselves as part of the parks' history, they became aware of the degree to which they themselves were manipulating nature in the parks.

There is certainly truth in this explanation for the ideological move away from interventionism, but it is not entirely satisfactory. Because of the continuing need to accommodate the public, the parks continued to be developed in the 1960s, casting doubt on the direct relationship between ideology and action. Nor is there an intrinsic reason to believe that one event, intervention, should produce an opposite reaction, a move away from intervention. Moreover, one could say that the coming decades actually proved staff's fears wrong: a park system supposedly overburdened with almost six million visitors in 1960 would prove capable of handling, without collapse, almost seventeen million in 1970. It may have been *reasonable* for the Parks Branch to develop a less interventionist philosophy in the 1960s, but there is nothing that made this decision obvious, necessary, or fated.

I believe the Branch's change in philosophy was related to its conception of itself as a purveyor of high culture. National parks had been championed throughout their history as safeguarding such ideals as beauty, love of nature, and patriotism. As custodian of the parks, the parks agency was also presumed to possess such qualities. But in postwar North America, it was more difficult for the Parks Branch to make such claims: the expansion and increasing wealth of the middle class made parks much more accessible, and the apparent wiping away of class difference made the promotion of elitist ideas less tolerable. The Parks Branch felt obliged to reconceptualize itself as an agency serving a single, mass clientele. The Atlantic parks themselves are eloquent testimony of this changing social vocation. Initial development at Prince Edward Island and Cape Breton Highlands had been premised on the assumption that each of these parks had to serve two distinct classes of tourists. Development at Fundy was centred on one location to serve the recreational needs of one class. This also signalled staff's belief that use by visitors was largely distinct from nature, and that, for the enjoyment of users and to meet the goals of preservation, people and nature should be kept separate.

But in the 1950s, the Parks Branch discovered that even in an age of recreational democracy, not all tourists were alike. A significant number of park visitors were shunning the facilities so elaborately and expensively prepared for them, preferring instead to get back to nature as much as possible. Staff were surprised to find that many people did not like the open, community-style campgrounds being developed. Similarly, the park system in the late 1950s experimented with a naturalist program, including trail walks and interpretive lectures, and staff were amazed at its great success. Field trips along the Bay of Fundy attracted two to three hundred people; eight hundred Fundy visitors sat in on a single campfire talk in 1962.[156] Considering that Parks Branch staff were part of the culture, and as caretakers of the park presumably in tune with the public's feelings about them, what was most remarkable was not that the Branch changed its operating philosophy, but that it was in many respects so slow to do so. It took years for the Parks Branch staff to recognize that many people were coming to parks primarily to experience nature, and not to experience recreation in close proximity to other people.

The Parks Branch's discovery of this constituency allowed it once again to speak of two classes of visitors, but this time the division between the classes was an intellectual rather than an economic one. Visitors were divided between those who understood and appreciated appropriate park recreations and those who did not. There were those who sincerely loved nature and those who were more interested in mass recreation (epitomized in the 1950s by the drive-in, whose potential inclusion at Banff created fierce controversy).[157] This was not a manufactured distinction; staff were honestly concerned about stress on the park system, and it seemed sensible that visitors who loved parks for their nature had the greatest right to their enjoyment. But, not coincidentally, this gave the Parks Branch a renewed opportunity to position parks as cultured places that bestowed on visitors – the right sort of visitors, experiencing the park in the right way – cultural capital.[158] It also let the Parks Branch claim to be the organization best situated to judge what was cultured in parks, and therefore obviously well cultured itself. In sum, the Parks Branch's ideological shift away from interventionism allowed it to reconceptualize a park elitism that had given it authority for most of its history, but which had disappeared after the war. Naturalist R.D. Muir provided this new elitism with what could have been its motto: "It is only the enlightened who see the order and pattern in Nature."[159]

Regardless of the motivations for advocating a less interventionist philosophy, the Parks Branch's move had repercussions in how parks were actually managed. Terra Nova National Park exhibited signs of this new naturalism, though some changes were as much about style as substance. Terra Nova was welcomed as a new sort of park for Eastern Canada and a return to the wilderness ideal. The only real development in the park was its cabins, and even these were relatively Spartan. The Branch blanched at the concessionaire's suggestion that they be named "Terra Nova Tourist Cabins," because it was trying to disassociate itself from tourism. H.S. Robinson, chief of the Education and Information Section, explained, "we have deleted phraseology such as 'Canada's National Playgrounds,' 'Tourist Accommodations,' 'Tourist Resort,' 'Tourist Attractions,' from all our publications and releases."[160] The more neutral "Terra Nova Park Bungalows" was chosen instead. But the park cabins were not even to be the centre-piece of park accommodations; the 417-unit Newman Sound Campground was where "the real park experience is available." Within a few years of Terra Nova's initial development, staff would also carve out seventy "primitive" sites six miles away from headquarters, along the Bonavista coast, "for those anxious to really rough it."[161]

Since the Parks Branch staff hoped Terra Nova would become the first Eastern wilderness park, it was all the more disappointing that, when creating it, they lost its wood supply to potential timber purposes and the Terra Nova River watershed to hydroelectric purposes. Without the watershed, the park had no good salmon streams and no place to reintroduce caribou. On top of this, once the park was established it was found to have a disappointing quantity and variety of wildlife. Though it was said that moose numbers were booming across Newfoundland, the wardens at Terra Nova reported seeing only a few, plus some foxes, rabbits, and other small animals. Superintendent Atkinson reported, "Quite candidly I miss the animals found in the Western Parks, as without them it is like being in an uninhabited piece of country."[162] On and off for the next decade, Ottawa asked Newfoundland for "a section of typical caribou country and some typical salmon fishing streams,"[163] but without success. The Parks Branch continued seeking to expand the park even when its wildlife numbers began rising in the 1960s. For a time, the government of Newfoundland considered donating Pitts Pond, outside the southwest corner of the park, but continually fretted about losing its potential economic benefit. Park staff pulled out all stops in trying to convince the province. Chief parks naturalist George Stirrett noted that the red pine to be found at Pitts Pond "occurs in only three locations in Newfoundland and the stand at Pitts Pond is the most easterly part of the range of the species in Canada. This red pine stand should be a talking point in our proposals to have the area of Pitts Pond included in TNNP."[164] Certainly this reference to the most easterly pines in Canada was a gimmick, but it nevertheless shows the lengths to which staff would seek justification for expanding a park to what was thought to be its necessary natural size. And more than that, it showed appreciation for the distinctiveness and inherent value of an Atlantic Canadian species – and a plant species at that. Such appreciation had never before been observed in three decades of national parks in the region.

Of all Canada's national parks, none was affected more by both the strain of park use and the subsequent defence of preservation than Prince Edward Island's. One of the smallest parks, at just seven square miles, it was also one of the busiest: by 1961, almost one million visitors were recorded as visitors to the park. Throughout the 1950s, the Parks Branch worked continually to meet demand in the park, building a bungalow camp, a recreation hall, campgrounds, and the Gulf Shore Parkway, which stretched along almost the entire park's length, broken only at Robinsons Island. When park ideology moved away from development in the early 1960s, it was as if staff had awakened from a dream, amazed at the decisions that had been made just a few years earlier. As they had at Fundy, staff disassociated themselves from PEI National Park's problems, by criticizing its very establishment and placing blame on decisions of early years. For example, Superintendent E.J. Kipping spoke freely at a 1962 public meeting of "mistakes of many years ago" now "coming home to roost."[165] Even Green Gables Golf Course, long the pride of the park, was now seen as an anachronism. The minister in charge of parks, Arthur Laing, refused a request by the PEI government for a second golf course by saying, "In the early days of National Parks' development a number of policies and practices were followed which seemed appropriate under the conditions then existing but which in the light of present day experience appear questionable and rather short-sighted. I do not think there is any doubt that golf courses fall into this category."[166] The park system would keep the golf courses it had, but it would not be designing more. Branch staff generally spoke in this era of the Island's national park as a failure. Parks engineer R.P. Malis believed that no matter what happened, it "will never be accepted by the people as a true National Park in the strict sense of the word" and agreed with a recommendation being floated to reclassify the park as a national seashore, like those in the US.[167] Though this did not happen, the fact that the Parks Branch was seriously contemplating the demotion of the second most visited member of the system shows how greatly the ecology movement was changing the conception of what a national park should be.

The Branch's ideological shift meant more to Prince Edward Island National Park, however, than just a denigration of its past and a pessimistic reading of its future. Staff worked to make the park a more natural experience. The greatest environmental and aesthetic threat was seen to be "ribbon development," private tourism development just outside the park's boundary. The park's border forests were relied on to block this out, and the Parks Branch even considered a tree-planting program to screen out the real world – a far cry from park practice less than a decade earlier to trim trees so that private tourist operations would have a better view of the water.[168] Superintendent Kipping believed that there was nothing to be done but buy up bordering land as it became available, and he even took to sending

Ottawa word about parcels that he learned were for sale. The Parks Branch initially expressed no interest, feeling that such a piecemeal expansion would accomplish nothing. Besides, there was no precedent for the federal government's purchasing land for a park.[169] But when Ottawa realized it could ask the Prince Edward Island government to cover the bill, the idea was reconsidered. Lobbying for support among Island politicians, Arthur Laing wrote, "It is now apparent that the land area originally set aside was inadequate and the best we can hope to do is to try and acquire land to round out the park at several key points."[170] The provincial government offered a site in Stanhope, outside the eastern wing of the park, on the condition that the Parks Branch build a golf course there; otherwise, the province had no interest in buying expensive property if the park planned to let it lie fallow.[171] The Parks Branch declined the offer. But the notion of buying buffer land was now firmly entrenched, and the Branch began lobbying its own government to buy land directly. In 1968 the Canadian government initiated a land acquisitions program for the park. Between 1969 and 1977 approximately 2400 acres of adjacent land were incorporated into the park.[172]

As the experience of Prince Edward Island National Park in the 1960s shows, the Parks Branch's move toward less interventionism affected how the agency thought about and actually managed the park. But that is not to say that development was curtailed. Throughout the decade, the Branch continued to satisfy the park's one million yearly visitors without limiting access by setting visitor capacities or even implementing entrance fees. Even in this era of less intervention, the Branch built a campground and day-use area on Robinsons Island, made accessible to the public only a few years earlier. The little park had no fewer than sixteen parking lots, including one at Dalvay for over twelve hundred cars, by the end of the decade. The idea of preservation had gained an ascendancy in the 1960s, but this had not relieved the Parks Branch of its responsibility to accommodate tourists.

The culminating moment for the Canadian park system in the 1960s was Arthur Laing's address on park policy to the House of Commons in September 1964. Park historians have called it "a very significant milestone" and one that "established the preservation of significant natural features in national parks as its 'most fundamental and important obligation.'"[173] The ingredients for drama were certainly there. A quiet revolution was already under way within the park system, as both staff and the public sought nature-friendly policies. The Parks Branch had been labouring for six years on a declaration explaining how it was attempting to reconcile preservation and use.

Now, the agency felt pressure to make this statement public: businessmen at Banff and Jasper were pressing their right to develop wherever they wanted in the townsites; a National and Provincial Parks Association of Canada had recently formed to protect the parks from development; and in 1963 *Maclean's* magazine published a scathing article, "Beauty and the Buck," that painted the park system as being on the verge of collapse.[174] Author Fred Bodsworth had pulled no punches: "We are losing them because a lax and indecisive parks policy, particularly toward business and political pressures, has allowed many national parks to deteriorate into commercialized, honky-tonk resorts where the major aim is no longer park preservation but rather separating tourists from their money."[175] It was hoped that Laing's speech would answer such criticisms by showing that the Parks Branch was already correcting past mistakes; it might also discourage developers from planning further incursions in the parks. Most of all, it would do away with the system of ad hoc, unwritten policy directions which had taken shape since 1930, replacing this with a single, decisive policy document.

It is surprising then that to today's reader, Laing's speech is not very decisive at all. The closest he comes to a pointed declaration of park priorities is to say:

National park policy cannot contribute to a solution of the crisis if it is based on one of the two extremes, maximum preservation on one hand, or maximum public use and development on the other. One would deprive the public of the benefits they receive from national parks; the other would destroy the special enjoyment and pleasure the public receives from lands kept in a near natural state. The objective of national park policy must be to help Canadians gain the greatest long term recreational benefits from their national parks and at the same time provide safeguards against excessive or unsuitable types of development and use.[176]

It helped, of course, for this position to be clarified for the Canadian public, and for Parks Branch staff to have such a ministerial statement from which to refer.[177] But rather than providing a solution, Laing's address simply restated the problem: that parks were always under a dual mandate of preservation and use. This was entirely appropriate. Though this was a period when park staff and the public leaned toward less interventionist policies – and, indeed, Laing's speech was interpreted as a victory for these beliefs – the Parks Branch could not and should not codify non-interventionism. A perfect balance between preservation and use must always be the park system's goal, even though it is a goal which will never be perfectly achieved.

9 Conclusion

While staff did the day-to-day work of maintaining Cape Breton Highlands, Prince Edward Island, Fundy, and Terra Nova national parks for the benefit of all Canadians, one group of Canadians paid special attention. Local residents, particularly those who had land expropriated, were the national parks' most observant critics. They knew that the parks were not in fact pristine wilderness and they perceptively drew out the inconsistencies of park policy. Of course, what made local residents so astute also made them less than objective. Many locals felt that the park belonged to them, either because they personally had lost land at the park's creation or because their community had given the park territory and deserved special consideration in return.

Locals' feelings revolved around issues of both preservation and use, and in both cases worsened when the park system worked to become less interventionist in the 1960s. Many were upset that park resources were no longer available to them. Trees could no longer be cut, deer no longer hunted, trout no longer fished without permit. This was considered especially onerous at Terra Nova and Cape Breton Highlands, where so much of the local land had been free for all to use. It was all the more galling when park staff committed the same acts in the name of management. Why were staff at Cape Breton Highlands allowed to shoot moose during periods of overpopulation, while locals were not? Why did the Parks Branch operate a sawmill in Fundy, when locals were not allowed to cut? How, for that matter, could the Parks Branch preach the sanctity of nature when its own road-building and development of facilities tore up the land? Such inconsistencies were made worse by the fact that through time the Parks Branch tended to grow less and less accommodating of locals' needs. Staff did not feel the same sense of obligation to nearby residents as park establishment receded further into the past. For example, through the years the Parks Branch stopped giving timber permits to those who had once cut on Terra Nova. As a result, not only did locals feel they had lost the right to an important resource promised them at park creation, but they interpreted the Branch's action as a deliberate, piecemeal process.[1] Staff also grew less accommodating because of the system-wide move away from natural resource use in the parks in the 1960s. Neil MacKinnon, who worked at Cape Breton Highlands for thirty years, remembers cutting trees for the flagpoles at warden stations; in later years, the wood had to be brought in from outside.[2] Don Spracklin told of getting in trouble with the staff at Terra Nova for pumping a thousand gallons of water out of Udells Cove Pond, for use in the Christmas Seal boat that went around the Newfoundland outports each year giving chest X-rays.[3] Others told me stories about run-ins people had with staff over the attempted removal of park sand, stones, and ice. Such stories are sometimes told with amusement, sometimes with disgust. In either case, they are used to demonstrate the Parks Branch's tenuous connection with real life, its willingness to put ideology before actual human needs.

Local residents also felt that the parks were not the economic salvation that politicians and park staff had promised they would be. The most consistent complaint I heard at all four parks was that, after the initial burst of park development had subsided, not enough people were ever hired. Those whose land had been expropriated felt especially deceived, in that many claimed they and their families had been promised jobs for life. A typical sentiment was that of Winnie Smith, who said, "You were moved out of the park, and now you can't be hired back in."[4] It was also felt that too many of the jobs that were created went to people from more distant communities, or, worse still, other provinces. These complaints were justified in some cases: it is clear that at Terra Nova, for example, the Smallwood government did promise jobs to locals, and then took to hiring more from some communities than others. The parks and their staff should not be blamed for this, however, or for the quite reasonable fact that most development took place within the first few years of park creation. Aware of the importance of park jobs to the local economy (and the importance of local harmony to the smooth running of the park) and recognizing that more appropriations meant the completion of more park projects, the Parks Branch did do its utmost to keep employment high[5] and to spread hiring throughout local communities.[6]

"How can we do things for people if people keep getting in our way!?" by
Bob Chambers, Halifax *Chronicle-Herald*, 12 May 1973. Republished with permission of
the estate of Bob Chambers.

Created just as the park system was turning from interventionism,
Terra Nova National Park drew less development than the other
parks, and consequently received more complaints from local residents. When would Terra Nova get its swimming pool, golf course,
and other such amenities?[7] Premier Joey Smallwood promised in the
late 1960s that if a second park were built on Newfoundland's western shore, as was being discussed, the province would demand
much more development: "Just taking so many hundred square
miles of all the wilderness we have, putting a boundary around it
and calling it a Park, well that's ridiculous! That's not what they
have across Canada. ... They have National Parks that are National
Parks in which untold millions of dollars of Canadian money was
spent to provide the people of the other provinces with these magnificent National Parks. ... But [to] settle for another wilderness park,
such as we already have at the Terra Nova ... would be intolerable
for us to accept."[8] The park system's ideology was moving in the

opposite direction, however. Parks in the future would be even less likely to satisfy local communities' requests for development.[9]

Because local residents had different interests from the Parks Branch when it came to issues of preservation and use, and more fundamentally because the park had been imposed on them and their community, many considered themselves adversaries of park staff. The parks themselves seemed set up to foster this opinion. Though the majority of staff were hired from the park vicinity, most of these worked at lower-rung jobs. Those in the most senior position, the superintendents, were always brought in from elsewhere. This was done not only in early years, when locals might not be expected to have the required knowledge of or experience in parks, but also later, when the Parks Branch feared that local staff might have conflicts of interest when making hiring and management decisions. Ottawa staff displayed in correspondence a constant wariness of local residents, and occasional outright distrust. When Superintendent Doak of Cape Breton Highlands asked in 1955 to be transferred after unstated problems, deputy minister Robertson noted that locals were "inclined to ride over anyone who is at all lenient, and they regard any concessions or favours as a sign of weakness." Though Robertson thought Doak should stay, if he were to go he was to be replaced "with a strong and decisive personality, who will not be intimidated or upset by the efforts that the local people will undoubtedly make to recapture the upper hand that they apparently had at one stage."[10] It became policy in this era to transfer superintendents from park to park every few years to improve their training; locals could not help but think that this was to keep supers from forming a close affiliation with any one park or its people. Indeed, an ex-staffer told me that when a park was facing a difficult time with local citizens, Ottawa would send out "an enforcer" with no links at all to the region.[11]

It is to some degree understandable if the Parks Branch slipped into an "us-versus-them" mindset. Local residents demanded more from the parks than anyone and in return were the most likely to violate preservationist policies. But for staff to give in to such thinking was counterproductive. Experience with poaching is a good example of this. There is some evidence of illegal hunting in all four parks' histories – it is sprinkled throughout the archival correspondence.[12] Perhaps the greater surprise is that there is not more. I expected to be told in interviews that poaching was a method of taking a stand against the park, but everyone who spoke of poaching insisted this was not the case. Those people who hunted in the park did so, it was said, because they needed the food or liked the sport, but not because they saw it as a political act.[13] Consequently, the most respected staff were those who made allowances for human need and tolerated occasional indiscretions.[14] On the other hand, people spoke disparagingly of staff who treated all locals as potential poachers – such as the warden at Terra Nova who would walk along the streets in the nearby village of Charlottetown and touch car engine bonnets, deducing who had just been out.[15] That this incident would be so clearly remembered decades after the fact shows just how precarious relations between staff and residents were. Sitting in a group around a kitchen in Cheticamp on the border of Cape Breton Highlands National Park, one man – forgetting I was there – started to talk about doing a little hunting in the park. Everyone laughed, and he turned to me, sputtering, "But that's my people's land back there!" It wasn't. His family had been expropriated from Cap Rouge, about 10 miles away. This was his people's land only if one thought of the park as a single unit and the people on its border as a single unit. In that sense, the park had created, where none had existed before, a community of residents who considered all of the park theirs.

The popularity of national parks in the 1960s ensured the creation of more parks and hence more interactions between locals and staff. The thought of having a tourist attraction which might attract a million visitors per year and which would be administered by the Canadian government encouraged provincial governments to lobby for parks to a degree unseen since the 1930s.[16] The Liberal federal government also approved of more parks, considering them popular signs of the national state and good investments for tourism creation. It seemed urgent to create more soon, before the best of Canadian recreational land was snapped up by developers, particularly American ones.[17] Moreover, the Parks Branch liked the idea of expanding the park system – for one thing, it would reduce visitor stress on existing parks. The primacy of the environmental justification for parks convinced staff of the need to preserve many more types of Canadian nature.

However, expropriation of land for parks in the late 1960s promised to be a different matter from what it had been even a decade earlier. The building of postwar public works projects such as highways and airports had highlighted countless difficulties with existing federal and provincial expropriation laws. There were procedural problems involving arbitrary and inadequate notification, and clear cases of insufficient compensation.[18] In the late 1960s, legislatures across Canada began modifying the laws to make them more uniform, and to change the basis of compensation to "market value," which was more likely to compensate the owner fairly.[19] This legal change reflected – and reinforced – changing Canadian attitudes concerning the state's responsibility to the public. Canadians no longer saw government as an immovable force whose will must be obeyed.

It might seem that these changes would make the Parks Branch wary about park system expansion, but the opposite seems to be true. As a part of the culture, the Branch understood it could not ride roughshod over citizens' concerns, and welcomed the chance to forge new and clean relationships with citizens at new parks. This was pragmatic thinking, and it also reflected a better ecological understanding that park lands had their own cultural history which had shaped their nature, and staff could not erase this history just by wiping away reference to past inhabitation. Staff in the Atlantic parks began in fact to grow interested in showcasing this cultural history. In discussing the proposed extension of Cape Breton Highlands to the northern tip of the island, Atantic regional director H.A. Johnson argued that though "planned communities should replace the scattered developments," there was nothing intrinsically wrong with having residents within the parks. He wrote:

Large numbers of the travelling public are interested in seeing typical fishing villages. Planning these communities will to some extent change their character but it is doubtful that they would lose their visitor appeal. If properly done, this planning will actually guarantee preservation of certain aspects of their character appeal, e.g. Peggy's Cove outside Halifax. … It is not a serious nor complete disadvantage to have these communities dependent to a degree on the park. They provide an essential labour supply and other services which, if properly controlled and planned, could provide a sound social and economic unit.[20]

In sum, there were legal, humanitarian, ecological, bureaucratic, and even aesthetic reasons to imagine that future parks should be characterized by more interaction between parks and local citizens.

In the late 1960s the Parks Branch began expanding the park system in earnest. Kejimkujik in the Nova Scotia interior, Kouchibouguac on the eastern shore of New Brunswick, and Gros Morne on the west coast of Newfoundland were established between 1968 and 1970, along with a park in British Columbia and two in Quebec.[21] Three parks were also being discussed for the Canadian North, as well as a third Nova Scotia park, and another for little Prince Edward Island. In the same period, the Parks Branch began releasing provisional master plans which documented how policy in each of the national parks, old and new, was to proceed. As proof of its commitment to good relations with the public, the Parks Branch announced in 1969 that it would host public meetings to discuss each of the master plans.[22]

Of the four parks studied here, only Cape Breton Highlands and Fundy were treated to public hearings. The Parks Branch set the format, one which seemed designed to keep things harmonious. The agency made itself moderator, with the responsibility of weighing the meaning and import of briefs. The sheer number of briefs – Fundy had sixty written ones and thirty-three oral ones, everything from the National and Provincial Parks Association of Canada to the New Brunswick Dance Teachers Association[23] – ensured a deadening of polemical opinions. The Parks Branch even solicited briefs from groups likely to share its views.[24] Finally, the Branch shaped the proceedings by setting the hearings' time and location. This was especially significant in the case of the Cape Breton Highlands hearing, held on a Wednesday in Sydney, almost a hundred miles from the park itself.[25] Although these factors could conceivably have made the public hearings sedate, they turned out to be quite spirited affairs. The transcripts of the Cape Breton Highlands and Fundy hearings suggest that citizens near the parks welcomed the chance to have their say about park policy. Just as important, the hearings were a harbinger of what the Parks Branch could expect of public opinion when creating new Atlantic Canadian parks in the 1970s.

There were many interests represented at the hearings, and even groups with similar views had quite different motivations. Some lobbied for more development, some less; some demanded expansion of the park, some that its size be frozen. Representatives from distant communities, business clubs, and tourist associations tended to speak up for development – figuring they would get some residual benefits from federal dollars and tourism, yet were far enough removed not to be directly affected by park policy.[26] Citizens and representatives of communities closer to the parks also made specific requests for more park facilities. These speakers reiterated the demand that residents be given greater consideration for hiring, as had been promised at the park's creation. The mayor of Alma, just outside Fundy, explained, "Local residents were led to believe by the politicians of the day that, lumbering having ended, they would supply the labour force for the development of the park. This was the case for many years; however, the passage of time has brought many changes, and at the present time there seems to be little concern for the local residents by those in positions of control."[27] A common complaint was that too much of a park was not in use, either for tourism development or resources. A Cape Breton MLA said that only one per cent of the park was developed and that 99 per cent was "absolutely nothing."[28] In the New

Brunswick hearing, there were calls to let Fundy's forests be lumbered, and economic, ecological, and aesthetic justifications were offered.[29] Others repeated the long-standing Atlantic Canadian claim that the region was owed more funding from the Parks Branch as "deferred credit" for the fifty years when only Western Canada had national parks.[30]

Both Cape Breton Highlands and Fundy were considering expansion in this period, and many briefs at both hearings enthusiastically approved. Cape Breton Highlands' provisional master plan talked of taking in the northwestern section of Cape Breton, leaving fishing communities intact.[31] Likewise, there was interest in expanding westward along the shore at Fundy National Park, creating a scenic Fundy Drive and making an easier approach to the park for tourists from Maine. Those who sought development at the parks approved wholeheartedly of the idea of park expansion – as long as it was not their land that would be needed. Charles Polley of the Moncton Fish and Game Association said of a westward Fundy expansion, "It would interfere with a few woods operations, but we are not too sympathetic in this line – right or wrong, we are not. The woods operations denude the land, so on and so on."[32]

There was also a school of speakers who opposed development or expansion of any kind. Naturalist groups wanted the parks largely left alone, but like pro-development business groups would not be directly affected by any decisions made.[33] Some local residents shared the naturalists' opinions, particularly people near Fundy who adamantly opposed the chemical spraying operation that was under way.[34] Other speakers decried any thought of increased park development because they already considered the park a detriment to local business. At Cape Breton Highlands, Maynard MacAskill, president of a citizens' group entitled the North Victoria Landowners' Protective Association, stated, "Presently you build subsidized campgrounds with our tax dollars and then you force us to compete with you. The national park is stifling private enterprise and it is unfair for the individual in our society to have to compete against his own tax dollar." As could be expected, MacAskill also opposed park expansion: "We lived the National Park for 34 years. Our stomachs are full of it. We will tolerate it, but we want no more of it."[35] There was not much talk at the Fundy hearing about expansion because it was still so hypothetical there, but those residents at Cape Breton Highlands who were potentially affected by the park's plans made the most of the hearing to voice their emphatic opposition. A seven-hundred-name petition against expansion was presented, as were the results of three public meetings held in northern Cape Breton: ten of twelve communities voted with large majorities (and some 100 per cent) against expansion, and the other two communities endorsed expansion only if no one would have to move.[36] This stand was unanimous among those who spoke at the Sydney hearing. The most vitriolic attack was voiced by W. Gwinn, an older resident of the threatened region. He described the establishment of Cape Breton Highlands: "They sent this Mr. Smart down there and he told more lies and falsehoods than you would find in the thousand tales of the Arabian Nights. All kinds of promises. As soon as they got things in their possession these promises were completely forgotten." After comparing the expropriation to the clearances of the Scottish Highlands, Gwinn's voice rose to a crescendo:

I think if you went down there and asked the majority of the people what their dearest wish would be, they'd answer you something like this: "We wish this cursed tiger was chained in the bottomless pit. We wish the National Park would destroy every vestige they have created down there. We wish they would burn every building they have ever erected. If they do that that will be the most glorious day that ever dawned in north Cape Breton." ... I'll tell you what would happen. The people would be out on the hills and the fields singing and rejoicing, "Glory, glory, hallelujah, the curse is lifted, the bloody tiger is chained in the bottomless pit, the Iron Curtain is torn down. Thank God we have freedom and liberty again." That's what you would hear. Thank you.[37]

The chair, introducing the next speaker, called Gwinn a tough act to follow.

It is difficult to know how the Parks Branch assessed these hearings. It released documents responding to public concerns and further explaining its policies – but did the hearings actually shape policy?[38] Even given the great diversity of opinions presented, the Parks Branch should have learned two lessons from the proceedings: first, that though parks were for all Canadians, they most directly affected those who lived next door to them; second, that expropriation for parks in the 1970s would bring passionate and unified opposition, much more so than in previous decades. Since even if a park was established expropriatees were likely to take residence nearby, the Parks Branch seemed destined for a future of difficult local relations.

Yet no matter how fair it promised to be in land acquisition, the park system still needed land if it was to make new parks. With the assistance of provincial governments, the Branch moved forward with parks at Kouchibouguac in New Brunswick and Gros Morne in Newfoundland, and announced its intent to create a park at Ship

Harbour in Nova Scotia. In all three cases, it was met with a hailstorm of opposition. The Ship Harbour park was cancelled after a sustained protest spearheaded by a citizens' group, the Association for the Preservation of the Eastern Shore.[39] Kouchibouguac was established, but its story was hardly one of success.[40] Beginning in 1970, the family of Jackie Vautour, one of about 225 families to be dispossessed, mounted a decade-long protest against the park. They returned the cheque the province had offered and stayed on their land. In 1976, they were forcibly removed and their home destroyed. Vautour moved his family back into the park in a tent, and subsequently built a more permanent shelter there. Following Vautour's loss of a Supreme Court challenge in 1980, he and his supporters organized two violent protests at the park headquarters. They promised that the park would never be left in peace.[41] Kouchibouguac was a watershed in Canadian national park history. The Parks Branch recognized that it could no longer pretend to be uninvolved in the process of park creation simply because the provincial governments were responsible for acquiring land; just as important, provincial governments recognized that expropriation for parks was politically inexpedient. The Newfoundland government conceived a new approach to park creation in the making of Gros Morne. Though originally 175 families were to be moved, it was decided in the face of public opposition that the park would be established around them.[42] By 1979, the Parks Branch officially adopted this method. In the future, more public input and support would occur before a park would even be considered, and no land would be expropriated – it would only be acquired if the owner was willing to sell. Though the process of national park creation was bound to take longer, it would at least be smoother.[43]

The protests at the newer Atlantic Canadian national parks did not much affect the first four parks of the region.[44] Nor did citizens' groups at the newer parks refer back to the expropriation experiences of the earlier parks: the submissiveness of landowners at Cape Breton Highlands, Prince Edward Island, Fundy, and Terra Nova was not something they wished to emulate. These first four Atlantic parks are still in existence today, of course. Cape Breton Highlands National Park did not expand to the north, and, thanks to a Consultative Committee with members from ten communities surrounding the park, there now exists a better relationship between staff and locals. Recently, the park has even placed a guide at Cap Rouge to tell visitors about the expropriation of the tiny settlement. Prince Edward Island National Park is still one of the most visited parks in Canada, and the threats of erosion on one side and tourism development on the other are as great as ever. There are more concessions to preservation, though, and in 1995 the park closed off part of the Gulf Shore Highway to permit dune migration across the road. Fundy National Park still has many of its 1950s trappings, but its long-ignored natural qualities have become central to visitors' enjoyment of the park. And the Fundy Trail is finally under way along coastal New Brunswick. Terra Nova National Park, which at its establishment was celebrated as a wilderness park, has been overshadowed in recent decades by the rugged, more mountainous beauty of Gros Morne National Park in western Newfoundland. Parks to the east have always had that problem.

During the development of Terra Nova in the late 1950s, the Parks Branch's new Planning Section believed that the park was missing something. It was, in the eyes of the planning report's authors, as yet only a park of typical Newfoundland: typical hills, typical bogs, typical inlets. In sum, "It is soon apparent that the park has no truly outstanding point of interest." Something was needed to capture the visitor's attention. The authors suggested portals at the park entrances: "These must be massive, rock masonry portals with the name of the park outlined with relief letters of black steel. The purpose of these portals would be psychological as well as informative. They should 'set the stage' for the visitor and place him in a receptive and appreciative mood for what is to come." This would impress tourists and remind local citizens that they were entering land where they were no longer free to hunt, fish, and cut timber. The planners believed that on approaching the portals, visitors would have "no doubt that they are entering a special area which has been set aside for a distinctive purpose."[45]

The portals were never built, but the Planning Section's call for them was astute. National parks always depended on such signs, though not usually literal ones. The forbidding of conduct such as hunting, cutting wood, or owning land was itself a sign that parks were different. Publicity photos and promotional brochures were likewise designed to reinforce the message that nature did not have the same sort of meaning in parks that it had elsewhere. Natural objects in parks were to possess neither economic nor moral significance, they just *were*. It was even hoped that, as the national park system gained its own history, its facilities such as resort hotels and golf courses, because they were associated with the parks, would themselves somehow signal a preference for the natural over the cultural here.

But portals suggest that there are limited points of access, and here the metaphor breaks down. The park and the outside world are not so different, and not only because they have the same *sort* of climate and geology, flora and fauna, but also because they share many of the very same *things*. Seeds float or are carried across borders, deer wander back and forth, and air circulates over the land equally. In growing recognition of this, much scientific research of national parks in recent years has studied them in terms of their relationship with the land outside. Of particular importance has been the application of concepts from island biogeography. Parks are seen as "islands," distinct in some ways from the land around them, yet with only a limited capability of maintaining species health on their own. Therefore, species need corridors to other islands if the parks are to fulfil their mandate of preservation.[46]

The cultural side of national parks needs to be studied in the same way. Humans create a park to be an island for nature, and in doing so prove the difficulty we have incorporating nature into our everyday world. But even when a park is created, we humans travel as effortlessly into and out of it as any species, carrying with us, in both directions, ideas about nature and about people's place in nature. The histories of Cape Breton Highlands, Prince Edward Island, Fundy, and Terra Nova national parks are for this reason not histories of just those places. In each case, the park that was selected, established, expropriated, and developed stands as an inviolable monument to the time when these events took place. Parks are very helpful, then, in serving to document how we have felt about and behaved toward nature in the past. As I hope I have shown, however, those attitudes and actions are never about nature alone, but also involve our own messy human aspirations for social, spiritual, and financial betterment. As such, parks serve to document how we have felt about and behaved toward one another.

Appendix 1
Departments in Charge of National Parks, and Senior Park Officials, 1930–70

Interior	1911–Nov. 1936
Mines and Resources	Dec. 1936–Jan. 1950
Resources and Development	Jan. 1950–Dec. 1953
Northern Affairs and National Resources	Dec. 1953–Sept. 1966
Indian Affairs and Northern Development	Oct. 1966–

MINISTERS

Charles Stewart	1926–Aug. 1930
Thomas Murphy	Aug. 1930–Oct. 1935
Thomas A. Crerar	Oct. 1935–April 1945
James Glen	April 1945–June 1948
James MacKinnon	June 1948–March 1949
Colin Gibson	April 1949–Jan. 1950
Robert Winters	Jan. 1950–Sept. 1953
Jean Lesage	Sept. 1953–June 1957
Douglas Harkness	June 1957–Aug. 1957
Alvin Hamilton	Aug. 1957–Oct. 1960
Walter Dinsdale	Oct. 1960–April 1963
Arthur Laing	April 1963–July 1968
Jean Chrétien	July 1968–

DEPUTY MINISTERS

W.W. Cory	1905–March 1931
H.H. Rowatt	April 1931–April 1934
James Wardle	Aug. 1935–Nov. 1936
Charles Camsell	Dec. 1936–Dec. 1945
Hugh Keenleyside	Jan. 1947–Sept. 1950
Hugh Young	Oct. 1950–Nov. 1953
R. Gordon Robertson	Nov. 1953–June 1963
Ernest Côté	July 1963–Feb. 1968
John A. MacDonald	March 1968–Jan. 1970
H. Basil Robinson	Jan. 1970–

SENIOR PARKS BRANCH OFFICERS

James Harkin (Commissioner)	1911–Nov. 1936
R.A. Gibson (Director)	Dec. 1936–Nov. 1950
James Smart (Director)	Dec. 1950–Feb. 1953
James Hutchison (Director)	March 1953–July 1957
J.R.B. Coleman (Director)	Aug. 1957–April 1968
J.I. Nicol (Director)	May 1968–

PARKS DIVISION HEADS (AFTER 1936)

F.H.H. Williamson (Controller)	1936–1941
James Smart (Controller)	1941–Nov. 1950
J.R.B. Coleman (Chief)	Dec. 1950–July 1957
B.I.M. Strong (Chief)	Oct. 1957–Nov. 1963
W.W. Mair (Chief)	Dec. 1963–Aug. 1966
W.W. Mair (Operations Chief)	Aug. 1966–Sept. 1966
Lloyd Brooks (Planning Chief)	Aug. 1966–May 1968
J.J.L. Charron (Operations Chief)	June 1967–Dec. 1968
Louis Lemieux (Operations Chief)	Aug. 1969–May 1970
H.K. Eidsvik (Planning Chief)	Oct. 1968–

Appendix 2
Attendance at Atlantic Canada National Parks, 1936–66

Ranking among all parks is in parentheses.

Year	Cape Breton Highlands	Prince Edward Island	Fundy	Terra Nova	All national parks
1936					908,161
1937	20,000 (9)	2500 (17)			1,008,690
1938	20,500 (10)	10,000 (15)			954,120
1939	22,035 (11)	35,488 (8)			995,270
1940	20,151 (11)	35,665 (9)			1,170,653
1941	23,694 (11)	40,470 (5)			1,000,563
1942	10,189 (12)	24,826 (6)			466,245
1943	17,612 (6)	25,963 (5)			415,351
1944	11,940 (12)	33,365 (5)			457,392
1945	18,863 (7)	48,068 (4)			602,409
1946	23,896 (10)	50,281 (6)			914,902
1947	27,507 (11)	67,508 (7)			1,154,699
1948	25,769 (13)	84,333 (5)			1,261,910
1949	31,508 (14)	95,623 (6)			1,688,367
1950	29,060 (14)	87,851 (6)	62,844 (10)		1,795,138

Year	Cape Breton Highlands		Prince Edward Island		Fundy		Terra Nova		All national parks
1951	31,903	(14)	107,961	(7)	81,064	(10)			2,016,797
1952	35,372	(15)	122,290	(7)	101,139	(10)			2,409,661
1953	33,610	(14)	146,827	(7)	107,793	(10)			2,857,268
1954	123,731	(8)	158,954	(6)	99,346	(12)			3,035,001
1955	75,310	(12)	172,884	(6)	105,487	(11)			3,305,149
1956	116,556	(11)	181,692	(8)	120,666	(10)			3,529,976
1957	128,397	(11)	200,748	(8)	143,662	(10)			3,940,711
1958	162,938	(11)	206,245	(9)	179,277	(10)			4,287,343
1959	193,684	(10)	224,781	(7)	199,777	(8)			4,600,434
1960	323,392	(8)	412,463	(5)	227,262	(9)	20,000	(30)	4,930,648
1961	371,686	(7)	775,583	(2)	280,006	(9)	29,710	(28)	5,491,663
1962	451,911	(6)	1,009,021	(2)	302,340	(12)	29,915	(31)	7,426,403
1963	615,133	(8)	1,019,104	(2)	494,157	(10)	55,926	(24)	9,426,857
1964	624,942	(8)	1,112,536	(2)	566,443	(10)	66,180	(23)	9,179,028
1965	729,443	(5)	967,372	(2)	679,406	(9)	108,738	(20)	9,845,283
1966	851,653	(6)	1,130,773	(2)	753,310	(9)	179,647	(17)	11,367,912

Rankings include National Historic Parks, but total attendance is for National Parks alone. All data are from the following year's departmental *Annual Report*.

Notes

CHAPTER ONE

1 The federal agency in charge of national parks in Canada has gone by a number of names. During the period under study, it was named the Dominion Parks Branch (1911–21), the Canadian National Parks Branch (1921–6), the National Parks Branch (1926–36), the Lands, Parks and Forests Branch (1936–47), the Lands and Development Services Branch (1947–50), the Development Services Branch (1950), the National Parks Branch (1953–65), the National and Historic Resources Branch (1965–6), the National and Historic Parks Branch (1966–73), and Parks Canada (1973 on). From 1936 on, within the Branch itself there was an agency involved solely with national park management. Its name was, by turns, the National Parks Bureau (1936–47), the National and Historic Sites Division (1950–5), and the National Park Service (1947–50 and 1955 on). For the sake of simplicity, I will refer to the National Parks Branch throughout.

This seems reasonable for three reasons: 1) most important, it was at the branch level that park policy was decided; 2) the parks agency began its existence at the branch level, and in 1973 became Parks Canada at the branch level; and 3) this name serves to differentiate the Canadian agency from the United States' National Park Service. A list of federal ministries in charge of parks, plus senior staff, may be found in appendix 1.

2 For the parks under discussion, the titles are Campbell, "A Report on the Human History of Cape Breton Highlands National Park"; Horne, *Human History: Prince Edward Island National Park*; Allardyce, "The Salt and the Fir: Report on the History of the Fundy Park Area"; and Major, *Terra Nova National Park: Human History Study.*

3 Bourdieu, *Distinction*, in particular chapter 1.

4 "[B]ecause the primary categories of culture have been the products of ideologies which were always subject to modifications and transformations, the perimeters of our cultural divisions have been permeable and shifting rather than fixed and immutable." Levine, *Highbrow/ Lowbrow,* 8.

5 Runte, *National Parks*, xix.

6 Runte, *Yosemite*, 2.

7 Nash, "The State of Environmental History"; Marsh, *Man and Nature*; Febvre, *A Geographical Introduction to History*; and Glacken, *Traces on the Rhodian Shore.*

8 Lorne Hammond's afterword in the second edition of Foster's *Working for Wildlife* is an excellent introduction to the field of Canadian environmental history.

9 Maclean, *A River Runs through It*, ix.

10 Parr, "Gender History and Historical Practice," 362.

11 Gaffield and Gaffield, "Introduction," *Consuming Canada*, 5.

12 See Worster's "Appendix: Doing Environmental History." Similarly, Barbara Leibhardt's model, in "Interpretation and Causal Analysis," breaks environmental history down into ecology, human economic relations, and cognition. Worster reintroduces his model in the 1990 *Journal of American History* "A Round Table: Environmental History." In response, Richard White calls it an unconscious homage to a model of base (natural history), structure (modes of production), and superstructure (culture and ideology). White believes that Worster fixates on the role capitalism plays in determining environmental change, to the degree that his model ignores other forms of environmental analysis. William Cronon accepts the tripartite model – which he simplifies as "Nature, political economy, and belief" – though he feels that environmental historians generally fail to include all three, and that most works are either materialist or idealist in character. Carolyn Merchant suggests that Worster needs to add a fourth issue, that of reproduction, but, interestingly, in an earlier context

she mimicked Worster's model with a list of the three questions environmental history poses: "1) What concepts describe the world? 2) What is the process by which change occurs?" and "3) How does a society know the natural world?" (see Merchant, "The Theoretical Structure of Ecological Revolutions").

13 Flores, "Place: An Argument for Bioregional History," 14.

14 White, "American Environmental History," 317.

15 Cronon, *Nature's Metropolis*, 18. On "nature," see Williams, "Ideas of Nature," and Evernden, *The Social Creation of Nature*.

16 Livingston, *Rogue Primate*, 82.

17 The two words that have been popularly used in "nature"'s stead in the last thirty years – "ecology" and "environment" – have done nothing to resolve confusion. "Ecology" can refer to a science, a movement, or a relationship between an organism and its surroundings. "Environment" has suffered much the same problem, serving as short form for a movement, as well as both a way of describing humans as part of their surroundings and a way of describing those surroundings themselves. For these reasons, I will avoid the use of "ecology" and "environment" throughout. On ecology and environment, see McIntosh, *The Background of Ecology*, especially 1–16; and Evernden, *The Natural Alien*. Meinig's introduction to *The Interpretation of Ordinary Landscapes*, 3–11, offers a helpful distinction between "nature," "scenery," "landscape," and "environment."

18 Cronon, "The Uses of Environmental History," 21n12.

19 Ecocriticism has blossomed from the pages of the *Interdisciplinary Studies in Literature and Environment*. As the following discussion will make clear, I have been influenced by Lawrence Buell's *The Environmental Imagination*, which may be called the first self-consciously ecocritical book, and by conversations on the Internet with members of the American Society for Literature and the Environment.

20 Buell, *The Environmental Imagination*, 102.

21 Ibid., 35. My enthusiasm for Buell's work does not mean that I believe ecocriticism, as a discipline, to be "better" than or even as mature as environmental history. Neither does my reliance on Buell here mean to suggest that he was the first to recognize that a description of nature involves both nature and observer. Just two examples of writers' discussion of this are Dearden, "Philosophy, Theory, and Method in Landscape Evaluation," and Lopez, *Arctic Dreams*. Dearden describes beauty as "an interaction between landscape and observer rather than specifically residing in one or the other" (264). Lopez writes, "What one thinks of any region, while traveling through, is the result of at least three things: what one knows, what one imagines, and how one is disposed" (271).

22 Turner, "The Significance of the Frontier in American History," and Webb, *The Great Plains*.

23 This field has grown a rich literature in the last twenty years. See Barrell, *The Dark Side of the Landscape*; Bermingham, *Landscape and Ideology*; Zukin, *Landscapes of Power*; Mitchell, *Landscape and Power*; Daniels, *Fields of Vision*; and Cosgrove and Daniels, *The Iconography of Landscape*.

24 Schama, *Landscape and Memory*, 61.

25 Buell, *The Environmental Imagination*, 13.

26 Marsh, *Man and Nature*, 51.

27 Canada, House of Commons, *National Parks Act*, 20–1 George V, chapter 33 (1930).

28 Brown, "The Doctrine of Usefulness," 107.

29 Ibid., 98.

30 Ibid., 97. See also Brown, *Canada's National Policy*.

31 Runte, *National Parks*. See also "The National Parks: A Forum on the 'Worthless Lands' Thesis." Runte was apparently not aware of Brown's (earlier) work.

32 Lothian, *A History*. See also Lothian, *A Brief History*.

33 W.F. Lothian to Gwendolyn Smart, 9 March 1984, James Smart papers, MG30 E545, NA.

34 Cited in McNamee, "From Wild Places to Endangered Spaces," 30.

35 Bella, *Parks for Profit*, 1.

36 Ibid., 2.

37 McNamee, "Fom Wild Places to Endangered Spaces," 28–30.

38 Brown, "The Doctrine of Usefulness," 107.

39 Canada, House of Commons, *Debates*, 3 May 1887, 233.

40 Brown, "The Doctrine of Usefulness," 107.

41 Canada, House of Commons, *Debates*, 3 May 1887, 233.

42 Ibid., 227.

43 Ibid., 29 April 1887, 196.

44 Brown, "The Doctrine of Usefulness," 103.

45 See LaForest and Roy, *The Kouchibouguac Affair*; and Thomas, "The Kouchibouguac National Park Controversy."

46 Todd, *The Law of Expropriation*, 1. Todd's book is a helpful introduction to Canadian expropriation law, as is Boyd, *Expropriation in Canada*. Discussion of rising Canadian opposition to expropriation may be found in Dacre, "Expropriation."

47 MacEachern, "No Island Is an Island."

48 See Hutcheon, *As Canadian As ... Possible* and *Irony's Edge*.

49 Cronon, "The Uses of Environmental History," 18.

CHAPTER TWO

1 Harkin, *The History and Meaning*, 5.

2 Williams, *Guardians of the Wild*, 3.

3 The Branch's creation story also exists in a slightly different version, one in which Harkin was given the option either to head the national parks or to oversee federal water power policies.

4 Foster, *Working for Wildlife*, 222; Lothian, *A History*, 2:16; Nicol, "The National Parks Movement in Canada," 39; Nash, "Wilderness and Man in North America," 77. Other useful in-depth studies of this period of park history include Van Kirk, "The Development of National Park Policy," and Johnson, "The Effect of Contemporary Thought."

5 Henderson, "James Bernard Harkin," 29. Henderson, it is true, is not the most neutral of biographers. He was the first executive director of the National and Provincial Parks Association of Canada (later the Canadian Parks and Wilderness Society) and a winner of the society's J.B. Harkin Conservation Award. Recently, Bill Waiser in *Park Prisoners* hints at a revisionist reading of Harkin's tenure, calling the commissioner a "hapless cheerleader" for park tourism in the 1920s (49).

6 The best biographical sketch is an unpublished one by W.F. Lothian to be found in Fergus Lothian research papers, M113 Acc. 1947, folder 6, Whyte Museum of the Canadian Rockies Archives, Banff, Alberta. Thanks to Pearl Anne Reichwein for passing this on to me.

7 Dorothy Barbour to Mabel Williams, 8 June (?), Dorothy Barbour papers, in possession of Robin Winks. Barbour worked in the Parks Branch from its beginning in 1911. Thanks to Robin Winks for sharing this material with me. There is some mystery as to why Harkin kept the Stefansson material. As Barbour wrote to Williams, "in speaking of it he always put us off saying two more have to die before the world knows – Steffansson and me." Harkin was concerned that the explorer was blackmailing the Canadians, asking for expedition money to proclaim Canadian sovereignty in the Far North, and obliquely threatening to go to the Danes or the Americans if they refused. William R. Hunt, in *Stef*, calls this interpretation ludicrous: "the Byzantine devices suggested by Harkin confound the mind" (282n21). Yet in 1949 Stefansson asked the deputy minister of mines and resources, Hugh Keenleyside, for all accounts of the proposed expedition to be assembled in the National Archives and declared classified until the deaths of all involved. Harkin took this as an insinuation that he was withholding files, which he called "absurd." It is possible, then, that Harkin's archival papers consist of information that he *was* withholding, or conversely that their existence proves he accommodated Stefansson's demand. For more on Arctic exploration in this period, and Stefansson and Harkin's place in it, see Zaslow, *The Northward Expansion of Canada*.

8 See RG10 vol. 8052, C11317, file 242,830, NA; February 1927, RG84 vol. 2161, file U346 vol. 1, NA; January 1931, RG84 vol. 483, file F2 vol. 2, NA; 17 September 1936, RG84 vol. 484, file F2 vol. 3, NA; and 22 September 1934, Munro 1932–7 file, Hoyes Lloyd papers, MG30 E441, vol. 12, NA. In

1921, Prime Minister Arthur Meighen referred to the forty-five-year-old Harkin in Parliament as "a very old civil servant." Canada, House of Commons, *Debates*, 29 March 1921, 1343. In a bit of rather circuitous research, I communicated through my friends Winnie and David Wake to Frances Girling, niece of James Harkin's secretary Mabel Williams. Girling remembers her aunt describing Harkin as a very active person, and wonders if perhaps Harkin's wife had a debilitating illness which made him miss work. Thanks to Frances Girling and the Wakes for their help on this.

9 MacEachern, "Rationality and Rationalization," 199–200.

10 This reminds us of the difficulty in attributing anything to anyone in a large bureaucracy. Throughout, I given authorial credit to the person whose signature is at the bottom of a document, unless there is specific mention that it was drafted by someone else.

11 This would seem to support the work of historian Richard Jarrell, who has found Britain to be of declining interest to Canadian scientists in the late nineteenth century. See Jarrell, "British Scientific Institutions."

12 See Foster, *Working for Wildlife*, 82; and Sellars, *Preserving Nature in National Parks*, 28 and 33. Comparative works on Canadian-American national park policy include Turner and Rees, "A Comparative Study," and Dubasek, *Wilderness Preservation*. See also Kline, *Beyond the Land Itself*, and Gillis and Roach, "The American Influence on Conservation in Canada."

13 In 1911, the US park system was granted $244,000, and attracted 230,000 visitors to twelve parks; Canada's was granted $223,000, and attracted 75,000 to six parks. In 1917, the US national parks had 487,000 visitors and a $515,000 appropriation, whereas the Canadian parks had about 68,000 visitors and an appropriation of $345,000. Canada, Department of the Interior, *Annual Reports* (1911–17). United States, Department of the Interior, *Annual Reports* (1911–20).

14 Harkin did also send information to the United States when asked, though this seldom occurred. In a famous letter outlining plans for the American park system, the US secretary of the interior, Franklin K. Lane, told National Parks Service director Stephen Mather, "In particular you should maintain close working relationship with the Dominion Parks Branch of the Canadian Department of the Interior, and assist in the solution of park problems of an international character." Lane, though, was born in Prince Edward Island. See Dilsaver, *America's National Park System*, 51.

15 Canada, Department of the Interior, *Annual Report*, Report of the Commissioner of National Parks (1915), 8.

16 Harkin, *The History and Meaning*, 7.

17 Harkin, Speech to Good Roads Association, Victoria BC, 1922, cited in Bella, *Parks for Profit*, 63. Defending a national parks grant increase of

$100,000 in 1920, Arthur Meighen had noted, "the returns per acre being figured by the officers of the department as being more for the rocks and waste lands of our parks than even for our wheat fields." Canada, House of Commons, *Debates*, 8 June 1920, 3283. On national parks and tourism, see Bella, *Parks for Profit*; and Hall and Shultis, "Railways, Tourism and Worthless Lands."

18 Harkin, *The History and Meaning*, 9, and "Our Need for National Parks," 98.

19 See Henderson, "James Bernard Harkin," 30; Nash, "Wilderness and Man," 77; McNamee, "From Wild Places to Endangered Spaces," 24; Foster, *Working for Wildlife*, 79.

20 Harkin, "Memorandum re: Dominion Parks 20 March 1914," James Harkin papers, MG30 E169 vol. 1, NA. See also Altmeyer, "Three Ideas of Nature in Canada," and Hewitt, *The Conservation of the Wildlife of Canada*.

21 Harkin, "Canadian National Parks and Playgrounds," 37.

22 Harkin, cited in Alderson and Marsh, "J.B. Harkin, National Parks and Roads," 10. Elsewhere, he proclaimed, "Each citizen of Canada is the owner of one share of stock in the National Parks. Our part is to see that the value of their holdings is kept up." Harkin, *The History and Meaning*, 13.

23 Canada, Department of the Interior, *Annual Report*, Report of the Commissioner of National Parks (1915), 5.

24 Ibid. (1916), 5.

25 Ibid. (1917), 9.

26 Ibid. (1919), 12.

27 Harkin, "Our Need for National Parks," 101. Harkin here was drawing upon a cauldron of societal concerns: the rise of cities, lack of physical fitness, immigration, gender confusion, and physical degeneration. See Boyer, *Urban Masses and Moral Order*, and Pick, *Faces of Degeneration*.

28 Williamson to Harkin, undated, RG84 vol. 103, file U36 pt. 1, NA. Harkin noted elsewhere that the results of modernity could be found "in the slums, the prisons, the asylums and hospitals of the land." Harkin, "Our Need for National Parks," 99. While little biographical detail is available on Harkin, there is even less on F.H.H. – Frank – Williamson. I have found that he was committed to the labour movement, serving as chair of the civil service sports committee in 1908, and the honourable secretary of the Professional Institute of the Civil Service in 1926. In 1910, a Frank Williamson, secretary of the Executive Board of the Hamilton Trades and Labor Council, wrote Prime Minister Laurier offering a resolution which endorsed the proposed bill to have an eight-hour work day. Included in the resolution was the statement that "a shorter workday provides more times for recreation and sociability, thereby causing the advance of civilization and moral attainments." If this was F.H.H. Williamson, it suggests

that the deputy commissioner saw his park work as a component of a broader social movement. See Williamson to Laurier, 1 September 1908, Laurier papers, MG26G, c866, 144441–4; Trades and Labor Council of Hamilton to Laurier, 23 December 1910, ibid., c897, 178672; Arthur Meighen to Williamson, 18 November 1926, Meighen papers, MG26I, 60485.

29 A useful essay on this is McCombs, "Therapeutic Rusticity." See also Gosling, *Before Freud*.

30 Harkin, "Our Need for National Parks," 101.

31 Canada, Department of the Interior, *Annual Report*, Report of the Commissioner of National Parks (1919), 12.

32 Harkin, "Canadian National Parks and Playgrounds," 37.

33 Canada, Department of the Interior, *Annual Report*, Report of the Commissioner of National Parks (1922), 16.

34 Ibid. (1917), 9.

35 Harkin, "Canadian National Parks and Playgrounds," 37.

36 Harkin, *The History and Meaning*, 13.

37 Harkin, "Our Need for National Parks," 98.

38 For a general introduction to the topic of nature and religion, see Albanese's *Nature Religion in America*. Schama's *Landscape and Memory* examines forests (214–42) and mountains (411–33) as holy places. In a different vein, Sears's *Sacred Places* describes how nineteenth-century Americans associated certain landscapes with religious ideals, and helped create tourist sites out of pilgrimages there.

39 Harkin, *The History and Meaning*, 13.

40 Harkin, "Our Need for National Parks," 107–8.

41 On mountains, see Nicolson, *Mountain Gloom and Mountain Glory*, and Sadler, "Mountains as Scenery."

42 Harkin, "Canadian National Parks and Playgrounds," 37, and Harkin, in Canada, Department of the Interior, *Annual Report*, Report of the Commissioner of National Parks, (1926–7), 88.

43 My discussion of Romanticism, the picturesque, and the sublime ignores the contradictions of and debates about their meanings. For instance, Romanticism may be seen as a sign of people either trying to come closer to God in an increasingly secular age (see Cronon, "The Trouble with Wilderness," 73) or seeking a replacement for God in nature (as Weiskel writes in *The Romantic Sublime*, "The Sublime revives as God withdraws" (3)). Likewise, it may be said that the formalizing of strict aesthetic conventions was a rationalist betrayal of the Romantic spirit (see Jasen, *Wild Things*, 11), yet Burke's work, which helped create the Romantic project, began as an attempt to find a law of aesthetics, a "theory of our passions" (*A Philosophical Enquiry*, 1). On the picturesque, see Hussey's classic *The Picturesque*. On the sublime, see Burke, *A Philosophical Enquiry*, and Kant,

Critique of Aesthetic Judgement. Other important work includes Nicolson, *Mountain Gloom and Mountain Glory,* and Weiskel, *The Romantic Sublime.* My understanding of how these related to North America, national parks, and tourism has been assisted by Demars, "Romanticism and American National Parks"; Jasen, *Wild Things,* especially 7–13; Nash, *Wilderness and the American Mind;* Runte, *National Parks;* and Cronon, "The Trouble with Wilderness."

44 Burke, *A Philosophical Enquiry,* 36.

45 Canada, Department of the Interior, *Annual Report,* Report of the Commissioner of National Parks (1916), 3; and Harkin, *The History and Meaning,* 12.

46 See Runte, *National Parks,* 11–47.

47 Demars, "Romanticism and American National Parks," 19.

48 On Banff's landscape, see Todhunter, "Banff and the Canadian National Park Idea."

49 Canada, Department of Interior, *Annual Report,* Report of the Commissioner of National Parks (1917), 9.

50 Ibid. (1918), 6. See also memo, 20 March 1914, MG30 E169, vol. 1, NA.

51 See, for example, Canada, Department of Interior, *Annual Report,* Report of the Commissioner of National Parks (1914), 10.

52 Harkin memo, 20 March 1914, MG30 E169, vol. 1, NA.

53 Williamson to Harkin, undated, RG84 vol. 103, file U36 pt. 1, NA.

54 Canada, Department of Interior, *Annual Report,* Report of the Commissioner of National Parks (1922), (1925), and (1927).

55 Ibid. (1926), 90. Harkin had commented on this the year earlier as well, ibid. (1925), 90. On autocamping in the United States, see Belasco, *Americans on the Road.*

56 Harkin, "Canadian National Parks and Playgrounds," 37; and Canada, Department of Interior, *Annual Report,* Report of the Commissioner of National Parks (1927), 88.

57 Ibid. (1929), 111.

58 Ibid. (1923), 112.

59 Harkin to Rev. C.R. Harris, 11 April 1925, RG84 vol. 983, file CBH2 vol. 1 pt. 3, NA; Harkin to W.B. McCoy, secretary of Department of Industries and Immigration, Nova Scotia, 4 March 1925, ibid.; and Harkin to A. McCall, general manager of Nova Scotia Steel and Coal Co., 19 October 1922, RG84 vol. 983, file CBH2 vol. 1 pt. 2, NA.

60 Ibid.

61 Harkin to James McKenna, Saint John *Telegraph Journal,* 31 October 1927, RG84 vol. 483, file F2 vol. 1, NA. The size Harkin quoted was rather arbitrary; four months earlier, he had suggested that a hundred square miles was sufficient. Harkin to Charles D. Richards, minister of mines and lands, New Brunswick, 6 July 1927, RG84 vol. 483, file F2 vol. 1, NA.

62 Harkin to A. McCall, general manager of Nova Scotia Steel and Coal Co., 19 October 1922, RG84 vol. 983, file CBH2 vol. 1 pt. 2, NA.

63 E.W. Robinson, MP for Kings, Nova Scotia, in Canada, House of Commons, *Debates*, 20 April 1925, 2237.

64 Arthur Newbery to Donald MacKinnon, 5 May 1923, RG84 vol. 1777, file PEI2 vol. 1 pt. 2, NA. On Newbery's work as a park designer, see Hennessey and MacDonald, "Arthur Newbery," 25–9.

65 Canada, Senate, *Reports and Proceedings of the Special Committee on Tourist Traffic* (1934), 179.

66 Prince Albert National Park was a gift from Prime Minister William Lyon Mackenzie King, who had parachuted into the constituency of Prince Albert in 1926. As a result, Manitoban politicians pressed hard to get a park for their own province. See Bella, *Parks for Profit*, 76–7; Lothian, *A Brief History*, 68–9 and 74–5; and Waiser, *Saskatchewan's Playground*.

67 Taylor, "Legislating Nature," 132–4.

68 Harkin to Richards, 6 July 1927, RG84 vol. 483, file F2 vol. 1, NA.

69 Harkin to Richards, 13 June 1927, ibid.

70 Cited in Taylor, "Legislating Nature," 134.

71 Canada, Department of Interior, *Annual Report*, Report of the Commissioner of National Parks (1913), 5. Though one can take from this that Harkin did not find the East as scenic as the West, more interesting is that Maine's popularity does not seem to have suggested to him that Canada's East Coast might also have highly profitable scenery.

72 On Maine tourism in this period, see Judd, "Reshaping Maine's Landscape," and chapter 6 of Brown, *Inventing New England*. The Parks Branch was aware that the idea of park standards more generally was undergoing change in the United States. In notes to prepare the minister of the interior for discussion of the 1930 National Parks Act, the Branch included an article in the American magazine *Parks and Recreation* entitled "National Park Standards," and highlighted the sentence "National Parks should differ as widely as possible from one another in their physical aspects, and the National Park system should represent a wide range of typical land forms of supreme quality." See "National Parks Act 1930. Explanations for Minister re Parliament," 48, RG84 vol. 1959, file U1. A, NA.

73 Albright to Harkin, 25 February 1929, RG84 vol. 483, file F2 vol. 1, NA.

74 Cautley to Harkin, 7 June 1929, RG84 vol. 1964, file U2.12.1, NA.

75 Taylor, "Legislating Nature," 128.

76 Ibid.

77 According to the Toronto *Mail and Empire*, "A national park maintained by the federal government in every province of the dominion is the aim of Hon. Charles Stewart, minister of the interior, he told a meeting of the Ottawa Women's Liberal association held here to-day." 10 January 1930.

The Vancouver *Daily Province* reported, "A chain of national parks from the Atlantic to the Pacific – this is the ideal toward which the parks branch at Ottawa is working, and in its work it needs the co-operation of the various provincial governments." 2 February 1930. See also 13 January 1930, *Toronto Star*.

78 Charles Stewart, in Canada, House of Commons, *Debates*, 26 May 1930, 3959.

79 Harkin to Col. G.S. Harrington, premier of Nova Scotia, 12 November 1931, RG84 vol. 983, file CBH2 vol. 1 pt. 2, NA; and Harkin to Harrington, 20 September 1933, RG84 vol. 983, file CBH2 vol. 1 pt. 3, NA.

80 Murphy, in Canada, House of Commons, *Debates*, 3 July 1931, 3376.

81 Taylor, *Negotiating the Past*, 110.

82 In discussion on the provisions of the National Parks Act in 1930, for instance, Bennett complained about the Parks Branch's involvement in Banff affairs. "You cannot have a game of baseball unless Mr. Harkin says so, and he is in Ottawa three thousand miles away. Do you want to go out and do a little fishing? Oh no." Canada, House of Commons, *Debates*, 9 May 1930, 1935. Correspondence concerning Banff runs through the Bennett papers, MG26K, NA.

83 On Bennett, see Gray, *R.B. Bennett*, and Glassfore, *Reaction and Reform*.

84 Lothian, *A History*, 2:17.

85 Taylor, *Negotiating the Past*, 110n12. Taylor does not explain why Bennett did not simply fire Harkin.

86 Per capita, Maritimers received one-third of the national average in all federal relief programs. See Forbes, "Cutting the Pie into Smaller Pieces."

87 Murphy, 18 November 1930, *Calgary Herald*.

88 The 1930s saw national park systems expand in a number of countries. In the United States, Shenandoah and Great Smokies National Parks were formed, transplanting communities in ways very similar to those soon to be pursued in the Maritime national parks. See Runte, *National Parks*, 106–37; Perdue, Jr. and Martin-Perdue, " 'To Build a Wall around These Mountains' "; and Reeder and Reeder, *Shenandoah Heritage*. In Mexico, only two national parks had been created before 1934; from 1935 to 1940, forty more were established. Simonian, *Defending the Land of the Jaguar*, 94. In German-occupied Poland, national forest reserves were created by emptying regions of people. Schama, *Landscape and Memory*, 70.

89 Ilsley, in Canada, House of Commons, *Debates*, 11 March 1935, 1611.

90 Thus we have speeches such as those of Nova Scotian MP William Duff, who offered a tour of northern Cape Breton: "From there I would take him along the celebrated Cabot trail over the tip of Cape North, up the side of a mountain the size of which is such that it makes even the Alps and the Rocky mountains dwindle, and on which cars go in low gear until they reach the top." Ibid., 1612.

91 MacEachern, "No Island Is an Island," 120–1. On the Tourist Committee, see March, *Red Line*, 233–6, and Canada, Senate, *Report and Proceedings of the Special Committee on Tourist Traffic* (1934).

92 Ibid., x.

93 Taylor, *Negotiating the Past*, 111.

94 Cautley report on New Brunswick park sites (1930), RG84 vol. 1964, file U2.13 vol. 1, NA.

CHAPTER THREE

1 Harkin to Cautley, 4 September 1934, RG84 vol. 983, file CBH2 vol. 1 pt. 2, NA.

2 Cautley to Harkin, 7 June 1929, RG84 vol. 1964, file U12.2.1, NA.

3 Cautley report to Harkin, December 1934, RG84 vol. 1964, file U2.12.2, NA. Cautley's "Report on Examination of Sites for a National Park in the Province of Nova Scotia" as it relates to the proposed Cape Breton site is in MacDonald, *Transportation in Northern Cape Breton*, 47–76. The page references to Cautley's report that follow are from MacDonald's book.

4 On Baldwin, see Parkin, *Bell and Baldwin*.

5 Cautley report on Nova Scotia sites, 50. Burke had noted that the ocean could produce sublime emotions in the viewer in a way that flat land could not (*A Philosophical Enquiry*, 53).

6 Cautley report on Nova Scotia sites, 50.

7 Cautley's unpublished memoirs are entitled "High Lights of Memory" and can be found at E/C/C31, British Columbia Archives. Cautley makes only passing mention of leading the investigations of Riding Mountain, Cape Breton Highlands, and Fundy National Parks, and no mention of his work at Prince Edward Island National Park.

8 Cautley report on Nova Scotia sites, 54 and 55.

9 Ibid., 56.

10 As Margaret Warner Morley wrote in 1900, "Ingonish is the end of the tourists' explorations as a rule. Few find their way there, still fewer go north of there." *Down North and Up Along*, 253.

11 *Guide Book to Cape Breton*, 28 and 30.

12 According to MacMillan, the dream came after a day spent walking from Pleasant Bay to Cheticamp (twenty-seven miles of very difficult terrain), driving to Baddeck (fifty more difficult miles) and "retiring early." MacMillan, "A Dream Come True," 2 and 66–70.

13 Gordon Brinley's *Away to Cape Breton* quotes the warning (86). Brinley's travel account is a good example of how the Cabot Trail became a test of toughness. After describing his and his wife's conquest of the route, he tells of a man they met whose spirit has been broken trying to drive it.

14 Warner, *Baddeck and That Sort of Thing*, 108.
15 Harkin to Cautley, 1 September 1934, RG84 vol. 983, file CBH2 vol. 1 pt. 2, NA.
16 Ibid.
17 Harkin to Gibson, 14 December 1934, RG84 vol. 983, file CBH2 vol. 1 pt. 1, NA.
18 Harkin referred to "the only serious deficiency of the northern area, namely, – its lack of recreational facilities." Harkin to Gibson, 6 February 1935, RG84 vol. 983, file CBH2 vol. 1 pt. 1, NA.
19 Harkin to Gibson, 14 December 1934, ibid.
20 Harkin to J.M. Wardle, deputy minister of the interior, 19 December 1935, RG84 vol. 985, file CBH2-Cap Rouge vol. 2 pt. 3, NA.
21 See the original draft letter, Thomas Crerar to Angus L. Macdonald, 18 March 1936, RG84 vol. 985, file CBH2-Cap Rouge vol. 2 pt. 2, NA; and the final draft, 25 March 1936, ibid. Wardle told Harkin, "It was the Minister's wish, however, that for the present no mention of the Bras d'Or Lakes area be made." 27 March 1936, ibid.
22 Harkin to Wardle, 6 April 1936, RG84 vol. 985, file CBH2-Cap Rouge vol. 2 pt. 2, NA.
23 Cited in Cautley report on Nova Scotia sites, 49.
24 Harkin to Cautley, 4 September 1934, RG84 vol. 983, file CBH2 vol. 1 pt. 2, NA.
25 Cautley report on Nova Scotia sites, 61.
26 Ibid.
27 Ibid., 63.
28 Ibid., 66.
29 Halifax *Herald*, 30 June 1937. Cap Rouge's size is also given in Williamson to Wardle, 4 November 1936, RG84 vol. 984, file CBH2 vol. 3 pt. 2, NA.
30 Brinley, *Away to Cape Breton*, 94.
31 Cautley report on Nova Scotia sites, 59 and 68.
32 On Smart's career, see RG32 vol. 237, file 1888.02.29, NA.
33 Smart to Harkin, 14 August 1936, RG84 vol. 984, file CBH2 vol. 3 pt. 2, NA.
34 Ibid.
35 Nova Scotia, General Assembly, *An Act to Provide for Establishing a National Park in Nova Scotia* (1935) stated that the park could be no more than 256,000 acres.
36 Williamson later remembered this agreement as basically a land swap. Williamson to Gibson, 30 June 1938, RG84 vol. 984, file CBH2 vol. 4 pt. 3, NA.
37 Interview with Angus Leblanc, Cheticamp, 2 August 1994.
38 Sydney *Post-Record*, 25 April 1936.
39 Cautley to Harkin, 16 June 1933, RG84 vol. 1964, file U2.12.1, NA.
40 Cautley report on Nova Scotia sites, 60.

41 Williamson to Gibson, 21 September 1937, RG84 vol. 139, file CB28, NA.
42 Hector Moore, letter to the editor, *Victoria-Inverness Bulletin*, 27 November 1936. M. MacLean, MP for Sydney Mines, wrote Crerar that he was receiving numerous complaints about park staff trespassing on private property. 30 August 1938. RG84 vol. 984, file CBH2, vol. 4 pt. 2, NA.
43 Fred Mathers to F.H.H. Williamson, 19 August 1934, RG84 vol. 984, file CBH2 vol. 4 pt. 2, NA.
44 Notice of the boundary changes was published in the *Victoria-Inverness Bulletin* on 5 and 12 August 1938.
45 Creighton, *Forestkeeping*, 58. Creighton heard that his predecessors had failed because they were drunk all the time, and even tried to take a cut from settlements. This might be true, but rumours about the government negotiators' duplicity were legion. I was told during one interview that Creighton himself earned a "commission" for striking low settlements with landowners.
46 Creighton with Donovan, "Wilfred Creighton and the Expropriations," 13. This price seems reasonable. Cautley had declared in 1934, "I was informed that the transfer of land between themselves is usually made at a rate of from three to four dollars an acre." Landowners probably expected more from the government than they would have from another buyer. A listing of expropriation settlements may be found in Nova Scotia, *Public Accounts* (1936–45).
47 Creighton, *Forestkeeping*, 60.
48 Creighton with Donovan, "Wilfred Creighton and the Expropriations," 9.
49 Interview with Wilf Aucoin, Cheticamp Island, 1 August 1994.
50 Maureen Scobie interview, in "Stories from the Clyburn Valley." Also, material from an interview with Maurice and Emma Donovan, Ingonish Beach, 5 August 1994.
51 Interview with Angus Leblanc, Cheticamp, 2 August 1994.
52 Interview with Gordon Doucette, Ingonish Beach, 5 August 1994.
53 See letters in RG84 vol. 986, file CBH16.1 vol. 1 pt. 2 and 3, NA.
54 Phoebe Folger Jordan poem, in RG84 vol. 986, file CBH16.1 vol. 1 pt. 3, NA.
55 Interview with Eva Deveau, Cheticamp, 2 August 1994.
56 See Sandberg, "Forest Policy in Nova Scotia." On American pulp and paper companies in the Maritimes, see Parenteau and Sandberg, "Conservation and the Gospel of Economic Nationalism."
57 In his report, Cautley made note that most of the proposed park site was under lease to the Oxford Paper Company, but the province was fully aware of this and apparently not concerned. Cautley report on Nova Scotia sites, 61.
58 *Victoria-Inverness Bulletin*, 18 April 1936.
59 Cautley report on Nova Scotia sites, 50.

60 Evidence of Wilfred Creighton, Oxford Paper Co. arbitration case, RG10 series B, vol. 204 file 11, 2384 and 2398, PANS.

61 One of the men Nova Scotia hired to evaluate the land, F.T. Jenkins, wrote T.D. McDonald, deputy attorney-general of Nova Scotia, "In my opinion the Oxford Paper Co. has received generous compensation, but I believe that, taking everything into consideration, the Nova Scotia government has not done too badly." 8 January 1941. Oxford Paper Co. arbitration case, RG10 Series B, vol. 202 file 4, PANS.

62 A.S. MacMillan stated that "in lieu of taxes" Oxford was paying $6000 per year toward the maintenance of roads in Inverness and Victoria Counties. Could MacMillan have been referring to the lease payment, which was $6000 per year? If so, Oxford Paper was not only paying a very low lease rate, it was in effect given the chance to ensure that its money would be spent by government in a way directly of use to the company. MacMillan, "A Dream Come True."

63 Harkin to Cautley, 4 September 1934, RG84 vol. 983, file CBH2 vol. 1 pt. 2, NA.

64 Cautley report on Nova Scotia sites, 58. As proof of the caribou's past numbers, and to show that northern Cape Breton was once home to many animals, Cautley cited a letter from the 1780s.

65 Ibid.

66 Harkin to Gibson, 14 December 1934, RG84 vol. 983, file CBH2 vol. 1 pt. 1, NA.

67 Cautley to Harkin, 7 June 1929, RG84 vol. 1964, file U2.12.1, NA.

68 This point was made to me by two interview subjects.

69 I was asked not to identify the source of this anecdote.

70 Cautley report on Nova Scotia sites, 72.

71 On the last day of the legislative sitting in 1939, opposition member G.Y. Thomas did complain, "I still say that they might as well hand over the whole government to you and let you run the whole thing. A company of yours gets contracts from the Federal government – will you deny that?" MacMillan replied, "The company that I have an interest in." Halifax *Chronicle*, 17 April 1939. MacMillan suggests that he was not closely involved with Fundy Construction, but in fact he wrote the Parks Branch on behalf of Fundy Construction during road-building as "Hon. A.S. Mac-Millan, Fundy Construction Company, Halifax." RG84 vol. 520, file CBH200 pt. 1, NA. On MacMillan, see Joudrey, "The Public Life of A.S. MacMillan."

72 W.H. Stuart, acting superintendent, to M. MacLean, MP, 3 September 1938, RG84 vol. 72, file CBH313, NA.

73 Grand Etang happenings, *Victoria-Inverness Bulletin*, 2 April 1937. This quotation is as it was in the original.

74 Canada, Department of Mines and Resources, *Annual Report* (1938), 88; (1939), 98; and (1940), 86.

75 D.A. Cameron to Crerar, 2 April 1937, RG84 vol. 984, file CBH2 vol. 4 pt. 3, NA.

76 A.S. MacMillan to J.L. Ilsley, minister of national revenue, 24 February 1938, RG84 vol. 990, file CBH16.112 pt. 1, NA.

77 Harkin to Cautley, 4 September 1934, RG84 vol. 983, file CBH2 vol. 1 pt. 2, NA.

78 Cautley report on Nova Scotia sites, 75.

79 Smart to T.S. Mills, chief engineer, Surveys and Engineering Branch, 6 July 1938, RG84 vol. 984, file CBH2 vol. 4 pt. 2, NA. Also, see Williamson to Wardle, 30 July 1938, RG84 vol. 139, file CBH28, NA.

80 Lothian, "James Bernard Harkin," M113 Acc. 1947, Whyte Museum of the Canadian Rockies Archives. See also Huxley, "Golf Courses in National Parks," 14.

81 Barclay, *Golf in Canada*, 365–72.

82 Thompson to Gibson, 4 June 1938, RG84 vol. 72, file CBH313, NA.

83 Thompson, as stated by Gibson to Williamson, 18 October 1938, RG84 vol. 72, file CBH313.7, NA.

84 Ibid.

85 Local workers recognized that their enjoyment of even short-term benefits would be short-term. A song by Lubie Chiasson of Cheticamp, written while working on the Cabot Trail, concludes (translated from the French): "At the end of the song / You will be well able to understand / That we work for the government. / At the end of two weeks / Or at the end of the month, / We will draw a small cheque, / And we will go home. / The money will all go / To the small merchants." Chiasson, *Cheticamp*.

86 Gibson to Williamson, 18 October 1938, RG84 vol. 72, file CBH313.7, NA.

87 In regard to operating the Lodge at a loss, see Harold Connolly, minister of industry and publicity for Nova Scotia, to R.J.C. Stead, 17 April 1942, RG84 vol. 139, file CBH113.200, NA.

88 McKay, "Tartanism Triumphant." On Macdonald, see also Hawkins, *The Life and Times of Angus L.*

89 Bourinot, "Notes of a Ramble through Cape Breton," 91.

90 Browne, "A Great Sanctuary in Nova Scotia," and Walworth, *Cape Breton*, 37.

91 *Victoria-Inverness Bulletin*, 16 June 1934.

92 McKay, "Tartanism Triumphant," 33–4. McKay overestimates the importance of this grant when he writes, "Once the Province of Nova Scotia accepted the gift, the campaign to establish a national park in Cape Breton was strengthened immeasurably" (34). The parcel of land was so small and isolated that it had little influence on the National Parks Branch and Nova Scotia government's decision on where to create a park.

93 Smart to Williamson, 27 November 1936, RG84 vol. 984, file CBH2 vol. 3 pt. 1, NA.

94 Smart to Gibson, 30 July 1938, RG84 vol. 22, file PEI109, NA.

95 Gibson to Williamson, 18 July 1938, RG84 vol. 139, file CBH28, NA.

96 Course designer Stanley Thompson and Loudon Hamilton, "a Scotchman, widely travelled and a student of Scotch lore," spent a winter evening coming up with the names. Stanley Thompson to James Smart, 27 October 1943, RG84 vol. 141, file CBH313 pt. 2, NA. The names they chose were a pastiche of Scottish, Gaelic, and Scottish-sounding names. As early as 1943, the names were losing what little original meaning they had. What Thompson remembered, for example, as "Canny Slap" because it called for "a tricky shot, one had to be canny," was now known as "Candy Slap" in reference to "a small opening or slap in a hedge or fence." See Dennis Sutherland to Superintendent J.P. MacMillan, October 1943, ibid.

97 Macdonald to Crerar, 5 June 1936, RG84 vol. 985, file CBH2–Cap Rouge vol. 2 pt. 1, NA. "Isle Royale" was rejected immediately by the Parks Branch, because the American Park Service had established Isle Royale National Park in Michigan in 1931.

98 Smart to Williamson, 27 November 1936, RG84 vol. 984, file CBH2 vol. 3 pt. 1, NA.

99 Smart to T.S. Mills, chief engineer, 3 January 1942, RG84 vol. 990, file CBH16.112 pt. 1, NA.

100 Canada, Dominion Bureau of Statistics, *Census of Canada* (1931), 2:336.

101 Bishop, "Cape Breton," in *A Cold Spring*. On Bishop in Nova Scotia, see Barry, "The Art of Remembering."

CHAPTER FOUR

1 R.W. Cautley to R.A. Gibson, 30 November 1936, RG84 vol. 1777, file PEI2 vol. 2, NA. Other evidence of this is found in a report by Smart to Mills, 30 June 1938, RG84 vol. 1778, file PEI2 vol. 11, NA: "The roof should be painted and the outside walls redecorated with white paint or white-washed again. The trim of the building and around the gables should be painted green and all the windows should have green shutters – the roof painted green." See also Anita Webb, interview with Fred Horne, PEINP files.

2 Canada, Parks Canada, *Resource Inventory and Analysis, Prince Edward Island National Park* (1977).

3 Donald MacKinnon to Harkin, 2 April 1923, RG84 vol. 1777, file PEI2 vol. 1, NA.

4 Stewart to R.A. Gibson, 27 June 1923, ibid.

5 See Taylor, *Negotiating the Past*, 28–31.

6 Stewart to MacLean, 17 May 1930, RG84 vol. 1777, file PEI2 vol. 1, NA.

7 Harkin to Rowatt, 19 April 1933, RG84 vol. 483, file F2 vol. 2, NA.

8 The 1935 election had been won with Walter Lea officially as leader. Lea was a sentimental choice but was ill throughout the campaign and died less than five months after assuming office. It was understood during the campaign that a vote for Lea's leadership was a vote for Campbell's.

9 Campbell, 31 March 1936, from the Charlottetown *Patriot*, speeches of the legislature, RG10 vol. 102, PAPEI. The date cited here, as in future citations to this source, refers to when the quotation was made, not when it appeared in the newspaper.

10 Harkin to MacDonald, 18 March 1936, RG84 vol. 985, file CBH-Cap Rouge vol. 2 pt. 3, NA.

11 Campbell to Howe, 8 February 1936, RG84 vol. 1777, file PEI2 vol. 1, NA; and Prince Edward Island, Executive Council, *Minutes*, 17 March 1936, RG7 series 3, box 12 no. 2640, PAPEI.

12 B.W. LePage, 14 April 1937, cited in Charlottetown *Patriot*, speeches of the legislature, RG10 vol. 102, PAPEI.

13 Campbell, 16 April 1936, cited in Charlottetown *Guardian*, speeches of the legislature, RG10 vol. 102, PAPEI.

14 Canada, House of Commons, *Debates*, 20 June 1936, 3994.

15 Crerar to Campbell, 31 March 1936, RG84 vol. 1777, file PEI2 vol. 1, NA.

16 For the public reaction to the park as seen in the local press, see Horne, *Human History*, 136–58.

17 Malcolm MacNeill, Charlottetown *Guardian*, 7 July 1936, and "Observer," ibid., 9 July 1936.

18 Ibid., 15 July 1936.

19 Ibid., 8 June 1936.

20 Charlottetown *Patriot*, 18 October 1937.

21 Ibid., 7 July 1936.

22 W.S. Stewart, ibid., 26 October 1937.

23 Charlottetown *Guardian*, 23 October 1937, and ibid., 13 October 1937. This has traditionally been a common compliment to pay an area while trying to end discussion of a proposed national park. Thomas Crerar, for example, explained in Parliament that he would not be creating a park on Vancouver Island because "the island itself is one big park." Canada, House of Commons, *Debates*, 20 May 1938, 3117. James Overton, in *Making a World of Difference*, 2, says the same was often said of Newfoundland.

24 F.H.H. Williamson memo, 23 March 1936, RG84 vol. 1777, file PEI2 vol. 1, NA.

25 Ibid.

26 Ibid.

27 Williamson to Harkin, undated [1936], ibid. See Sackett, "Inhaling the Salubrious Air," for discussion on the purposeful creation of an earlier Maritime seaside resort, St Andrews. Sackett notes that the seaside resort was an anti-modern escape from mass, urban, modern existence; the PEI national park was an attempt to incorporate the masses into this ideal.

28 Cromarty was an architect and planner trained in England, who had joined the Parks Branch in 1921. See Taylor, *Negotiating the Past*, 111.

29 The diary of Harry T. Holman, a prominent Island businessman, provides a glimpse at local lobbying for a park location. Holman originally favoured a Dunk River site for the park, later switching his allegiance to nearby Holman's Island when it was clear his first suggestion would not fly. Much of his energy was devoted to defaming the Dalvay proposal, which he suspected had already been approved. The day before he was finally to meet the two park inspectors, Holman wrote, "Will be very glad when it is over as it has entailed a lot of work and worry." See especially 28 February, 21 March, 24 March, 2 April, 10 April, 9 May, 2 June, 17 June 1936 of the diary, Harry T. Holman papers, Acc. 4420 vol. 7, PAPEI.

30 Williamson and Cromarty report on PEI sites, 28 July 1936, RG84 vol. 1777, file PEI2 vol. 1, NA.

31 Ibid.

32 L.M. Montgomery, 23 September 1928, in Rubio and Waterston, *The Selected Journals of L.M. Montgomery*, 3:358.

33 See Horne, *Human History*, especially 129–30. Horne credits Montgomery's books with singlehandedly taking Cavendish "out of the tourism doldrums" (129).

34 Williamson memo, 23 March 1936, RG84 vol. 1777, file PEI2 vol. 1 pt. 2, NA.

35 "Notes re PEI park" – unsigned [Harkin (?) to Wardle (?)], RG84 vol. 1777, file PEI2 vol. 1 pt. 2, NA.

36 Williamson to Harkin, undated [1936], RG84 vol. 1777, file PEI2 vol. 1, NA.

37 Levine, *Highbrow/Lowbrow*, 68.

38 Prince Edward Island, Executive Council, *Minutes*, 21 September 1936, RG7 series 3, box 12 no. 2640, PAPEI.

39 Williamson to Wardle, 5 October 1936, RG84 vol. 1777, file PEI2 vol. 2, NA.

40 Crerar to Campbell, 19 April 1937, and W.J.F. Pratt to Campbell, 20 May 1937, "C-1937" file, Thane Campbell papers, RG25 vol. 32, PAPEI.

41 An illegible signature on a memo of 7 August 1937 prevents us from knowing who officially christened the park. It was not the Branch's new head, F.H.H. Williamson. On 13 August 1937, publicity director Robert J.C. Stead defended the park's name, saying that "When a manufacturer is placing a new product on the market great importance is attached to the name" and that it must be easy to spell, pronounce, and remember. Williamson wrote beside this, "I think you might add: 4. Not too long. 'PEI Natl. Pk.' and 'CBH Natl. Pk.' are rather mouthfuls." RG84 vol. 1777, file PEI2 vol. 3, NA.

42 Hungerford to C.D. Howe, 18 January 1937, RG84 vol. 1777, file PEI2 vol. 2, NA.

43 Joseph Van Wyck to Howe, 15 March 1937, RG84 vol. 1777, file PEI2 vol. 3, NA.

44 Webb to Williamson, 11 March 1937, RG84 vol. 1777, file PEI2 vol. 3, NA.
45 Campbell to Crerar, 12 November 1936, RG84 vol. 1777, file PEI2 vol. 2, NA.
46 Cautley to Williamson, 31 October 1936, RG84 vol. 1777, file PEI2 vol. 3, NA.
47 Cautley to Williamson, 11 June 1937, ibid.
48 Public notice, 31 May 1937, ibid.
49 The complete range: extra good land, $65 per acre; good land, $40-$50; fair land, $25–35; light land, $20–30; sand dunes $6. See Provincial Secretary fonds, Prince Edward Island National Park files, RG7 series 14, box 30 subseries 2, file 753, PAPEI.
50 Ibid.
51 Charlottetown *Guardian*, 23 August 1937.
52 It was perhaps inevitable that the Island's Liberals would be compared to some famous fascists of the day. Jeremiah Simpson, the most cantankerous and unyielding of dispossessed landowners, wrote to "T.A. Carar" to complain. Having stated his case, he asked the minister to handle the matter personally, suggesting, "You need not send this to the Campbell's or [president of the Executive Council Bradford] Lepage in Charlottetown as they have been trying to imitate Musolini and Hitler." Crerar replied that this was a provincial matter, and that he had passed Simpson's letter over to Premier Campbell. 24 February 1938, RG84 vol. 1777, file PEI2 vol. 5, NA. ("Carar" and "Musolini" are as in the original.)
53 Provincial Secretary fonds, Prince Edward Island National Park files, RG7 series 14, box 30 file 748, PAPEI.
54 Charlottetown *Patriot*, 30 September 1937.
55 Charlottetown *Guardian*, 18 February 1938.
56 See Committee of Dispossessed Landowners' petition (at RG7 series 14, box 30 file 748, PAPEI) and the final settlement list (at RG7 Series 3, box 12 file 205, PAPEI). Certainly partisan politics was a standard part of park creation. In 1938 Premier Campbell wrote Crerar with local Liberals' complaints that a known Conservative had been hired. He asked that Crerar get a list of acceptable candidates from the four PEI members of Parliament, all Liberal. Campbell to Crerar, 2 August 1938, Thane Campbell papers, RG25 vol. 32, file "C-1938," PAPEI.
57 Canada, House of Commons, *Debates*, 15 February 1938, 487.
58 MacKinnon, 8 April 1938; MacIntyre, 23 March 1938; and McKay, 12 April 1938, from the Charlottetown *Patriot*, speeches of the legislature, RG10 vol. 102, PAPEI.
59 Campbell, 30 March 1938, ibid.
60 Landowner Sydney Ranicar accused the government of telling owners in Cavendish that those in Brackley and Tracadie had already sold, and vice versa. Charlottetown *Guardian*, 21 September 1937.

61 Landowner Wendell Kelly later stated that he might as well sell, since tourists were already treating his land like a park, pulling down his fences to get to the shore. Kelly interview with Fred Horne, 8 June 1978, PEINP files.

62 Horne, *Human History*, 148.

63 Charlottetown *Patriot*, 25 August 1937.

64 Campbell, 21 March 1939, from the Charlottetown *Patriot*, speeches of the legislature, RG10 vol. 102, PAPEI.

65 Canada, House of Commons, *Debates*, 15 February 1938, 490.

66 Williamson to Cautley, 21 June 1937, RG84 vol. 1784, file PEI16.112.1, NA.

67 Cautley to Williamson, 25 June 1937, ibid.

68 James Smart to R.A. Gibson, 19 July 1941, RG84 vol. 150, file PEI313 vol. 2, NA.

69 Interview with Ernest Smith, Tea Hill, 4 August 1995. See also RG84 vol. 23, file PEI336, NA.

70 Williamson to Superintendent Ernest Smith, 11 May 1940, ibid.

71 In the legislature, Premier Campbell decried the *Star*'s "unfortunate propaganda." He argued that development was necessary so that the apparently inevitable tourists would not destroy "the natural amenities, woods, and the streams, and such-like." 22 March 1939, Charlottetown *Patriot*, speeches of the legislature, RG10 vol. 102, PAPEI. During development, only geodetic survey engineer W.C. Murdie questioned the logic of attempting any work on Green Gables. He wrote that it was an old house and "can really never be anything else. ... Would it not be worthy of consideration to retain Green Gables as an unoccupied building and only keep it in sufficiently good repair to retain its original appearance and atmosphere to enable it to be used as a museum and a point of historical interest ...?" Murdie to Mills, 2 August 1938, box 3, Green Gables golf course file, PEINP files. This part of Murdie's letter was ignored in response.

72 Gibson to D. Roy Cameron, chief forester, Department of Mines and Resources, 16 June 1939, RG84 vol. 23, file PEI182.3, NA. Smith had graduated from the University of New Brunswick's Department of Forestry in 1934, and worked for the Anglo-Canadian Pulp and Paper Company in Quebec. He was the choice of J. Walter Jones, an Island politician who would soon be premier. Smith and Jones were from the same community, Pownal. Interview with Ernest Smith, Tea Hill, 4 August 1995.

73 Ernest Webb to Superintendent Smith, 11 January 1939, box 1, Yankee Gale monument file, PEINP files. ("Crearer" is as in the original.)

74 As well as disputes involving individuals losing rights to the park, there were also cases in which individuals lost sole right to what had become a common property by way of the park. According to Superintendent Ernest Smith, Mike Doyle complained that other farmers were using "his"

cove for Irish moss harvesting. Smith located the deed in the land record office and found that Doyle did have a special right to the cove "for one peppercorn per year." Doyle admitted that he had not been paying it, so Smith told him that he had lost his right. Smith recalls, "I never heard another word about it." Interview with Ernest Smith, Tea Hill, 4 August 1995.

75 RG84 vol. 1777, file PEI2 vol. 5, NA.
76 Simpson to Crerar, 24 February 1938, RG84 vol. 1777, PEI2 vol. 5, NA.
77 Ibid., 24 February 1938 and 19 January 1939.
78 Smith to Williamson, 20 June 1940, ibid.
79 Statistics are from Canada, *Annual Departmental Reports* (1936–43).
80 Stead, *Canada's Maritime Playgrounds*.

CHAPTER FIVE

1 This description is actually from Wilson's memoir, *Naturalist*, 361. Here, Wilson is reviewing his earlier *Biophilia* and provides a more exact description of humans' ideal habitat.
2 Wilson, *Biophilia*, 115.
3 Allan McAvity, New Brunswick Fish and Game Association, to Harkin, 20 January 1927, RG84 vol. 483, file F2 vol. 1, NA.
4 Saint John *Telegraph Journal*, 12 September 1928.
5 F.M. Sclanders, commissioner of the Saint John Board of Trade, to Harkin, 15 February 1929, RG84 vol. 483, file F2 vol. 1, NA. Also, see Sclanders to Harkin, 28 February 1929, ibid. On published reports on progress in having a park established, see Saint John *Telegraph Journal*, 13 February and 27 September 1929, and 27 January 1930.
6 Harkin to Cory, 19 September 1929, RG84 vol. 483, file F2 vol. 1, NA.
7 Harkin to Sclanders, 21 September 1929, ibid.
8 Saint John *Telegraph Journal*, 14 January 1930 and 27 March 1930.
9 Cautley report on New Brunswick sites (1930), 1 and 2, RG84 vol. 1964, file U2.13 vol. 1, NA.
10 Ibid., 3.
11 Ibid., 4 and 12.
12 Ibid., 13. Colpitts, "Alma, New Brunswick," 184.
13 Cautley report on New Brunswick sites (1930), 13, 17, and 12.
14 Ganong believed Mount Carleton was the most beautiful point in the province, and felt the fact that part of the area was currently under timber lease was not a problem: "All that would be necessary would be to arrange that the lumbering be so done as not to damage attractive natural features." Ganong to W.H. Davidson, secretary of the Newcastle Branch of the New Brunswick Fish and Game Association, 2 January 1929, RG84 vol. 483, file F2 vol. 1, NA.

15 Cautley report on New Brunswick sites (1930), 27. On Mount Carleton, see Shaw, *Mount Carleton Wilderness*.

16 Cautley report on New Brunswick sites (1930), 17. Cautley admitted that the 50 for practicability of purchase was arbitrary, since he did not know how expensive and easy it would be to reclaim alienated Crown lands.

17 New Brunswick's patience can be seen in the correspondence of RG84 vol. 484, file F2 vol. 2, NA.

18 Cited in Cautley to Harkin, 22 December 1934, RG84 vol. 484, file F2 vol. 3, NA.

19 It is not clear whether Bennett himself got involved with bringing the park to Albert County, or whether his supporters wished to put it there as a memorial (just as William Lyon Mackenzie King helped bring Prince Albert National Park to Saskatchewan).

20 Cautley to Gibson, 22 November 1934, RG84 vol. 484, file F2 vol. 3, NA.

21 This was a name given to the Albert County plans, as mentioned by R.B. Hanson, MP for York-Sunbury, in Canada, House of Commons, *Debates*, 26 May 1939, 4649.

22 Minister of lands and mines F.W. Pirie to Crerar, 28 January 1936, cited by Fred Squires, in New Brunswick, Legislative Assembly, *Synoptic Report* (1937), 21.

23 Cited by Squires, ibid., 20–21.

24 Cautley to Harkin, 6 March 1936, RG84 vol. 484, file F2 vol. 3, NA.

25 Cautley to Harkin, 24 June 1936, ibid.

26 Squires, in New Brunswick, Legislative Assembly, *Synoptic Report* (1937), 21 and 22.

27 Cautley report on New Brunswick sites (1936), 14–15, 3, 9–10, 5, RG84 vol. 1964, file U2.13 vol. 1, NA.

28 When asked why Prince Edward Island and Nova Scotia had parks and New Brunswick did not, Crerar tried to put the best face on matters: "There is some difference of opinion as to the best location for a park in that province. I would point out something that I think is obvious to everyone acquainted with the situation. A park located in Nova Scotia or in Prince Edward Island is bound to be of some benefit to New Brunswick because of the fact that tourists could reach a park in Nova Scotia only by travelling through New Brunswick." This response did not, of course, satisfy New Brunswickers. Canada, House of Commons, *Debates*, 13 May 1936, 2789 and 2791.

29 Cited by Gibson to Williamson, 13 April 1937, RG84 vol. 484, file F2 vol. 3, NA. The competition between sites appears to have led to accusations of impropriety, though I found no reference to this in either the *Synoptic Reports* or the Saint John *Telegraph Journal* at the time. But the 15 April 1936 *Island Farmer*, in an editorial on the proposed Prince Edward Island park, noted, "There are 2 things that can ruin the park project from the very

start. The first of these is the least suspicion of racketeering, which has come up in our sister province of New Brunswick in connection with their National Park; the other is the choosing of an unsuitable location." Nine years later, at the end of the war, Ian Sclanders wrote a piece for the *Telegraph Journal* entitled "Get Ready for Tourists!" in which he spoke of the "bitter controversy about the proper location for the project" in 1936. It had been "political dynamite," and a "case of sectionalism and lack of co-operation" (18 December 1945). My own research found very few references to the New Brunswick national park in political and editorial discourse in 1936 – surprising especially since New Brunswickers saw Prince Edward Island and Nova Scotia successfully obtaining parks.

30 Canada, House of Commons, *Debates*, 1 March 1937, 1384.

31 Cautley to Harkin, 6 March 1936, RG84 vol. 484, file F2 vol. 3, NA.

32 Smart report on New Brunswick sites (1937), 10, RG84 vol. 1023, file F2 pt. 1, NA.

33 Crerar to Pirie, 16 July 1937, RG84 vol. 1023, file F2 vol. 4 pt. 2, NA.

34 Pirie, in New Brunswick, Legislative Assembly, *Synoptic Report* (1938), 32.

35 Canada, House of Commons, *Debates*, 26 May 1939, 4649.

36 On parks during the war, see Waiser, *Park Prisoners*; and Bella, *Parks for Profit*, 94–103. It was accepted during the war that parks were peacetime luxuries. Member of Parliament Rev. Alexander Nicholson stated, for example, in discussion of increasing Banff's Lake Minnewanka hydro development, "in time of war those who prize these natural beauty spots must be prepared to see them sacrificed if power is necessary." R.B. Hanson of York-Sunbury noted, "If this will have the effect of changing the economy of that part of Alberta, then I think we are justified in voting for this measure, even though we may be violating some other tenets which we hold with regard to national parks." Canada, House of Commons, *Debates*, 4 June 1941, 3484 and 3491.

37 The amounts were: 1938, $1.41 million; 1939, $1.29 million; 1940, $1.17 million; 1941, $1.19 million; 1942, $1.00 million; 1943, $.87 million. Canada, Department of Mines and Resources, *Annual Report* (1938–43).

38 Canada, House of Commons, *Debates*, 30 July 1942, 5033.

39 Ibid., 20 March 1941, 1740.

40 Crerar to Pirie, 10 August 1943, RG84 vol. 1023, file F2 vol. 4 pt. 2, NA.

41 The report stated, "New Brunswick is the only province which has no National Park. It is urgent that such a park be created, and it is recommended that the Provincial Government select one or more National Park sites now." New Brunswick, Legislative Assembly, *Report of the New Brunswick Committee on Reconstruction* (1944), 35. Actually, Quebec did not have a national park either. See also Colpitts, "Alma, New Brunswick," 195; and Young, "'and the people will sink into despair.'"

42 New Brunswick, Legislative Assembly, *Synoptic Report* (1946), 253; and (1937), 21.
43 Colpitts, "Alma, New Brunswick," 196. Kings County voted Conservative candidates into the provincial legislature in every election from 1908 through the 1970s, except for 1935.
44 Albert County elected Liberal members in every provincial election from 1930 until 1952. Notably, in that election – the first since completion of Fundy National Park – the riding went Conservative, and continued to do so through the 1970s.
45 Smart memos, 26 April 1947 and 21 June 1947, RG84 vol. 1023, file F2 vol. 4 pt. 1, NA.
46 Saint John *Telegraph Journal*, 28 July 1947, 1 and 4. The official announcements of park creation were made in the Canadian Order in Council #3211 and the New Brunswick Order in Council #47–538.
47 See Saint John *Telegraph Journal*, 4 August 1947; New Brunswick, Legislative Assembly, *Synoptic Report* (1948), 77–8; and (1949). As early as 7 October 1949, the *Telegraph Journal* would report, "No matter how much is spent on the park, it will not achieve its full purpose until the 'Fundy Trail' is a reality." On 26 January 1950, it announced, "In fact, it is doubtful whether Fundy National Park can be regarded as adequately planned and exploited until the seaside trail is run through."
48 Canada, Department of Mines and Resources, *Annual Report* (1943–9).
49 Allardyce, "'The Vexed Question of Sawdust,'" 120, and "The Salt and the Fir."
50 Majka, *Fundy National Park*, 30–1.
51 Bella, *Parks for Profit*, 129.
52 Colpitts, "Alma, New Brunswick," especially chapter 4, "The National Park: Sawmilling Community to Government Support System," 168–221. Also, in "Sawmills to National Park," Colpitts seeks to disprove the notion held by historians A.R.M. Lower and S.A. Saunders that the New Brunswick sawmilling industry in general was in decline in the first half of this century.
53 Colpitts, "Sawmills to National Park," 109.
54 Interview with Winnie Smith, Riverside-Albert, 31 August 1994.
55 Cited in Cooper and Clay, *History of Logging and River Driving in Fundy National Park*, 5.
56 Interview with Leo Burns, Alma, 1 September 1994.
57 Cautley report on New Brunswick sites (1930), 14 and 12, RG84 vol. 1964, file U2.13 vol. 1, NA.
58 Cautley report on New Brunswick sites (1936), 14 and 15, ibid.
59 Smart report on New Brunswick sites (1937), 5 and 6, RG84 vol. 1023, file F2 pt. 1, NA.
60 Smart to Gibson, 14 January 1948, RG84 vol. 1039, file F200 vol. 1, NA.

61 New Brunswick, Department of Lands and Mines, *Annual Report* (1949), 24. Deputy minister of lands and mines G.H. Prince reminded Smart that according to his notes of one of their earlier meetings, "It was intimated that the Dominion would give consideration to developing a well managed forest in the area with the likelihood that good forest management would be adopted." Prince to Smart, 6 October 1947, RG84 vol. 1023, file F2 vol. 4 pt. 1, NA.

62 Colpitts, "Alma, New Brunswick," 197.

63 W.P. Keirstead to Smart, 4 November 1949, RG84 vol. 1025, file F16.112.1 vol. 1 pt. 2 (1949–52), NA.

64 Interview with Winnie Smith, Riverside-Albert, 31 August 1994.

65 This is not to suggest that the New Brunswick government was in any way reacting to the expropriations at the first two parks. In fact, there is no mention of Cape Breton Highlands or Prince Edward Island parks during the establishment of Fundy.

66 The arbitration proceedings are explained by Gill in New Brunswick, Legislative Assembly, *Synoptic Report* (1949), appendix, 16.

67 New Brunswick, Department of Lands and Mines, *Annual Report* (1947), 121; and (1949), 21.

68 In January of 1948, Smart mentioned in a memo that seventy of the seventy-nine and a half square miles of the park had already been settled. This means that settlement with Hollingsworth and Whitney had already been made, though no other owners had even been contacted yet. Smart to Gibson, 24 January 1948, RG84 vol. 1024, file F2 vol. 5, NA.

69 Saint John *Telegraph Journal*, 28 August 1947.

70 See Colpitts, "Alma, New Brunswick," 204–5.

71 Ibid.

72 K.B. Brown, in Fred Colpitts file, Fundy National Park Land Assembly Records, Department of Natural Resources, RG10 RS145, PANB.

73 Cleveland, in Judson Cleveland file, ibid.

74 For example, in answering Fred Colpitts's complaints, K.B. Brown suggested that residents were lucky to be dealing with the government: "If Hollingsworth and Whitney had decided to convert all wood to pulpwood, as they had told us they planned to do, would you have had a claim against them?" See Colpitts file, ibid. Colpitts, "Alma, New Brunswick," 201.

75 Keirstead was eventually granted $200. Here is a good example of how difficult it is to judge settlement offers today. In the expropriation records, Bob Keirstead's property is said to be twenty-four acres, not thirty, and his brother is said to have been given $539, not $600, for thirty-eight acres, not thirty. Are these the final settlements in each case? Are the surveyed sizes inaccurate? Is Keirstead's memory of the properties or of the settlements

wrong? See Fundy National Park Land Assembly Records, Department of Natural Resources, RG10 RS145, PANB.

76 The assessor, Brown, feared that granting Colpitts this concession would spark trouble with other landowners. See Colpitts, "Alma, New Brunswick," 203.

77 Smart to Brown, 29 May 1948, RG84 vol. 1024, file F2 vol. 5, NA.

78 Smart to Gibson, 10 December 1952, RG22 vol. 239, file 33.6.1 pt. 3, NA.

79 See RG84 vol. 1024, file F16.1, NA.

80 Baizley file, Fundy National Park Land Assembly Records, Department of Natural Resources, RG10 RS145, PANB.

81 Ian Sclanders, "New Brunswick Parade," Saint John *Telegraph Journal*, 26 July 1948.

82 Interview with Leo Burns, Alma, 1 September 1994.

83 Interview with Pearl Sinclair, Alma, 2 September 1994.

84 Allardyce, "The Salt and the Fir," 5.

85 Interview with Winnie Smith, Riverside-Albert, 31 August 1994.

86 Canada, Department of Resources and Development, *Fundy National Park*, opening ceremonies booklet, 1950.

87 Canada, Department of Mines and Resources, *Annual Reports* (1937–48). See appendix 2.

88 Ibid. (1945–50).

89 In 1948, $540,000 was spent; in 1949, $870,000; in 1950, $835,000. It should be noted that in the same period $2.6 million was spent in rebuilding sections of Cape Breton Highland's Cabot Trail. RG22 vol. 474, file 33.9.1 pt. 4, NA.

90 Shortly after the park was created, deputy minister of lands and mines G.H. Prince offered James Smart his notes on their previous meeting about park development. His memory was perhaps imperfect – he recalled, for example, that Smart had agreed to ski runs. Prince to Smart, 6 October 1947, RG22 vol. 366, file 304.73 pt. 1, NA.

91 Maxwell, '*Round New Brunswick Roads*, 104–5.

92 Editorial, Saint John *Telegraph Journal*, 5 January 1950.

93 Lothian, *A Brief History*, 111.

94 *Moncton Times*, undated. From clipping collection of Greta Geldart Elliott.

95 From clipping collection of Greta Geldart Elliott. It is possible that these writers were relying on the same Parks Branch press release.

96 R.W. Murphy. From clipping collection of Greta Geldart Elliott.

97 From clipping collection of Greta Geldart Elliott.

98 Ian Sclanders, "New Brunswick Parade," Saint John *Telegraph Journal*, 26 July 1948.

99 Smart to Esther Clark Wright, 20 December 1950, RG84 vol. 46, file F109, NA.

100 Gordon L. Scott to J.R.B. Coleman, chief, National Parks Branch, 31 April 1951, RG84 vol. 140, file F28.1, NA.
101 Park administrators were just beginning to question the wisdom of investing so heavily in golf courses. Referring to work at Fundy, H.S. Robinson, assistant chief of the Historical and Information Section of the Department of Mines and Resources, wryly noted, "As something less than 4% of park visitors make use of park golfing facilities, we believe that it would create better relations with about 96% of the public if we were able to publicize the availability of extensive up-to-date campground facilities." Robinson to Gibson, 8 December 1949, RG84 vol. 484, file F36 pt. 1, NA.
102 Lawson and Sweet, *This Is New Brunswick*, 123.
103 Smart to Moore, 24 February 1949, RG84 vol. 142, file F317, NA.
104 James Hutchison to Smart, 20 February 1953, RG22 vol. 239, file 33.6.1 pt. 3, NA.
105 Smart to Gibson, 17 April 1952, ibid.
106 New Brunswick, Legislative Assembly, *Synoptic Report* (1946), 221.
107 See New Brunswick, Department of Lands and Mines, *Annual Report* (1950), 18.
108 And as deputy minister of mines and resources Hugh Keenleyside later said when defending cost overruns on the house, "the construction of appropriate departmental headquarters promptly is a token of the intention of the Administration to undertake worthwhile development of the Park." Keenleyside to R.B. Bryce, Treasury Board, 16 June 1949, RG84 vol. 141, file 56.1, NA.
109 Smart to Gibson, 4 May 1949, ibid.
110 R.A. Gibson told the assistant controller, J.A. Wood, "Surely everyone associated with the National Park Service has realized that the reported cost of this structure has been criticized by Treasury Board and we had to make a special explanation." Gibson to Wood, 12 July 1949, ibid.
111 Wood to Gibson, 13 July 1949, ibid.
112 Interview with Bob Keirstead, Alma, 29 August 1994.
113 Interview with Leo Burns, Alma, 1 September 1994.

CHAPTER SIX

1 Lee Wulff, in Newfoundland Tourism Board, *Annual Report* (1942), Reid Company papers, W. Angus Reid files, MG17 pt. 3, box 3, PANL. On tourism in Newfoundland, see Overton, *Making a World of Difference*.
2 Hiller and Harrington, *The Newfoundland National Convention*, 2:148.
3 See editorials, St John's *Evening Telegram*, 1 June 1957, and 21 May 1959.
4 Newfoundland, *Report on Meetings between Delegates from the National Convention*, summary of proceedings and appendices, 25 June– 29 September 1947, 20 and 122.

5 Cited in J.R. Baldwin, Canadian Privy Council Office, to Keenleyside, 4 September 1947, RG84 vol. 1942, file TN2 vol. 1, NA.

6 Smart report on Newfoundland sites (1950), 2–3, RG84 vol. 1942, file TN2 vol. 1, NA. In 1947, Smart had believed that a Newfoundland park should be five hundred to a thousand square miles, but by 1949 he was ready to state, "we would not expect a very big area to be set aside." See Smart to Gibson, 31 May 1947, ibid.; and Smart to Gibson, 11 January 1949, ibid.

7 Smart report on Newfoundland sites (1950), 9, ibid.

8 Ibid., 17–18.

9 Ibid.

10 For attendance figures, see appendix 2. For expenditures, see Canada, Department of Resources and Development, *Annual Reports* (1951–4), and Department of Northern Affairs and National Resources, *Annual Reports* (1955–7). On delaying the park because of defence expenditures, see St Laurent to Smallwood, 3 February 1954, RG22 vol. 177, file 33.11.1, NA.

11 See Smallwood to St Laurent, 24 December 1953, ibid.

12 Smallwood, *I Chose Canada*, 359. On Smallwood, also see Gwyn, *Smallwood*, and Rowe, *The Smallwood Era*.

13 On Pickersgill's place in the politics of the day, see Granatstein, *The Ottawa Men*, especially 207–25. Pickersgill describes the Ministry of Northern Affairs and National Resources shuffle in *My Years with Louis St. Laurent*, 207, and his courtship by Smallwood, 181 on. Peter Neary notes that Smallwood saw the fact that the Newfoundland representative in Cabinet was "from away" as an advantage: Pickersgill would not be a competitor for provincial leadership and would rely on Smallwood for patronage. Neary, "Party Politics in Newfoundland," 218.

14 See Pickersgill, *My Years with Louis St. Laurent*, 229–30. But as we have seen, Smallwood had this in mind by the winter of 1953.

15 Smallwood to St Laurent, 24 December 1953, RG22 vol. 177, file 33.11.1, NA; Jackson to Robertson, 11 January 1954, ibid.; and Robertson to Lesage, 11 January 1954, ibid.

16 Lesage to St Laurent, 26 January 1954, ibid.

17 Kennedy, Cameron, and Goodyear, *Report of the Royal Commission on Forestry* (1955), 47.

18 In *I Chose Canada*, Smallwood lists a dozen international companies – some with no experience in pulp and paper – that he pursued (438).

19 P.J. Murray to Robertson, 1 October 1954, RG22 vol. 177, file 33.11.1, NA.

20 Hutchison to Robertson, 15 October 1954, ibid.

21 D.A. MacDonald to Jackson, 21 October 1954, ibid.

22 Robertson to Murray, 26 October 1954, ibid.

23 Kennedy, Cameron, and Goodyear, *Report of the Royal Commission on Forestry*, 186.

24 Corner Brook *Western Star*, 21 March 1955.

25 See Lesage to Pickersgill, 8 September 1955, RG22 vol. 177, file 33.11.1, NA.

26 "Report of Proceedings of Provincial Tourism Conference, 28 February–
   1 March, 1955," 12, Reid Company papers, W. Angus Reid files, MG17
   pt. 3 box 3, file "Tourism 1953–1955," PANL.

27 Hutchison to the chief, 2 May 1955, RG84 vol. 1942, file TN2 vol. 2, NA.

28 Scott to Hutchison and Coleman, 14 May 1955, RG22 vol. 177,
   file 33.11.1, NA.

29 Scott report to Coleman, 21 May 1955, ibid., 15.

30 As Scott noted, "the caribou is the Provincial emblem and it would be
   somewhat ridiculous if we did not have a herd of these animals in the
   Park." Ibid., 5.

31 Scott to Hutchison and Coleman, 14 May 1955, ibid.; and Scott report to
   Coleman, 21 May 1955, ibid., 16.

32 Hutchison to Robertson, 2 June 1955, ibid.

33 Smallwood to St Laurent, 31 August 1955, cited in RG84 vol. 1954,
   file TN200 vol. 4, NA.

34 As the premier told the legislature, "the Government of Canada will
   reimburse us every cent we will have spent on that road and then proceed
   to pave it, then they will provide the maintenance at their own expense
   for ever, if it should be designated a National Park." Smallwood, in
   Newfoundland, House of Assembly, *Proceedings* (1956), 1080.

35 Jenkins, *Report on Forest Survey.*

36 Smallwood to Pickersgill, 28 December 1955, RG22 vol. 474, file 33.11.1
   pt. 2, NA. It is worth noting that this letter is in the park files: Pickersgill
   was sharing with the federal ministry information and strategy being sent
   him by Smallwood.

37 Robertson to Jackson, 3 October 1956, RG22 vol. 474, file 33.11.1 pt. 2, NA;
   and Lesage to Pickersgill, 17 October 1956, ibid. This was not a new equa-
   tion: Smallwood mentioned it as early as 28 December 1955 in a letter to
   Pickersgill, ibid.

38 Lesage to Pickersgill, 17 October 1956, ibid.

39 Pickersgill to Smallwood, 22 October 1956, 3.25.00, J.R. Smallwood
   Collection, Centre for Newfoundland Studies Archives; and Smallwood
   to Lesage, 13 November 1956, RG22 vol. 474, file 33.11.1 pt. 2, NA.

40 Robertson to Jackson, 3 October 1956, ibid.

41 Ibid.

42 D.J. Learmouth, forestry engineer, Department of Northern Affairs
   and Natural Resources, to Coleman, 28 December 1956, RG22 vol. 474,
   file 33.11.1 pt. 2, NA.

43 MacDonald to Côté, 7 February 1956, ibid.

44 C.R. Granger, Pickersgill's private secretary and executive assistant, to
   Murray, 22 January 1957, ibid.

45 Lesage to Pickersgill, 28 January 1957, ibid.

46 See F.A.G. Carter memo, 4 February 1957, ibid.

47 Harold Horwood column, St John's *Evening Telegram*, 27 December 1956, 4. Pickersgill was also fêted for having unemployment insurance extended to fishermen.

48 See Lesage memo, 5 April 1957, RG84 vol. 1942, file TN2 vol. 3, NA; and Lothian, *A Brief History*, 118.

49 Lesage to the new Newfoundland minister of mines and resources, W.J. Keough, 28 January 1957, RG22 vol. 474, file 33.11.1 pt. 2, NA. On whether to include the North West Brook watershed, see Murray to Hutchison, 17 December 1956, RG84 vol. 1942, file TN2 vol. 2, NA; and Robertson to Murray, 4 January 1957, ibid.

50 Scott to Hutchison, 8 November 1955, ibid.

51 Scott, for example, wrote a note to Hutchison that began, "Since time is obviously a factor in proclaiming a park area …" Scott to Hutchison, RG22 vol. 474, file 33.11.1 pt. 2, NA. The Liberals' sudden haste to finalize park plans was not lost on the opposition. Lesage noted in Parliament, "the boundaries have now been defined, and I expect that a formal agreement will be signed between this government and the government of Newfoundland within the next 10 or 15 days." An opposition member asked, "Before the election?" and Lesage retorted, "We work on this side, we don't lose time trying to play tricks." Canada, House of Commons, *Debates*, 8 April 1957, 3234.

52 Pickersgill, *My Years with Louis St. Laurent*, 230.

53 There was no debate about the name; the Newfoundland government christened the park the day it was proclaimed. Canada, House of Commons, *Debates*, 12 April 1957, 3467.

54 St John's *Evening Telegram*, 12 April 1957.

55 Lothian, *A Brief History*, 118.

56 The Newfoundland government did not give a breakdown of park expropriation costs in its yearly public accounts, as Nova Scotia, New Brunswick, and Prince Edward Island had done to varying degrees. In the 1958 *Annual Report* of the Department of Mines and Resources, there is reference to $4338.42 spent on national park land assembly (175), in 1959 another $200 spent on the same item (162), and in 1960 $22,000 on expropriation (98). This final figure is for expropriation in the entire province, and it is unclear how much, if any, concerns the park.

57 See Major, *Terra Nova National Park*, 49–61.

58 W.C. Wilton and H.S. Lewis, "Forestry Problems of the Bonavista Peninsula Newfoundland," Forest Research Division of the Department of Northern Affairs and National Development, Technical Note #26, 1956, RG84 vol. 1942, file TN2 vol. 2, NA.

59 Murray to Robertson, 2 October 1957, RG22 vol. 474, file 33.11.1 pt. 3, NA.

60 (?) to Flanagan, 9 November 1967, RG84 vol. 1954, file TN200 vol. 3, NA. Don Spracklin, who operated a sawmill in Bread Cove, estimates that

there were fourteen operations working out of Charlottetown alone (though many of these would be getting their wood from outside the park area). Interview with Don Spracklin, Charlottetown, 18 August 1994.

61 Sutherland's "The Men Went to Work by the Stars" and "We Are Only Loggers"are excellent social histories of Newfoundland logging.

62 Interview with Clayton and Mildred King, Sandy Cove, 19 August 1994.

63 Interview with Mark Lane, Eastport, 17 August 1994.

64 Interview with Clayton King, Sandy Cove, 19 August 1994.

65 Scott to Hutchison, 8 November 1955, RG84 vol. 1942, file TN2 vol. 2, NA. The Parks Branch also was aware that the Jenkins report had found that "Practically all parts of the proposed Park area have been logged for saw-timber, and sometimes for poles or pit-props. Those sections close to settlements have been heavily cut to small diameter for firewood." The Parks Branch copy of the Wilton and Lewis report highlights the passage, "Devastating forest fires in the past and reckless exploitation by man had caused a decrease in the area on which merchantable timber could be expected to develop even under intensive silvicultural care. The productive forest lands remaining have been subjected to such intensive cutting for sawlogs, fuelwood, and other forest products that most stands are in deplorable condition" (3). This last quotation refers specifically to the eastern end of the Bonavista peninsula, not the western end of which the park area could be said to be a part.

66 J.D.B. Harrison, director, Forestry Branch, to deputy minister, 29 November 1957, RG22 vol. 474, file 33.11.1 pt. 3, NA. "Mature" is defined as over eighty years of age, "young growth" as forty-one to eighty years.

67 Hutchison to Robertson, 12 April 1957, RG84 vol. 1944, file TN28 pt. 1, NA.

68 Coleman report, 7 August 1957, RG22 vol. 474, file 33.11.1 pt. 3, NA; and Robertson to Coleman, 7 October 1960, RG22 vol. 1097, file 311.1 pt. 1, NA.

69 Scott to Hutchison, 28 May 1957, RG84 vol. 1944, file TN28 pt. 1, NA.

70 See, for example, Coleman report, 7 August 1957, RG22 vol. 474, file 33.11.1 pt. 3, NA.

71 Robertson to Hutchison, 17 April 1957, ibid.

72 Cited in Scott to Hutchison, 28 May 1957, RG84 vol. 1944, file TN28 pt. 1, NA.

73 Coleman to G.B. Williams, chief engineer, Development and Engineering Branch, Department of Public Works, Canada, RG84 vol. 1944, file TN28 pt. 2, NA.

74 Coleman to Robertson, 24 December 1957 and 11 February 1958, RG84 vol. 1943, file TN16.112, NA.

75 See Robertson to Charles Granger, deputy minister of highways for Newfoundland, 17 September 1957, RG22 vol. 474, file 33.11.1 pt. 3, NA. That same winter, Newfoundland sent the Parks Branch a bill for $15,797.47 for fire protection at Terra Nova, as they had agreed. The

Branch responded seeking a few clarifications, but an internal memo notes, "This letter is really a polite stall. … [T]here was no money provided for this work – we were depending on a surplus developing and none has! We can't pay till the new fiscal year starts." F.A.G. Carter to (?), 12 March 1958, RG22 vol. 1097, file 311.1 pt. 1, NA.

76 See editorial, St John's *Evening Telegram*, 8 July 1960; and "Resident of Glovertown," 7 February 1962.

77 E.P. Henley, president, Gander Chamber of Commerce, to Dinsdale, 23 January 1962, RG84 vol. 1097, file 311.1 pt. 1, NA.

78 Robertson to Coleman, 7 October 1960, ibid.

79 Interim Annual Report, Atkinson to chief, 21 January 1960, RG84 vol. 1952, file TN112 year vol. 1, NA.

80 For instance, twenty-five carpenters were employed in the middle of winter to build houses for park staff. J.E. Wilkins to Scott, 19 June 1958, RG84 vol. 1944, file TN28 pt. 1, NA.

81 Pickersgill told the House of Commons in 1961 that hiring locally had been agreed to by all parties at establishment. "The provincial government asked for this assurance because they felt they were more or less taking away the livelihood of these people in establishing the park. They wanted to be sure that no hardship was effected." Canada, House of Commons, *Debates*, 10 May 1961, 4627.

82 St John's *Evening Telegram*, 16 December 1957.

83 Coleman report, 7 August 1957, RG22 vol. 474, file 33.11.1 pt. 3, NA.

84 See Canada, House of Commons, *Debates*, 2 July 1958, 1825; 14 August 1958, 3536 and 3558; 18 February 1959, 1149.

85 See Doak to Coleman, 9 September 1963, RG84 vol. 1954, file TN200 vol. 3, NA. To his credit, when Pickersgill was given a list of "our key men that we would like to keep," he told Doak that "some of them were opponents of his, but regardless they should not be laid off or replaced."

86 (?) to (?), 31 August 1959, RG84 vol. 1944, file TN28 pt. 2, NA.

87 Strong to Doak, 9 March 1960, ibid.

88 Interim Annual Report, Atkinson to B.I.M. Strong, 21 January 1960, RG84 vol. 1952, file TN112 year vol. 1, NA.

89 Carter to Robertson, 4 December 1957, RG84 vol. 1944, file TN28 pt. 1, NA.

90 Selby Moss, Allied Aviation Ltd, letter to the editor, St John's *Evening Telegram*, 18 December 1957.

91 Harry Horwood editorial, St John's *Evening Telegram*, 28 December 1957.

92 Interview with Clayton King, Sandy Cove, 19 August 1994.

93 Interim Annual Report, Atkinson to B.I.M. Strong, 21 January 1960, RG84 vol. 1952, file TN112 year vol. 1, NA.

94 See Robertson to Granger, 17 September 1957, RG22 vol. 474, file 33.11.1 pt. 3, NA.

95 Robertson to Murray, 28 October 1957, ibid.

96 Warden H.T. Cooper to Lothian, 11 October 1957, RG84 vol. 1954, file TN200 vol. 1, NA.

97 See Annual Report (1959), Atkinson to Strong, 1 June 1959, RG84 vol. 1952, file TN112 year vol. 1, NA; Interim Annual Report (1960), Doak to Strong, 23 January 1961, ibid; and Annual Report (1961), Doak to Strong, 14 April 1961, ibid. The later estimation is given in (?) to Flanagan (?), 9 November 1967, RG84 vol. 1954, file TN200, vol. 3, NA. One thousand fbm scaled by the International 1/4" Log Rule of logs cut from 8" trees form a pile of stacked wood measuring three to five cords. Therefore, the most wood cut from the park was somewhere between 120 and 200 cords, in 1957.

98 Ibid.

99 See Sutherland, "'We Are Only Loggers.'"

100 See Robertson to Murray, 28 October 1957, RG22 vol. 474, file 33.11.1 pt. 3, NA.

101 Coleman to Robertson, 2 January 1963, RG84 vol. 1954, file TN200 vol. 3, NA.

102 There is a large literature on underdevelopment and modernization theory in Atlantic Canada. See Burrill and McKay, *People, Resources, and Power*, and Brym and Sacouman, *Underdevelopment and Social Movements*.

103 Coleman to Robertson, 30 April 1963, RG84 vol. 1954, file TN200 vol. 3, NA.

104 Ibid.

105 Interview with Don Spracklin, Charlottetown, 18 August 1994.

106 Doak to W.W. Mair, chief, National Parks Branch, 18 January 1964, RG84 vol. 1952, file TN132, NA.

107 Ibid. This continued to be the Newfoundland position up to at least 1967. See (?) to Flanagan (?), 9 November 1967, RG84 vol. 1954, file TN200 vol. 3, NA, which states, "the Newfoundland Minister of Mines & Resources has stated he wishes to keep logging and sawmilling by small enterprise out of the Park and considers the timber as a growing asset for the third pulp mill. The Premier of Newfoundland has stated in public that the timber reserves in the Park are available to the Province for this purpose."

CHAPTER SEVEN

1 Canada, House of Commons, *Debates*, 9 May 1930, 1943.

2 There has never been a need for a well-defined and publicly accepted philosophy for the Department of Agriculture, for example.

3 Some readers may believe that analysis of preservation and use has become of only academic interest: preservation would seem to have won, now that Parks Canada has declared (in 1988 amendments to the National Parks Act and in its 1994 Guiding Principles and Operational

Policies) that its primary commitment is to ecological integrity. I disagree. While these declarations have been symbolically significant, and are sure to make future policy more preservationist-minded, the tension between preservation and use remains. Parks Canada has shifted the point at which the needs of use outweigh the needs of preservation, but staff will still have to make the decision as to when that point has been reached. There is still a twin mandate. It is as if Parks Canada has proclaimed the old softball adage that the tie goes to the runner, when any ump will tell you that there actually *are* no ties.

4 Bella, *Parks for Profit*, 128.

5 Canada, House of Commons, *Debates*, 3 May 1887, 233.

6 Of all the parks created before the 1960s, only Wood Buffalo National Park on the border of Alberta and the Northwest Territories could be said to have no real potential for attracting tourists.

7 "Fundy National Park."

8 Saint John *Telegraph Journal*, 31 July 1950.

9 It was well known within the Parks Branch that attendance figures were crude and calculated in different ways throughout the system. In 1955, the director, Hutchison, noted, "At Fundy Park, vehicles are stopped at the Lakeview and Alma entrances to the Park and recorded together with the number of passengers. ... At Cape Breton Highlands National Park, vehicles are recorded at the Ingonish and Cheticamp entrances and the number of passengers estimated at 3 persons per car. ... At Prince Edward Island National Park, attendance figures are compiled from registrations at Dalvay House and Green Gables plus the estimated number of visitors to the Cavendish, Brackley, and Stanhope sections of the Park." Hutchison to Jackson, 28 October 1955, RG22 vol. 470, file 33.2.43 pt. 1, NA. The Parks Branch thought that statistics for Prince Edward Island National Park in particular were grossly inflated, because they included so much day use by Island beach-goers. From the Branch's point of view this was not "real" tourism, since it brought so little money into the economy. It should be noted that although local visitors had less of an impact on cultural features such as hotels and campgrounds, they put just as much strain on the park's natural features as anyone else.

10 Annual Reports, PEINP files.

11 Saint John *Telegraph Journal*, 6 December 1961. For Fundy tourism statistics, see RG84 vol. 1036, file F121.13 vol. 2, NA.

12 Doak to Strong, 5 October 1960, RG84 vol. 1944, file TN36, NA.

13 For Harkin's and Smart's love of golf, see Lothian, "James Bernard Harkin," 10; and Lothian, *A History*, 4:20, respectively. Crerar's correspondence with ex–Nova Scotia premier Angus L. Macdonald in the 1950s consisted almost entirely of golf talk. See Thomas Crerar papers, acc. 2117, series 3, box 122, 1945–50 file, Queen's University Archives.

14 This is a quite well documented period in the history of North American tourism. See, for example, Wilson, *The Culture of Nature*, especially 28–33. In "Gender, Recreation, and the Welfare State in Ontario," Tillotson shows that recreation was an unheralded but important part of the Canadian state in the 1950s.

15 Canada, Department of Resources and Development, *Playgrounds of Eastern Canada*, no date.

16 Dilsaver, "Stemming the Flow," 251.

17 *Playgrounds of Eastern Canada*, 15.

18 Wilson, *The Culture of Nature*, 131.

19 See Dunlap, *Saving America's Wildlife*, 79–82; Runte, *National Parks*, xix.

20 See Dunlap, *Saving America's Wildlife*, 99–102; Worster, *Nature's Economy*, 342–53. National parks were not seen as just a figurative escape from nuclear war. In 1957, member of Parliament Lloyd Crouse of Queens-Lunenburg lobbied for a second Nova Scotia national park on the grounds that "In the event of a nuclear war, these national parks would be invaluable. People in thickly settled areas could be evacuated to these camp sites where water and sewage facilities are provided. Because of their location these facilities would not likely be contaminated by fall-out so I would urge the government to give some thought to this proposal." Canada, House of Commons, *Debates*, 13 November 1957, 1078.

21 On camping's renaissance in the 1950s, see Wall and Wallis, "Camping for Fun." Runte, *National Parks*, 157, notes that autocamping, or "sagebrushing," was popular in the 1920s as a reaction to the extravagances of park resort tourism of the day. It suggests something about the relationship between our appreciation of nature and class consciousness that camping came into fashion in the 1920s and 1950s, but was out of fashion in the 1930s when some North Americans were camping out of necessity.

22 See RG84 vol. 103, file U36 vol. 3; and Canada, Department of Northern Affairs and National Resources, *Annual Reports* (1953–64).

23 J.B. McKee Arthur, New Jersey, to Alan Field, Director, Canadian Government Travel Bureau, 16 December 1958, RG84 vol. 1036, file F121 vol. 3, NA.

24 R.G. Talpey, no date, ibid.

25 Meeker, "Red, White, and Black"; and Sax, *Mountains without Handrails*.

26 A 1975 study of Point Pelee National Park found that 74 per cent of adult visitors had at least some university training, and about half had a university degree. Grant and Wall, "Visitors to Point Pelee National Park."

27 Meeker, "Red, White and Black," 203 and 200. In a 1995 interview, black actress Halle Berry said, "Recently, there was a role of a park ranger that I really wanted. We kept pushing and pushing until finally the studio called my agent and said, 'We don't know if a park ranger would be black.'" *New York Times*, 12 March 1995.

28 Smart to Superintendent Smith, 25 August 1942, RG84 vol. 1781, file PEI16.2 vol. 3, NA.

29 Smith to Smart, 4 September 1942, ibid.

30 Spero, assistant controller, to H.S. Robinson, supervisor, Parks and Resources Information, 4 May 1948, RG84 vol. 1781, file PEI16.1 vol. 4, NA.

31 Gibson to Nason, 12 May 1948, ibid.

32 Elsie M. Kieran, Women's Voluntary Services, Quebec, to Robinson, 15 June 1949, ibid. Saul Hayes, national executive director of the Canadian Jewish Congress, to Keenleyside, deputy minister, 12 September 1949, RG22 vol. 242, file 33.21.1 pt. 2, NA.

33 Harold DeWolf to Robert Friars, 1 May 1960, RG84 vol. 1025, file F16.112.1 vol. 3, NA.

34 Friars to DeWolf, 24 June 1960, ibid.

35 The following pages are based on my "Why Martin Luther King Didn't Spend His Summer Vacation in Canada."

36 DeWolf's role as advisor on King's Ph.D. has been closely scrutinized since the 1990 discovery that parts of the dissertation had been plagiarized, not least from a doctoral thesis that DeWolf had supervised only three years earlier. See "Becoming Martin Luther King, Jr."

37 Oates, *Let the Trumpets Sound*, 101.

38 DeWolf to National Parks Branch, 21 July 1960, RG84 vol. 1025 file F16.112.1 vol. 3, NA.

39 Strong to MacFarlane, 29 July 1960, ibid.

40 MacFarlane to Strong, 3 August 1960, ibid.

41 Strong (?) to a Mr Kelly, 5 August 1960, ibid.

42 Strong to DeWolf, 12 August 1960, ibid.

43 See Winks, *The Blacks in Canada*, 464–5; and Walker, *Racial Discrimination in Canada*, 16–24.

44 Cautley report on Nova Scotia sites, in MacDonald, *Transportation in Northern Cape Breton*, 74.

45 N. Milton Browne, Cape Breton Tourist Association, to Smart, 30 March 1939, RG84 vol. 990, file CBH16.112 vol. 1 pt. 3 (1937–49), NA.

46 Smart to Becker, 1 October 1937, ibid.

47 Smart to Browne, 2 May 1939, ibid.

48 Smart to A.S. MacMillan, minister of highways, Nova Scotia, 21 March 1938, RG84 vol. 990, file CBH16.112 pt. 1, NA. (The grammar is as in the original.)

49 J.D. MacKenzie, minister of highways, Nova Scotia, to Crerar, 19 September 1941, RG84 vol. 986, file CBH16.1 vol. 2 pt. 2, NA.

50 Cautley to Cromarty, 22 December 1936, RG84 vol. 1784, file PEI16.112.1, NA.

51 Cautley to Williamson, 16 June 1937, ibid.

52 Katherine Wyand, "Open Letter to Members of the Legislature," Charlottetown *Guardian*, 1 April 1938.

53 Blake Sinclair, Dalvay House, to Williamson, 4 October 1939, RG84 vol. 1784, file PEI16.112.1, NA.

54 Smart to Williamson, 4 June 1940, ibid.

55 Gibson to Smart, 19 January 1948, RG22 vol. 242, file 33.21.1 pt. 2, NA.

56 (?) to Smart, 9 January 1946, RG84 vol. 103, file U36 vol. 2, NA. See also Smart to all superintendents, 17 October 1945, ibid. Cape Breton Highlands was targeted by the Parks Branch, the Canada Youth Commission, and the Nova Scotia Department of Health as the potential home of "family camps" for local communities. It was thought that "such a project would have more prospects of success if the idea were to originate from amongst the people most likely to benefit e.g. the mining communities of Nova Scotia." Robinson to Smart, 2 December 1946, ibid.

57 From clipping collection of Greta Geldart Elliot.

58 Smart to Hugh Young, deputy minister, 24 February 1953, RG22 vol. 239, file 33.6.1 pt. 3, NA.

59 Lesage to MacEachen, 28 January 1954, RG22 vol. 473, file 33.9.1 pt. 4, NA.

60 See, for example, Charlottetown *Patriot* and Charlottetown *Guardian*, 25 September 1947.

61 Hutchison to Young, 23 October 1952, RG84 vol. 1025, file F16.112.1 vol. 1 pt. 1 (1952–1955), NA.

62 See Jackson to Lesage, 24 October 1956, RG84 vol. 1025, file F16.112.1 vol. 2, NA; and Coleman to Robertson, 3 February 1960, RG22 vol. 898, file 307.1, NA. The Parks Branch had intended to sell the cabins itself, and had in fact done so, when it was discovered that this would also mean selling the land under them. The "sales" became "leases." See Lothian to Coleman, 6 April 1964, RG84 vol. 1943, file TN16.112.1 vol. 3, NA.

63 Lesage to Neil Matheson, MP for Queen's, 22 August 1956, RG84 vol. 1785, file PEI16.112.2 vol. 3, NA. Maritimers occasionally voiced this concern themselves. W.W. Reid, supervisor of the PEI Travel Bureau, told Smart, "It would appear to me that our people down here are a bit too conservative [in that] they are quite keen to make something once the Government backs the project, and at the same time they do not appear to be too anxious to share any profit with the Government." 8 May 1948, RG84 vol. 1797, file PEI56.2 vol. 2, NA.

64 Smart to Gibson, 10 December 1952, RG22 vol. 239, file 33.6.1 pt. 3, NA.

65 Coleman was concerned that even the playground might "give a shoddy appearance." Robertson disagreed with Coleman: "there is a serious need for some recreational amenities of a modest kind and of an outdoor character in the park." Coleman to Robertson, 11 May 1960, RG84 vol. 1943, file TN16.112.1 vol. 1, NA; and Robertson to Coleman, 7 October 1960, RG22 vol. 1097, file 311.1 pt. 1, NA.

66 Doak to Strong, 5 June 1961, RG84 vol. 1943, file TN16.112.1 vol. 1, NA.

67 Doak to Coleman, 20 March 1964, RG84 vol. 1943, file TN16.112.1 vol. 3, NA.

68 Memo, 15 September 1961, RG84 vol. 1943, file TN16.112.1 vol. 1, NA.

69 See Bella, *Parks for Profit*, chapter 2.

70 Taylor, "Legislating Nature," 132. The Spray Lakes episode is quite similar to the more famous US case in which Yosemite National Park's Hetch Hetchy Valley was dammed in the 1910s. The great difference is that a widespread opposition arose to the Hetch Hetchy case, resulting in public support for the national park idea; no such support arose in Canada to save Spray Lakes.

71 See Bankes, "Constitutional Problems." H. Ian Rounthwaite in "The National Parks of Canada" suggests that parks might be able to win the right to permanent protection through common law public trust, though this has not been attempted in the courts.

72 Premier G.F. Harrington, cited in Cautley to Harkin, 7 June 1929, RG84 vol. 1964, file U2.12.1, NA. The greatest proof that the Parks Branch's message was understood was the manner in which the provinces tried to sidestep inviolability at park establishment. New Brunswick tried to include a clause in the agreement that created Fundy National Park which stated, "It is noted, however, that this grant reserves to the Province all coals and also all gold and silver and other mines and minerals." The Parks Branch explained why this would not do, and the province conceded the point. J. Allison Glen, minister of mines and resources, Canada, to R.J. Gill, minister of lands and mines, New Brunswick, 21 February 1948, and Gill to Glen, 4 March 1948, RG84 vol. lo24, file F2 vol. 5, NA.

73 Crerar to J.H. MacQuarrie, minister of lands and forests, Nova Scotia, 9 June 1936, cited in Lesage to George Prudham, minister of mines and technical surveys, Canada, 4 May 1954, RG84 vol. 516, file CBH31 pt. 1, NA.

74 Premier Angus L. Macdonald to Crerar, 31 March 1936, RG84 vol. 985, file CBH2 Cap Rouge, vol. 2 pt. 2, NA.

75 Wardle to W. Stuart Edwards, deputy minister of justice, 4 April 1936, RG84 vol. 985, file CBH2–Cap Rouge, vol. 2 pt. 2, NA; and Edwards to Wardle, 11 April 1936, ibid.

76 Cited in Harkin to Daly, departmental solicitor, Department of the Interior, 17 June 1936, RG84 vol. 985, file CBH2 –Cap Rouge, vol. 2 pt. 1, NA.

77 Cited in Gibson to Camsell, 29 November 1938, RG84 vol. 516, file CBH31 pt. 1, NA.

78 Gibson to Williamson, 1 December 1938, ibid.

79 Camsell (?) to Gibson, 30 November 1938, ibid.

80 This episode is taken from the correspondence of RG22 vol. 366, file 304.73 pt. 1, NA.

81 Lothian, *A Brief History*, 114.

82 *Victoria-Inverness Bulletin*, 24 March 1954.

83 M.A. Patterson, minister of mines, Nova Scotia, to Hon. George Prudham, minister of mines and technical surveys, Canada, 22 April 1954, RG84 vol. 516, file CBH31 pt. 1, NA.

84 The Parks Branch may very well have been wrong. According to Bankes, an 1899 ruling makes clear that in a general land transfer from province to Canada, mineral rights are not transferred from provincial authority unless stated explicitly. And since Eastern parks like Cape Breton Highlands and Terra Nova were created by a simple land conveyance, "Without more, one must conclude in each case that the precious metals have remained vested in the provincial Crown" ("Constitutional Problems," 227). My point is not to dispute what the national parks did or did not own, but only to show that inviolability does not have a firm legal foundation.

85 Robertson to Boyer, deputy minister of mines, Nova Scotia, 25 May 1954, RG84 vol. 516, file CBH31 pt. 1, NA. See also Halifax *Chronicle-Herald*, 19 and 20 May 1954.

86 W.T. Dauphinee, minister of mines, Nova Scotia, to Prudham, 2 August 1955, RG84 vol. 516, file CBH31 pt. 1, NA.

87 The *Chronicle-Herald*, 28 April 1956, spoke highly of the federal decision to remove the section from the park.

88 G.D. Mader, Nova Scotia Power Commission, to Superintendent Doak, 22 October 1956, RG84 vol. 998, file CBH68 vol. 1, NA; and Coleman to Doak, 2 November 1956, ibid.

89 See Nova Scotia Power Commission, minutes of meetings, in H.D. Hicks papers, MG2 vol. 1240, files 1 and 3, PANS.

90 Morley Taylor, Nova Scotia Power Commission, to Côté, 2 November 1956, RG84 vol. 998, file CBH68 vol. 1, NA.

91 Scott to Hutchison, 22 November 1956, ibid.

92 W.W. Mair, chief, Canadian Wildlife Service, 7 November 1956, RG22 vol. 366, file 307.16.2 pt. 1, NA.

93 Coleman to Hutchison, 7 November 1956, ibid.

94 Hutchison to Robertson, 7 November 1956, ibid.

95 Côté to Taylor, 8 November 1956, ibid.

96 Coleman to superintendent, 23 November 1956, RG84 vol. 998, file CBH68 vol. 1, NA. See also Lesage to Governor General in Council, 29 November 1956, RG22 vol. 366, file 307.16.2 pt. 1, NA.

97 In the 13 November minutes of the Nova Scotia Power Commission, Chairman Hicks mentions that he had been talking to Robert Winters, the federal minister from Nova Scotia, about the Parks Branch's recalcitrance, and thought the matter would be settled soon. H.D. Hicks papers, MG2 vol. 1240, file 3, PANS.

98 Simon P. Boudreau, secretary, Cheticamp Board of Trade, to Hamilton, 25 September 1957, RG84 vol. 998, file CBH68 vol. 1, NA.

99 Hamilton to Boudreau, 8 October 1957, ibid.
100 Coleman to Robertson, 16 October 1957, RG22 vol. 366 file 307.16.2 pt. 1, NA.
101 Robertson to Dinsdale, 21 October 1957, ibid.
102 See Robertson to Coleman, 6 November 1957, ibid.
103 On the Wreck Cove debate in the 1970s, see "Wreck Cove Hydro-Electric Investigation;" and *4th Estate*, 28 November 1974, 7 May 1975, 16 June and 1 December 1976, and 17 February 1977.
104 Bush, *A History of Hydro-Electric Development*, 114.
105 Federal-provincial agreement, 12 March 1957, cited in Bankes, "Constitutional Problems," 225. See also Lothian, *A Brief History*, 118.
106 Bankes, "Constitutional Problems," 225.
107 Canada, Department of Northern Affairs and National Resources, *Annual Reports* (1954–60); and Lothian, *A History*, 4:21. The parks also benefited from Prime Minister John Diefenbaker's federal-provincial winter works program, designed to foster employment during the recession of the late 1950s. Killan, *Protected Places*, 107.
108 Runte, *National Parks*, 107. Results of the Mission 66 program included 2800 miles of new and rebuilt roads, 936 miles of trail, and 575 campgrounds. Wright, *Wildlife Research*, 23.
109 Runte, *National Parks*, 107.
110 Canada, Department of Northern Affairs and National Resources, *Annual Report* (1959), 41.
111 Though the Higgs commissioners recommended excluding the island, they also suggested prices of $2900 for Cleve Robinson and $3200 for Percy MacAusland, the two owners. Robinson negotiated a $5000 settlement. MacAusland, though, refused to settle, and only accepted $5000 in 1942 with a promise from Premier Thane Campbell that this would not keep him from seeking more. Provincial secretary, PEINP files, RG7 series 14, box 30, subseries 1 file 749, and subseries 2 file 753, PAPEI. MacAusland argued the matter on and off for the next thirty years. See, for example, "Asks Queen's Help for Hearing," *Eastern Graphic*, 2 June 1976.
112 John William Robinson had bought the island in 1838, and the family owned all or part of it for the next century. Over the years it was known by a number of names, but atlases of the late nineteenth and early twentieth centuries call it Robinsons Island, and it went by that name during park expropriation. Recent lobbying by the Robinson family convinced Parks Canada to restore the name Robinsons Island in 1998. Charlottetown *Guardian*, 14 July 1998.
113 On the provincial tourism industry's desire to make access to the park easier, see PEI Innkeepers Association to Lesage, 30 May 1957, RG22 vol. 476, file 33.21.6 pt. 1, NA.

114 See Charlottetown *Patriot*, 14 February 1949, 30 July 1951, and 16 and
    17 August 1951.
115 Dr S.S. Masur, "Rustico Estuary Investigation, 1952–1953" report, cited
    in W.S. Veale, district director, Department of Public Works, to regional
    director, Department of Public Works, 14 May 1971, box 5, PEINP files.
116 The two positions are discussed in deputy minister Young to Winters,
    29 April 1953, RG22 vol. 476, file 33.21.6 pt. 1, NA; and W.S. Veale, "The
    Fisherman's Case," Veale to regional director, Department of Public
    Works, 14 May 1971, box 5, PEINP files. Of course, the fishermen's opin-
    ions were only given any credence in terms of fishing; the decision on
    the project's feasibility was the engineers'. Or as one Rustico fishermen
    put it bluntly, "No use talking to engineers, might as well talk to a gull."
    Beecher Court interview with Fred Horne, S2.290, PEINP files.
117 Hutchison to Jackson, 21 March 1955, RG22 vol. 476, file 33.21.6 pt. 1, NA.
118 Robertson report to Hutchison, 10 August 1955, RG22 vol. 317, file 33.21.1
    pt. 3, NA.
119 See Hutchison to Robertson, 27 June 1957, RG22 vol. 476, file 33.21.6 pt. 1,
    NA.
120 Lothian, *A History*, 4:21.
121 Lloyd Brooks, "Planning Considerations, Prince Edward Island National
    Park," January 1960, box 17, PEINP files. The long-term environmental
    effects of the Gulf Shore Parkway are discussed in Keith, *The Cumulative
    Effects*, 22–5.
122 Robertson report to Coleman, 6 August 1960, RG22 vol. 899, file 316.1 vol.
    1, NA.
123 Superintendent Kipping to Strong, 25 August 1960, box 17, PEINP files.
124 These figures are from assorted letters, maps, and photos, ibid. For re-
    ports on the erosion, see Charlottetown *Guardian*, "Robinsons Island
    Loses Another Section of Causeway," 30 March 1976; and "Fishermen
    Lay Blame on Man," 17 April 1976.
125 Charlottetown *Patriot*, 23 December 1971; *Journal-Pioneer*, 24 and
    27 December 1971.
126 Canada, House of Commons, *Debates*, 14 February 1970, 4048.

CHAPTER EIGHT

1 McNamee, "From Wild Places to Endangered Spaces," 28.
2 As early as 1938, Commissioner F.H.H. Williamson noted, "In Canadian
  Parks possibly the development has reached a point where we should
  call a halt on new work and confine ourselves simply to the improvement, completion, and maintenance of existing works." Williamson to
  Gibson, 4 January 1938, RG84 vol. 2101, file U172 (6), NA.

3 Power Bratton, "National Park Management," 119 and 126–7.

4 I acknowledge a danger in extending my argument into the United States, and even – as will be discussed in the following paragraph – locating my own work in relation to two historians writing on American parks. It is true that I am using the American case to learn more about Canada, and using the Canadian case to suggest something about the American one. However, I believe this can be defended. My research indicates many confluences between the American and Canadian situations in everything from predator policy to post–Second World War budget growth. On a related point, Canadian biologists in this period were generally trained in the States, and the Canadian Parks Branch depended on its American sister agency for advice in establishing and implementing policies. For secondary material, I was forced to depend on American sources because so little was available on Canadian national parks, wildlife, or ecological science from the 1930s to the 1960s. This book was written too late to incorporate Thomas Dunlap's history of environmental ideas in Canada, the US, New Zealand, and Australia, or Alexander Burnett's history of the Canadian Wildlife Service. For more on the relation between American and Canadian science in national parks, see my "Rationality and Rationalization."

5 Dunlap, "Wildlife, Science, and the National Parks," and "Ecology, Nature and Canadian National Park Policy."

6 Sellars, *Preserving Nature*.

7 Cautley report on Nova Scotia sites, in MacDonald, *Transportation in Northern Cape Breton*, 57, 58, and 76.

8 Williamson and Cromarty report on PEI sites, 28 July 1936, RG84 vol. 1777, file PEI2 vol. 1 pt. 2, NA.

9 Cautley report on New Brunswick sites (1930), 15 and 17, RG84 vol. 1964, file U2.13 vol. 1, NA.

10 There is a large literature on human perceptions of animals. A standard text, though defined more by its subtitle than its title, is Thomas, *Man and the Natural World: Changing Attitudes in England 1500–1800*. On animal conservation, see Livingston, *The Fallacy of Wildlife Conservation*; on animal rights, Singer, *Animal Liberation*; and on the history of environmental ethics, Mighetto, *Wild Animals and American Environmental Ethics*.

11 For the Wildlife Division and the origin of wildlife research in Canada, see Lothian, *A History*, 4:55–9; Pimlott, Kerswill, and Bider, *Scientific Activities*, 112; Cowan, "A Naturalist-Scientist's Attitudes," 93–6; and Lotenburg, "Wildlife Management Trends."

12 Lloyd's formal training was as a chemist, but he came to the Wildlife Division because of his interest in ornithology. On Lloyd, see Foster, *Working for Wildlife*, 159–61.

13 MacEachern, "Rationality and Rationalization," 204–5.

14 For ecology in this period, see Worster, *Nature's Economy.* On ecology in the national parks, see Sellars, *Preserving Nature,* and Dunlap, "Wildlife, Science, and the National Parks."

15 North American biologists were greatly affected by the lessons learned in the Kaibab National Forest in Arizona. US Biological Survey staff had cleansed this game preserve of all predators by 1920, and in subsequent years the deer population exploded. All foliage was soon picked clean, and by 1925 starvation wiped out much of the herd. In *Discordant Harmonies,* Botkin offers a valuable critique of the simplistic lessons learned from the Kaibab. Essentially, he points out that deer and predator populations do not simply move up and down in relation to one another. This ignores all other variables in their environment, and factors their "value" strictly by their absence or presence in the system. See also Dunlap, "That Kaibab Myth."

16 More research is needed on the effect that the Depression had on Canadian wildlife preservation policies. In the United States, it has been said of Franklin Roosevelt's 1930s that "No other decade or administration did so much to save wildlife." Worster, *An Unsettled Country,* 77. See also Cart, "A 'New Deal' for Wildlife."

17 Parks commissioner James Harkin was seen as Canada's greatest advocate of sanctuaries, in national parks as well as in provincial and public reserves. For Harkin's interest in sanctuaries, see Foster, *Working for Wildlife,* especially 88, 198, and 206, as well as his own "Wildlife Sanctuaries." On sanctuaries in general, see Gabrielson, *Wildlife Refuges;* and Worster, *Nature's Economy,* 259–60; and *An Unsettled Country,* 76–7.

18 Some hunters mistakenly believed that parks were sanctuaries *for* hunting rather than sanctuaries *from* them. In 1942, Ernest Smith, superintendent of Prince Edward Island National Park, thought this belief so prevalent that he asked his superiors if he could advertise in local papers that hunting was forbidden in the park. RG84 vol. 23, file PEI300, NA. When Fundy was established, Egbert Elliott of Alma wrote the National Parks Branch asking if he could set up a tourist business that would cater to hunters. RG84 vol. 1024, file F16, NA.

19 Cautley report on New Brunswick sites (1930), 1, RG84 vol. 1964, file U2.13 vol. 1, NA.

20 November 1939 report, RG84 vol. 1002, file CBH300 vol. 1 pt. 3, NA.

21 Following traditional notions about "good" and "bad" animals, special notice was made of predators in the parks. From Cape Breton Highlands, the August 1939 report states, "The predatory animals are bound to increase with the increase in game. Foxes and black bears seem to be increasing. There does not seem to be any necessity for controlling predators as long as wildlife is on the increase." Ibid. Much the same was

said in following months, but no predator eradication policies were introduced, or even suggested by the superintendent.

22 Superintendent J.P. MacMillan to Smart, 27 October 1941, ibid.

23 See RG84 vol. 140, file CBH272, NA for the beaver reintroduction. The
moose transfer is discussed in Williamson to Sarty, RG84 vol. 1002,
file CBH300 vol. 1 pt. 3, NA.

24 Sellars, *Preserving Nature*, 96–8.

25 Reintroduction of species killed off by locals in the past was also a way
of asserting the legitimacy of the park's takeover of the land and its
resources.

26 C.H.D. Clarke report, 23 March 1942, 5, RG84 vol. 1002, file CBH300 pt. 3,
NA.

27 Ibid., 7. In another case of rejecting exotic but otherwise desirable species
during this period, the Parks Branch declined the offers of Pheasants Unlimited on PEI and the provincial fish and game association of Nova
Scotia to let pheasants loose in the respective parks because they were not
native birds. Harrison Lewis of the Wildlife Division even suggested that
if pheasants began to arrive from outside the parks and adversely affected
native wildlife, they should be controlled. Lewis to Spero, 3 May 1946,
RG84 vol. 182, file PEI301, NA. Smart to Frank Nolan, president, Fish and
Game Protective Association, 3 December 1948, RG84 vol. 139,
file CBH301, NA.

28 Smart to Superintendent Smith, 16 September 1942, RG84 vol. 1802,
file PEI282, NA.

29 Smart to Smith, 16 November 1942, ibid. Smart did, however, indicate that
killing skunks might be justified in the future if they destroyed park property or were a serious nuisance.

30 Smart to Gibson, 23 September 1949, RG84 vol. 23, file PEI300, NA.

31 Canada, Departments of Resources and Development, Northern Affairs
and National Resources, and Indian Affairs and Northern Development,
Annual Reports (1948–69). Since there was a departmental reorganization
in 1947, and more items were targeted by this expenditure from then on, it
is difficult to know exactly how much of an increase this was.

32 Golley's *A History of the Ecosystem Concept* is helpful in demonstrating
how the size and budgets of scientific studies can shape the direction that
their work takes.

33 Smart to superintendents, 6 January 1948, RG84 vol. 2102, file U172 vol. 7,
NA.

34 Pimlott, Kerswill, and Bider, *Scientific Activities*, 112.

35 Press release, Deputy Minister Hugh Keenleyside, 11 June 1949, RG22
vol. 153, file 5.0.1.35 vol. 6, NA.

36 Lotenburg, "Wildlife Management Trends." Discussing the Canadian
situation generally, Stephen Bocking writes in "A Vision of Nature and

Society" that "most Canadian ecological research has been tied more or less closely to immediate resource management concerns" (16).

37 See chapter 14 of Worster, *Nature's Economy*, and Bocking, "A Vision of Nature and Society."

38 Lothian, *A History*, 4:59. Because the Parks Branch was ultimately responsible for management decisions, it is credited with the policies discussed in the remainder of this chapter. Of course, it should be noted that the Canadian Wildlife Service often played a critical role in decision-making on park wildlife issues, just as the Forestry Service did in park forestry matters – and politicians did to any matter in which they had an interest.

39 H.F. Lewis to George J. Keltie, president, Western Canada-Yukon Fish and Game Council, 31 March 1949, RG84 vol. 39, file U300 vol. 16, NA. See also Lewis to Smart, 4 March 1949, ibid.

40 Slaughters of elk, moose, and buffalo had been taking place in Western parks since the 1930s, and became annual affairs during the Second World War. These were justified (time and again) in local, practical terms; not until the 1950s did wildlife managers speak of overpopulation as the inevitable result of a park's existence.

41 "A Policy Statement Respecting Wildlife in the National Parks of Canada," Coleman to Hutchison, 21 January 1957, RG84 vol. 2140, file U300 pt. 18, NA.

42 John P. Kelsall report, "Mammal and Bird Survey, New Brunswick National Park, June 4 to July 4, 1948," to Smart, 29 October 1948, RG84 vol. 141, file F300, NA.

43 Edwards and Fowle's "The Concept of Carrying Capacity" discusses how a supposedly scientific term as this can lose precise meaning owing to its popularity and apparent universal applicability. In 1955 they were already writing that "most definitions of carrying capacity are vague and that some are almost meaningless" (589).

44 A.W.F. Banfield, "Fundy National Park Wildlife Investigations, March 12–17, 1951," RG84 vol. 1002, file CBH300 vol. 1 pt. 2, NA. Banfield did not yet recommend moose reduction.

45 Wright, *Wildlife Research*, 69.

46 J.S. Tener report, passed on by W.D. Taylor to Coleman, 20 August 1953, RG84 vol. 1039, file F300 vol. 1, NA.

47 J.A. Hutchison to Nason, legal advisor, 22 October 1953, RG84 vol. 1024, file F2 vol. 5, NA.

48 See RG84 vol. 486, file F216 pt. 1, NA.

49 "Moose reduction program, Fundy National Park, January 1955," ibid.

50 A.W.F. Banfield to Coleman, 28 May 1954, RG84 vol. 520, file CBH217 pt. 1, NA.

51 See R.A. Morrison, secretary, Cape Breton Fish and Game Association, to (?), 8 May 1954, ibid.; and Sydney *Post-Record*, 26 January 1955.

52 Tener to Lewis, 4 March 1952, RG84 vol. 1002, file CBH300 pt. 2 (1942–52), NA.
53 Banfield report, 2 November 1954, RG84 vol. 520, file CBH217 pt. 1, NA.
54 Doak to Coleman, 28 January 1955, ibid.
55 Research in the US has questioned wildlife managers' ability to judge browse quantity in the first place; what might seem to be overbrowsing may be quite natural for an area. Wright, *Wildlife Research*, 80.
56 There was discussion of this in the 1939 *Transactions of the Provincial-Dominion Wildlife Conference*.
57 F.A.G. Carter to Robertson, 23 September 1957, RG22 vol. 476, file 33.21.1 pt. 4, NA.
58 Superintendent F.C. Browning to Strong, 7 May 1958, RG84 vol. 1778, file PEI2 vol. 8, NA.
59 Many locals may have disliked this concession. In 1960, Superintendent Kipping wrote that "indeed there was local derision directed at the Park staff because of their inability to cope with the hunting." Cited by Coleman to Côté, 4 August 1964, RG84 vol. 1778, file PEI2 vol. 10, NA.
60 Smart to superintendent, 9 May 1951, RG84 vol. 141, file F300, NA; and Strong to superintendent, 15 December 1959, RG84 vol. 487, file F281 pt. 2, NA.
61 C.O. Bartlett, wildlife biologist, Canadian Wildlife Service, cited by Superintendent Kipping to Strong, 16 January 1963, RG84 vol. 1802, file PEI279, NA.
62 John P. Kelsall to Harrison Lewis, 9 October 1950, RG84 vol. 1802, file PEI275, NA.
63 Superintendent Kipping to "Gentleman and Olga" of CBC "Long Shot," 2 September 1959, box 17, whales and seals file, PEINP files.
64 Kipping to Strong, 13 November 1962, ibid.
65 Strong to superintendent, 15 December 1959, RG84 vol. 487, file F281 pt. 2, NA; and Smart to superintendent, 9 May 1951, RG84 vol. 141, file F300, NA.
66 Dinsdale to Ralph H. Olive, 8 June 1962, RG22 vol. 1083, file 300.52, NA.
67 See RG84 vol. 986, file CBH16.1 vol. 2 pt. 1, NA.
68 Sellars, *Preserving Nature*, 123. See also Runte, *Yosemite*, 65–6.
69 See Johnstone, *The Aquatic Explorers*.
70 Harkin to Gibson, 14 December 1934, RG84 vol. 983, file CBH2 vol. 1 no. 1, NA.
71 Cautley report on Nova Scotia sites, 58, in MacDonald, *Transportation in Northern Cape Breton*; and Cautley to Harkin, 16 June 1936, RG84 vol. 520, file CBH296 pt. 1, NA.
72 Cromarty and Williamson report on PEI sites, 11–12, RG84 vol. 1777, file PEI2 vol. 1 pt. 1, NA. They noted that this reserve should not include Rustico fishing grounds.
73 Wardle to William A. Found, deputy minister of fisheries, 6 November 1936, RG84 vol. 1802, file PEI296 vol. 1, NA.

74 Dr A.H. Leim, director, Atlantic Biological Station, 30 August 1939, RG84 vol. 1802, file PEI296 vol. 1, NA.

75 Canada, Departments of Mines and Resources, and Resources and Development, *Annual Reports* (1939–54). Salmon was the main fish stocked in early years, with Eastern brook trout increasingly stocked by the late 1940s. None of the Eastern parks had their own hatcheries, but the Branch's philosophy on stocking can be seen in Solman, Cuerrier, and Cable's "Why Have Fish Hatcheries."

76 Canada, Departments of Resources and Development, and Northern Affairs and National Resources, *Annual Reports* (1953–4).

77 J.V. Skiff's "Is There a Place for Stocking in Game Management?" (1948), delivered at the North American Wildlife Conference, discusses the evolution of conservation: from game protection, to game breeding and stocking, to game management, to wildlife management within a larger sphere of conservation of land and water coupled with good agricultural and forestry practices. Skiff argues that American managers had reached this final stage. In contrast, it would appear that Canadian policy of the period had only attained the stage of game management, in which it was believed all wildlife could be effectively managed.

78 Solman, "Limnological Investigations in Cape Breton Highlands National Park, Nova Scotia, 1947," RG84 vol. 140, file CBH296.12, NA.

79 In the 1950 film *Father of the Bride*, Elizabeth Taylor's character breaks off her engagement because her fiancé wants a Nova Scotia honeymoon, complete with fishing. This nicely alludes to Nova Scotia's popularity with sport fishermen, while at the same time suggesting that the province might need to remarket itself if it hoped to attract postwar American tourism.

80 Cautley report on New Brunswick sites (1930), RG84 vol. 1964, file U2.13 vol. 1, NA.

81 Solman to Lewis, 4 February 1948, RG84 vol. 1802, file PEI296 vol. 1, NA.

82 Ibid. See also *Hunting and Fishing*, November 1952.

83 Solman to Lewis, 11 February 1948, RG84 vol. 140, file CBH296.12, NA.

84 Williamson to MacKay, 28 March 1938, RG84 vol. 1802, file PEI296 vol. 1, NA. See also MacKay to Williamson, 10 June 1938, ibid.

85 Superintendent MacFarlane to Smart, 12 June 1952, RG84 vol. 200, file CBH3.1.1, NA. In this case, the decision backfired. A group of Americans arrived early the following summer, and were very upset that the fishing season was two weeks away. In the parks studied here, the fishing seasons constantly changed as the Parks Branch tried to find a period which would attract the most tourists and yet correspond closely to the provincial season.

86 See Lothian, *A History*, 4:60, 66–7; and Pimlott, Kerswill, and Bider, *Scientific Activities*, 62, for administrative changes in this period. See Bocking, "A Vision of Nature and Society," for changes in ecology.

87 MacKay to Williamson, 15 March 1938, RG84 vol. 1802, file PEI296 vol. 1, NA.

88 See Krumholz, "The Use of Rotenone."

89 Press release, 25 November 1954, RG22 vol. 156, file 5.0.1.35 vol. 14, NA.

90 Jean Lesage, "Canadian Wildlife Service," speech to the Canadian Institute of Surveying and Photogrammetry, Ottawa, 2 February 1956, RG22 vol. 159, file 5.0.1.39, NA.

91 This was the same phrase used in 1953 by J.C. Ward (biologist) and Jean-Paul Cuerrier (assistant limnologist) and in 1956 by Parks Branch director James Hutchison to describe the proposed use of rotenone in PEI National Park lakes. Yet in both cases, the speakers favoured the poisoning. See Ward report, 1953, RG84 vol. 1802, file PEI296.1, NA; and Hutchison to Pritchard, director, Department of Fisheries, 24 January 1956, ibid.

92 Ward report, 1953, ibid.

93 Cuerrier to Strong, 4 August 1960, RG84 vol. 520, file CBH296 pt. 2, NA.

94 Carter report, 1 August 1956, RG22 vol. 472, file 33.6.1 pt. 4, NA. According to the Wildlife Service, poisoning had not been considered. Coleman, 23 August 1956, ibid.

95 MacFarlane to Strong, 5 June 1958; and Strong to MacFarlane, 17 June 1958, RG84 vol. 487, file F296 pt. 1, NA.

96 In *A Walk in the Woods*, Bill Bryson tells the story of the US National Park Service's use of rotenone at Great Smokey Mountains National Park in 1957. Staff killed everything in Abram's Creek, including a fish thought extinct – which now was (97).

97 Strong to superintendent, 5 June 1959, RG84 vol. 1802, file PEI296 vol. 2, NA.

98 Roderick Nash's *The Rights of Nature* discusses the evolution of rights over time. Though animal rights are quite well established, the idea of rights for other living things is in its infancy. See Christopher D. Stone's influential "Should Trees Have Standing?"

99 In bringing Smart to the Parks Branch in 1930, deputy minister of mines and resources W.W. Cory wrote, "with the growing needs and responsibilities for the protection of the Parks' forests and development of silviculture methods involved in the growth of the Parks system, it becomes a necessity to have a fully qualified Forester appointed to this Branch." Cory to William Foran, secretary, Civil Service Commission, 10 April 1930, RG32 file 237, file 1888.02.29, NA.

100 Smart to Gibson, 14 January 1948, RG84 vol. 1039, file F200 vol. 1, NA.

101 Williamson also warned that "there shall be no repetition in Cape Breton Highlands National Park of the timber disposal complications experienced at Riding Mountain Park" in Manitoba. Williamson to Gibson, 4 December 1939, RG84 vol. 520, file CBH200 pt. 1, NA. When Riding Mountain was created in 1930, it was decided that traditional use of

park forests would be honoured, and sawmills were allowed to continue operation in the park. James Smart was the superintendent there in this period. Lothian, *A Brief History,* 78.

102 Smart to Williamson, 30 January 1940, RG84 vol. 520, file CBH200 pt. 1, NA.

103 W. Robinson to Smart, 15 October 1941, ibid.

104 Smart to superintendent, 19 February 1941, ibid. Also, D. Roy Cameron, Dominion forester, to J.C. Venness, Dominion Forestry Service, 4 February 1941, ibid.

105 Balsam have been known to take over areas wiped out by spruce budworm, and this may have been what Smart was referring to. He had to know, though, that this succession process was unusual. On foresters' poor understanding of succession in the early twentieth century, see Nancy Langston, *Forest Dreams, Forest Nightmares,* 122–34.

106 Smart to Gibson, 14 January 1948, RG84 vol. 1039, file F200 vol. 1, NA.

107 H.L. Holman to Smart, 1 December 1949, RG84 vol. 1802, file PEI181 vol. 1, NA.

108 Ibid.

109 Ibid.

110 Smart to Gibson, 3 December 1949, ibid. As a result of Holman's report and Smart's reply to it, the PEI nursery was abandoned. The Christmas tree business never materialized. J.C. Goodison, acting superintendent, to Coleman, 26 September 1955, ibid.

111 Robertson to Coleman, 20 August 1957, RG84 vol. 1786, file PEI28 vol. 1, NA. Complaints about park trees may be seen in North Shore Tourist Resort Operators to Department of Northern Affairs and Natural Resources, 8 October 1955, RG22 vol. 317, file 33.21.l pt. 3, NA; and acting superintendent H.A. Veber to Coleman, 30 June 1954, RG84 vol. 1786, file PEI28 vol. 1, NA.

112 Gibson to Timm, Dominion Forest Service, 17 May 1948, RG84 vol. 1039, file F200 vol. 1, NA; and Smart to Superintendent Saunders, 27 June 1952, ibid.

113 See Superintendent MacFarlane to Coleman, 25 May 1956, ibid.; Mullen, *Literature Review,* 71; and Burzynski, "Man and Fundy: Story Component Plan," Parks Canada, 1987, 43, internal document, Fundy National Park files. Those in charge of implementing the cut squeezed scientific benefit from even the most prosaic of forestry tasks. For example, at Bennett Brook "It is proposed to lop and scatter the brush on part of the area and pile and burn on the remainder to provide cost data and esthetic comparisons between the two brush disposal methods." D.J. Learmouth, forestry engineer, to Coleman, 20 October 1955, RG84 vol. 1039, file F200 vol. 1, NA.

114 Côté to Jackson, 11 September 1956, RG84 vol. 472, file 33.6.1 pt. 4, NA.

115 On spruce budworm in the Maritimes, see Lie, "The Spruce Budworm Controversy"; Restino, "The Cape Breton Island Spruce Budworm Infestation"; and May, *Budworm Battles*.

116 C.C. Smith report, December 1957, RG84 vol. 520, file CBH181.1 pt. 3, NA.

117 Learmouth report to Coleman, 24 October 1955, RG84 vol. 520, file CBH180.3 pt. 1, NA.

118 Learmouth report "A Fire Access Trail Program for Cape Breton Highlands National Park," 1956, RG84 vol. 8, file CBH62, NA.

119 Robertson to Hutchison, 28 December 1956, RG22 vol. 474, file 33.9.1 pt. 5, NA.

120 Fire was (and is) another troublesome forest management concern in national parks. By the time Robertson was writing, Cape Breton Highlands – its large, inaccessible interior susceptible to lightning strikes – had reported fires of varying sizes in 1939, 1943, 1945, 1947, 1949, and 1950. The 1947 fires burned 4330 acres, cost $21,000 to suppress, led to the evacuation of Pleasant Bay, and destroyed twenty homes there. And yet, by the 1950s plant biologists understood that the constant suppression of fires allowed forest litter and small brush to build up, making the next fire more likely to be a major one. Also, fires opened up areas for wildlife and young trees – by 1955, six-foot high spruce, fir, poplar, birch, and maples were taking over the Pleasant Bay burn area. The Wildlife Service explained all this to the Parks Branch in the hopes of showing how park forest management was interfering with wildlife management, but policy did not change. The Parks Branch had to show its commitment to the protection of local communities, and could not wait for fires to burn themselves out. More to the point, staff did not want to let fires burn: fires were seen as destroying forest beauty and utility, and had to be stopped. Harrison Lewis to Smart, 1 March 1952, RG84 vol. 2102, file U172 vol. 8, NA; and Smart to Hutchison, 13 March 1952, ibid. See Canadian Parks Service, *Keepers of the Flame*; and Wein, *Historical and Prehistorical Role of Fire*. For the history of North American attitudes to wildland fire, see Pyne, *Fire in America*.

121 This was a struggle faced by foresters across North America. See Langston, *Forest Dreams, Forest Nightmares*, 114–56.

122 Scribbler re Dalvay House information, PEINP files. For insect abatement in US national parks, see Runte, *Yosemite*, 176–8.

123 For pesticide use in postwar America, see Dunlap, *DDT*. On the relationship between war and pesticide, see Russell, "'Speaking of Annihilation.'"

124 Learmouth to Coleman, 17 October 1956, RG22 vol. 474, file 33.9.1 pt. 5, NA.

125 Coleman to Hutchison, 8 December 1955, RG22 vol. 473, file 33.9.1 pt. 4, NA.

126 Dunlap, *DDT*, 37; and Russell, "Speaking of Annihilation," 1525. In a 1946 issue of the *Journal of Wildlife Management* devoted to DDT, a leading biologist of the time made the pronouncement, "No chemical substance has excited greater general interest than DDT." Storer, "DDT and Wildlife," 181.

127 Cottam and Higgins, "DDT." See Smart to superintendent, 30 October 1946, RG84 vol. 1002, file CBH300 pt. 2 (1942–52), NA; and RG84 vol. 23 file PEI300, NA. Even prior to the civilian release of the chemical, the Canadian Provincial-Dominion Wildlife Conference passed a resolution stating, "Therefore, be it resolved, that this conference recommends a programme of scientific investigation of the complicated problems relating to DDT and wildlife before any large quantities of DDT are used for any large-scale commercial purpose, and also recommends that the use of DDT be kept under strict governmental control." February 1945, RG22 vol. 4 file 13, NA.

128 Smart to Harold Furst, manager, Seignory Club, Montebello, Quebec, 8 November 1948, RG84 vol. 39, file U300 vol. 16, NA. Smart wrote to find out how the Seignory Club kept insects away.

129 L.W. Ford to Coleman, 23 February 1955, RG84 vol. 1786, file PEI28 vol. 1, NA; and Canada, Department of Mines and Resources, *Annual Report* (1949), 22.

130 Doak told Coleman, "If you should decide to carry out an aerial spray probably the Province of Nova Scotia would co-operate and have the entire Island sprayed." 30 July 1956, RG84 vol. 520, file CBH181.1 pt. 3, NA. In this letter, he advocated the use of DDT, even though he had just received a letter from a New Brunswick forestry engineer expressly recommending the use of another spray, if any. C.C. Smith, Forest Biology Lab, Fredericton, to Doak, 20 July 1956, ibid.

131 Jackson to Hutchison, 1 August 1956, ibid.; and Hutchison to Coleman, 8 August 1956, ibid.

132 There was as yet no firm evidence that DDT posed a threat to humans. However, it was certainly suspected. Dunlap, *DDT*, 88–9. There was constant discussion of the risks and benefits of DDT in the North American Wildlife Conferences of the 1950s. It is telling that in a 1952 defence of the chemical, David G. Hall of the US Bureau of Entomology felt compelled to assure his readers, "DDT is not the human killer many people think it is." Hall, "Our Food Supply," 30.

133 Lothian, assistant chief, to Hutchison, 3 December 1956, rg22 vol. 474, file 33.9.1 pt. 5, NA.

134 Doak to Strong, 1 August 1957, RG84 vol. 520, file CBH181.1 pt. 3, NA.

135 See, for example, ibid.

136 Superintendent Browning to Strong, 23 July 1959, ibid. That staff had been expecting a mite outbreak can be seen in C.C. Smith, Forest Biology Lab, Fredericton, to Superintendent Doak, 20 July 1956, ibid.

137 Superintendent MacFarlane to Strong, 19 October 1961, RG84 vol. 1036, file F150.5 vol. 2, NA.

138 R.D. Muir, cited in Strong to superintendent, 30 July 1962, RG84 vol. 1802, file PEI181.1. vol. 2, NA.

139 Interview with Neil MacKinnon, Pleasant Bay, 4 August 1994. Three ex-staffers from Maritime parks described DDT use to me, but asked not to be quoted.

140 Don Morris, "Park Paradise Beckons," St John's *Evening Telegram*, 6 August 1960.

141 Carson, *Silent Spring*, 134–8.

142 Joseph G. Strauch Jr, Corvallis, Oregon, to David Munro, Canadian Wildlife Service, 12 September 1962, RG84 vol. 1002, file CBH301 pt. 2, NA.

143 Superintendent McCarron to Strong, 22 November 1962, ibid.

144 Narrator on "Diary of a Warden," CBC program "20/20", 1965. Thanks to Barb MacDonald of Prince Edward Island National Park for making this available to me.

145 Burzynski, "Man and Fundy," 50.

146 Sign at Point Wolfe Beach, Fundy National Park. The collapse of the North American peregrine falcon population was critical in convincing the US to ban DDT in 1972. See Dunlap, *DDT*, 130–2.

147 On science's cultural authority, see Dunlap's work, especially "Wildlife, Science, and the National Parks."

148 No one better exemplifies ecology's transformation than the American ecologist Aldo Leopold. His 1933 book *Game Management* taught a generation of administrators how to manage nature, but as early as the mid-1930s he had lost faith in humans' ability to do so. Until his death in 1948, Leopold sought to create a less intrusive relationship with nature. His final book, *A Sand County Almanac*, is a seminal text in the creation of the ecology movement.

149 Cited in Carter to Robertson, 4 December 1957, RG84 vol. 1944, file TN28 pt. 1, NA.

150 Robertson to Coleman, 7 October 1960, RG22 vol. 1097, file 311.1 pt. 1, NA.

151 Harold Eidsvik, planning officer, "Fundy National Park – Planning Considerations," February 1960, 1, FNP files. Assistant deputy minister C.W. Jackson noted, "At the present time, while the park is named for the Bay of Fundy, there are very few points within the park from which a view of the Bay can be obtained." Jackson to minister Lesage, 14 September 1956, RG22 vol. 472, file 33.6.1 pt. 4, NA.

152 Coleman "Addenda" to Eidsvik, "Fundy National Park – Planning Considerations," February 1960, FNP files. Coleman also said, "At one time it was considered good policy to make expenditures in the park readily apparent to the public; hence the prominent siting of the Superintendents' residences in some parks."

153 Robertson to Coleman, 19 October 1960, RG22 vol. 897, file 304.1 vol. 1, NA.

154 Muir to Stirrett, 20 January 1964, "Report on Field Trips, 1963," 84.

155 Canada, Northern Affairs and National Resources, *Annual Report* (1964), 17.

156 Ibid. (1963), 22. Appropriately enough, park recreation directors were replaced by naturalists in the early 1960s.

157 Concessionaires at Banff tried throughout the mid-1950s to win the right to build drive-in theatres within the townsite, but Parks Branch staff were adamant in their opposition. Drive-ins were a far cry from golf courses and swimming pools, which, as deputy minister Robertson reasoned, "are enjoyments that can be achieved only in large, open areas; they are the kind of physical and outdoor activities that (among others) the Parks are to provide." This is reasonable (though one could argue that drive-ins also encourage physical, outdoor activity). But Robertson added that in creating developments such as golf courses, "where well-handled, the impairment is not serious – in fact the clearing and opening of certain areas often increases the beauty and adds to the possibility of visual enjoyment." In the 1950s, a golf course could still infuse a park's nature with aesthetic value, whereas a drive-in could not. In time, not even golf courses would be safely ensconced in high culture. Robertson to Lesage, 5 June 1956, RG22 vol. 469, file 33.2.2 pt. 4, NA.

158 See Bourdieu, *Distinction*, 12–14.

159 Muir to Stirrett, 20 January 1964, "Report on Field Trips, 1963," 3.

160 H.S. Robinson, chief, Education and Information Section, to Coleman, 15 February 1962, RG84 vol. 1943, file TN16.112.1 vol. 1, NA.

161 Heppes, "Terra Nova National Park," 17.

162 Superintendent J.H. Atkinson to Strong, 21 January 1960, RG84 vol. 1952, file TN112 Year vol. 1, NA. See also Atkinson to Strong, ibid., and Strong to Atkinson, 25 November 1959, RG84 vol. 1954, file TN216, NA.

163 Lesage to W.J. Keough, 28 January 1957, RG22 vol. 474, file 33.11.1 pt. 2, NA.

164 Chief park naturalist George M. Stirrett to Brooks, 24 January 1964, RG84 vol. 1942, file TN2 vol. 5, NA.

165 Superintendent E.J. Kipping speech to Resource Development Council, PEI, 6 February 1962, RG84 vol. 1786, file PEI28 vol. 2 pt. 3, NA.

166 Laing to Leo Rossiter, minister of industry and natural resources, PEI, 3 February 1964, RG84 vol. 1779, file PEI2A pt. 4, NA.

167 R.P. Malis, regional engineer, to Coleman, 27 May 1964, RG84 vol. 1786, file PEI28 vol. 2 pt. 1, NA.

168 See Coleman to "Dr. Fischer," 14 December 1964, RG84 vol. 1802, file PEI181 vol. 2, NA.

169 Strong to Kipping, 1 December 1961, RG84 vol. 1779, file PEI2A pt. 3, NA. See also Coleman to Robertson, 30 January 1962, ibid.

170 Laing to J. Watson MacNaught, 12 September 1963, RG22 vol. 1103, file 330.3 pt. 1, NA.

171 See RG84 vol. 1779, file PEI2A pt. 4, NA for negotiations on this. Also, Charlottetown *Guardian*, 3 February 1964.

172 Canada, Parks Canada, *Prince Edward Island National Park Preliminary Master Plan* (1977), no pagination.

173 Nicol, "The National Parks Movement," 47; and McNamee, "From Wild Places to Endangered Spaces," 30. It is not clear whom McNamee is quoting, since this phrase does not appear in Laing's policy statement.

174 On the park system in the early 1960s, see Bella, *Parks for Profit*, 113–16; and McNamee, "From Wild Places to Endangered Spaces," 29–30. On the formation of the parks association (now the Canadian Parks and Wilderness Society), see Dubasek, *Wilderness Preservation*, 72–8. Bodsworth, "Beauty and the Buck," 25 and 41–66. The Bodsworth article received a great deal of discussion within the publicly minded Parks Branch. RG84 vol. 2103, file U172 vol. 12, NA contains considerable correspondence on it.

175 Bodsworth, "Beauty and the Buck," 25.

176 Canada, House of Commons, *Debates*, 18 September 1964, 8192.

177 Staff often quoted or referred to Laing's speech in correspondence. For example, Coleman to Superintendent McAuley, 21 May 1965, RG84 vol. 1802, file PEI1181 vol. 2, NA; and letter for Laing to Louis Robichaud, premier, New Brunswick, 8 October 1964, RG84 vol. 1039, file F200 vol. 4, NA.

CHAPTER NINE

1 Interview with Ralph Ford and Dennis Chaulk, Charlottetown (Nfld.), 19 August 1994.

2 Interview with Neil MacKinnon, Pleasant Bay, 4 August 1994.

3 Interview with Don Spracklin, Charlottetown (Nfld.), 18 August 1994.

4 Interview with Winnie Smith, Riverside-Albert, 31 August 1994.

5 For example, the ministry in the late 1950s agreed to introduce "work rotation" at Cape Breton Highlands National Park, whereby local labourers were hired and then replaced when they had worked enough to qualify for unemployment insurance. Hamilton to Robert Muir, MP for Cape Breton North and Victoria County, 30 January 1958, RG22 vol. 898, file 307.1, NA.

6 Local politicians would not have allowed otherwise. Cape Breton Highlands National Park, for example, was part of both Inverness and Victoria counties. In the mid-1960s, member of Parliament Robert Muir periodically asked in Parliament how many were hired for the park from each county. In 1963, the answer was forty-eight from Victoria, forty-five from Inverness. In 1967, there was even greater parity: forty-nine from

Victoria, fifty from Inverness. Canada, House of Commons, *Debates*,
2 December 1963, 5318, and 6 February 1967, 12670.

7 See Bob Moss, St John's *Evening Telegram*, 12 July 1968, and 30 April 1965.

8 Joey Smallwood, radio interview, 8 April 1970, Terra Nova National Park
file, Newfoundland Collection, St John's Public Library. In similar fashion, an *Evening Telegram* editorial of 9 May 1968 entitled "Those Obstinate
Federal Chaps" suggested, "let's do some horse trading next time and be
sure we get the best we can out of the federal branch."

9 In discussion on a proposed second national park for Prince Edward
Island in the mid-1960s, Premier Walter Shaw noted, "I am further surprised at the criteria imposing standards of a very high quality on sites for
National Parks. The standards have been apparently raised from those in
effect when the main National Park on Prince Edward Island was considered." Walter Shaw to Arthur Laing, 22 February 1966, Proposed Second
Prince Edward Island National Park files, Acc. 2617/3, PAPEI.

10 Robertson to Hutchison, 10 August 1955, RG22 vol. 473, file 33.9.1 pt. 4,
NA. See Hampton, "Opposition to National Parks."

11 Interview with an ex-staffer who requested anonymity.

12 See especially any of the parks' wildlife files, labelled "300" (such as RG84
vol. 1002, file CBH300, NA). At no time in the archival records did staff at
any of the four parks studied here believe that there was an epidemic of
poaching activity. Cape Breton Highlands superintendent Tim Reynolds
believes that today the amount of poaching is normal for a rural area, and
former Prince Edward Island superintendent Ernest Smith thinks it was
never much of a problem. Personal communication with Tim Reynolds,
3 August 1994. Interview with Ernest Smith, Tea Hill, 4 August 1995.

13 Interviews, for example, with Don Spracklin, Charlottetown (Nfld.),
18 August 1994; Wilf Aucoin, Cheticamp, 1 August 1994; and an individual at Cape Breton Highlands who requested anonymity.

14 A former superintendent of PEI National Park, Ernest Smith, told of
catching a man that he knew hunting in the park. Smith made him swear
that he would not do it again, and let him go. Interview with Smith, Tea
Hill, 4 August 1995. There is no way of knowing whether some staff themselves poached fish and wildlife in the parks, as some did in Western
Canadian parks. Wardens were originally hired in part because of their
knowledge of park land, which for some of them probably came by hunting. I discuss this issue in "Rationality and Rationalization," 207–8. My
favourite poaching story from the parks studied here was told at second-
hand about a staff member working at the tree nursery in Prince Edward
Island park. He went home each day with a sapling nestled in his
lunchbox.

15 Interview with Don Spracklin, Charlottetown (Nfld.), 18 August 1994.

16 Bella, *Parks for Profit*, 129.

17 Chrétien, "Our Evolving National Parks System."

18 Dacre, "Expropriation."

19 Todd, *The Law of Expropriation*, 2. See also Boyd, *Expropriation in Canada*, 2. The new Federal Compensation Act 1970 set compensation to be the higher of market value or the cost of establishing a similar property elsewhere.

20 Johnson to director, 17 September 1965, re "Cape Breton Highlands National Park, Boundary Revisions and Interior Development, A Preliminary Report, Report #40, April 1965," RG84 vol. 991 file, CBH28 vol. 4, NA. Cape Breton Highlands staff had noticed since the park had opened that Acadian fishermen appealed to tourists. In 1938, Superintendent J.P. Macmillan lobbied successfully to have the fishing shacks at Cap Rouge remain standing. He wrote, "I am of the opinion that these shacks serve as an added attraction to the Park from a tourist standpoint." Macmillan to Smart, 23 February 1942, RG84 vol. 520, file CBH296 pt. 1, NA. Over Superintendent Doak's protestations, the Parks Branch opted to tear the shacks down in the 1950s because they were deemed unattractive. Doak to chief, 11 May 1955, RG84 vol. 984, file CBH2 vol. 6, NA.

21 See Lothian, *A Brief History*, 122–9, for the Atlantic Canadian parks.

22 Lothian, *A History*, 4:26.

23 Some of the oral briefs were based on written ones. See Canada, National and Historic Parks Branch, *Transcript of Proceedings Fundy National Park Public Hearing* (1970).

24 The Sierra Club's brief, written at the Branch's request for the Fundy hearing, shows no sign of knowing anything about Fundy.

25 In analysis of the establishment of Gros Morne National Park, James Overton notes that in such matters, "'participation' is predominantly defined by the new administrative power system. They attempt to control the time, place and form of participation and also the issues that the public will be invited to express opinions on" (*Making a World of Difference*, 186).

26 Canada, National and Historic Parks Branch, *Transcript of Proceedings Cape Breton Highlands National Park Public Hearing* (1970), 25, 91, and 93; *Transcript of Proceedings Fundy*, 47, 50, 52, and 86.

27 Audley Haslam in *Transcript of Proceedings Fundy*, 116. The chair of the proceedings, director of the National Parks Branch John Nicol, disputed Haslam's contention. He said that at present, only 13 of the 125 working at Fundy were from outside New Brunswick. Better than 65 per cent of full-timers and 75 per cent of seasonals were from Albert County (164).

28 Fisher Hudson, MLA for Victoria County, in *Transcript of Proceedings Cape Breton Highlands*, 91.

29 *Transcript of Proceedings Fundy*, 25, 55–8, and 96.

30 John Hirtle, Voluntary Planning Board, Bridgewater, in *Transcript of Proceedings Cape Breton Highlands*, 25.

31 A Miss Mary Barker asked the chair of the Cape Breton Highlands public hearing, senior assistant deputy minister J.H. Gordon, a simple question: what species, what nature in Cape Breton's north was considered so precious that the park had to be expanded? Gordon fumbled, explaining that an inventory of flora and fauna had not yet been written up. But he then answered in the most basic aesthetic terms: "As you go gradually north, the scenery tends generally to become more and more spectacular, more and more rugged." Land that had been removed from park plans in the 1930s and refused by the park in the 1950s (as a trade for the loss of Cheticamp Lake), because it was considered ugly, was now, because the aesthetic had changed, spectacular enough to demand park expansion. *Transcript of Proceedings Cape Breton Highlands*, 25.

32 *Transcript of Proceedings Fundy*, 44. See also 47, 50, 52, and *Transcript of Proceedings Cape Breton Highlands*, 5.

33 Ibid., 20, 39, and 99; *Transcript of Proceedings Fundy*, 18, 93, and 131.

34 Ibid., 27, 72, and 138.

35 Maynard MacAskill, president of the North Victoria Landowners' Protective Association, *Transcript of Proceedings Cape Breton Highlands*, 57.

36 Ibid., 82.

37 W. Gwinn, member of the North Victoria Landowners' Protective Association, ibid., 66–7.

38 Canada, National and Historic Parks Branch, *Decisions Resulting … Fundy*, (1971); and Canada, National and Historic Parks Branch, *Decisions Resulting … Cape Breton Highlands* (1971).

39 See Reilly, "Planning for New National Parks."

40 See LaForest and Roy, *The Kouchibouguac Affair*, and Thomas, "The Kouchibouguac National Park Controversy." On park expropriation in general, see Maclean, "Leaving Behind a Bitter Legacy," 38.

41 Edward Gaunce, co-ordinator of the Kouchibouguac Committee for Justice for the Expropriates, said, "Let me say it in plain words. We either get the land back or we destroy it completely. Fire in the woods, oil to pollute the rivers." Vautour himself stated, "If they want the land back, they'll have to carry bodies out of here." Cited in Thomas, "The Kouchibouguac National Park Controversy," 12.

42 On the establishment of Gros Morne, see Overton, *Making a World of Difference*, chapter 8.

43 McNamee, "From Wild Places to Endangered Spaces," 33.

44 The nearest approximation was the burning of five toll booths at Prince Edward Island National Park over Christmas in 1975. This was in apparent protest over the implementation of entrance fees in Atlantic Canadian parks. (Visitors had paid to enter Western parks for the previous forty years.) See Charlottetown *Guardian*, 27 and 29 December 1975. Earlier that year, the *Guardian* had organized a petition to Prime Minister Pierre

Trudeau opposing the fees, arguing that tourists already paid a ferry rate to visit the Island, and should not have to pay again upon entering the park. The newspaper received four thousand names within four days. Letters of support were published, such as one that read, "We are happy to add our names to your telegram, and feel most strongly that such a levy should not be charged. The point that we are only one of two provinces where a fee must be paid to arrive and depart is an excellent one. … Yours truly, Mr. and Mrs. Gordon MacEachern, Mr. and Mrs. Arnold MacEachern." My grandparents and parents.

45 Planning Section report "Terra Nova National Park, Portals and Information Centres" (1959) RG84 vol. 1944, file TN28 pt. 2, NA.

46 For an introduction to island biogeography and patch dynamics in parks, see Theberge, "Ecology, Conservation and Protected Areas"; or Freemuth, *Islands under Siege*.

Bibliography

PRIMARY SOURCES

MANUSCRIPT

National Archives of Canada (Ottawa)
Records of the Canadian Park Service. RG84
Records of the Department of Indian Affairs. RG10
Records of the Department of Indian and Northern Affairs. RG22
Records of the Public Service Commission. RG32
R.B. Bennett papers. MG26 K
James Bernard Harkin papers. MG30 E169
William Lyon Mackenzie King papers. MG26 J
Wilfrid Laurier papers. MG26 G
Hoyes Lloyd papers. MG30 E441
Arthur Meighen papers. MG26 I
James Smart papers. MG30 E545
Robert J.C. Stead papers. MG30 D74

Public Archives of Nova Scotia (Halifax)
H.D. Hicks papers. MG2 vol. 1240
Oxford Paper Co. arbitration case papers. RG10 series B, vols. 200–206

Cape Breton Highlands National Park (Ingonish)
Campbell, Judith V. "A Report on the Human History of Cape Breton Highlands National Park." 2 volumes. No date
Cape Breton Highlands National Park files

Public Archives of Prince Edward Island (Charlottetown)
Prince Edward Island. Executive Council. *Minutes.* RG7 series 3, box 12 no. 2640
– Provincial Secretary. Prince Edward Island National Park files. RG7 series 14, box 30
– Speeches of the Legislature, 1936–60. Legislative Library Material. RG10 vols. 102–104a
Proposed Second Prince Edward Island National Park files. Acc. 2617/3
Thane Campbell papers. RG25 vol. 32
Harry T. Holman papers. Acc. 4420 vol. 7
J. Walter Jones papers. RG25 vol. 33

Prince Edward Island National Park (Dalvay)
"Diary of a Warden." CBC program "20/20"
Prince Edward Island National Park files

Provincial Archives of New Brunswick (Fredericton)
New Brunswick. Department of Natural Resources. Fundy National Park Land Assembly Records. RG10 RS145

Fundy National Park (Fundy headquarters)
Allardyce, Gilbert. "The Salt and the Fir: Report on the History of the Fundy Park Area." Internal document 1969
Burzynski, Michael. "Man and Fundy: Story Component Plan." Internal document 1987
Eidsvik, H.K. "Fundy National Park – Planning Considerations." Internal document 1960

Provincial Archives of Newfoundland and Labrador (St John's)
Reid Company papers, W. Angus Reid files. MG17 pt. 3

Newfoundland Collection, St John's Public Library (St John's)
Terra Nova National Park files

Centre for Newfoundland Studies Archives (St John's)
J.R. Smallwood collection. Acc. 3.25.00

Queen's University Archives (Kingston)
Thomas A. Crerar papers. Acc. 2117

Whyte Museum of the Canadian Rockies Archives (Banff)
Fergus Lothian research papers. M113 Acc. 1947

British Columbia Archives (Victoria)
R.W. Cautley papers. E / C / C31

Private Collections
Dorothy Barbour papers. In the possession of Robin Winks, New Haven, Connecticut
Fundy National Park clipping collection of Greta Geldart Elliott. In the possession of Larry Hughes, Riverview, New Brunswick

PRINTED

Government Sources
Canada. *Annual Departmental Reports.* 1914–75
– Department of Indian Affairs and Northern Development. National and Historic Parks Branch. *Decisions Resulting from the Public Hearings on the Provisional Master Plan for Cape Breton Highlands National Park.* 1971
– Department of Indian Affairs and Northern Development. National and Historic Parks Branch. *Decisions Resulting from the Public Hearing on the Provisional Master Plan for Fundy National Park.* 1971
– Department of Indian Affairs and Northern Development. National and Historic Parks Branch. *Fundy Master Plan.* 1966
– Department of Indian Affairs and Northern Development. National and Historic Parks Branch. *Transcript of Proceedings Cape Breton Highlands National Park Public Hearing, June 24, 1970.* 1970
– Department of Indian Affairs and Northern Development. National and Historic Parks Branch. *Transcript of Proceedings Fundy National Park Public Hearing, October 24, 1970.* 1970
– Department of Indian Affairs and Northern Development. Parks Canada. *Prince Edward Island National Park Preliminary Master Plan.* 1977
– Department of Indian Affairs and Northern Development. Parks Canada. *Resource Inventory and Analysis, Prince Edward Island National Park.* 1977
– Department of Resources and Development. National Parks Branch. *Fundy National Park.* Opening ceremonies brochure. 1950
– Department of Resources and Development. National Parks Branch. *Playgrounds of Eastern Canada.* No date
– Dominion Bureau of Statistics. *Census of Canada.* 1931
– House of Commons. *Debates.* 1887–1975
– Parliament of Canada. *National Parks Act.* 20–21 George V. Chapter 33. 1930
– Senate. *Report and Proceedings of the Special Committee on Tourist Traffic.* 1934
New Brunswick. Department of Lands and Mines. *Annual Reports.* 1945–52
– Legislative Assembly. *Report of the Committee on Reconstruction.* 1944
– Legislative Assembly. *Synoptic Reports.* 1927–52

Newfoundland. Department of Mines and Resources. *Annual Reports*. 1955–60
– House of Assembly. *Proceedings*. 1955–8
– *Report on Meetings between Delegates from the National Convention of Newfoundland and Representatives of the Government of Canada*. Summary of proceedings and appendices, 25 June–29 September 1947
Nova Scotia. General Assembly. *Acts*. 1935
– General Assembly. *Public Accounts*. 1936–45
United States. Department of the Interior. *Annual Report*. 1911–20

Newspapers and Magazines
Bulletin (Victoria-Inverness). 1934–9
The Busy East of Canada (Sackville, NB). 1919–35
Chronicle (Halifax). 1934–9
Chronicle-Herald (Halifax). Scattered dates, 1954–6
Citizen (Ottawa). 28 January 1954
Daily Province (Vancouver). 2 February 1930
Eastern Graphic (Montague, Prince Edward Island). 2 June 1976
Evening Telegram (St John's). 1955–62
4th Estate (Halifax). 28 November 1974, 7 May 1975, 16 June and 1 December 1976, and 17 February 1977
Guardian (Charlottetown). 1935–1939, scattered dates 1961–76
Herald (Calgary). 18 November 1930
Herald (Halifax). 1934–9
Hunting and Fishing. November 1952
Island Farmer (Charlottetown). 15 April 1936
Journal (Ottawa). 11 June 1923
Journal-Pioneer (Summerside, Prince Edward Island). 24 and 27 December 1971
Mail and Empire (Toronto). 10 January 1930
Patriot (Charlottetown). 1934–9, scattered dates 1961–76
Post-Record (Sydney). 1936–8, and 26 January 1955
Star (Toronto). 13 January 1930
Telegraph Journal (Saint John). 1929–31, 1935–7, 1947–50
Times (New York). 12 March 1995
Western Star (Corner Brook). 21 March 1955

Other Primary Sources
Bishop, Elizabeth. "Cape Breton." *A Cold Spring*. Reprinted in *Elizabeth Bishop: The Complete Poems 1927–1979*. New York: Noonday Press of Farrar, Straus, and Giroux 1979: 67–8
Bodsworth, Fred. "Beauty and the Buck: A Holiday through Our Magnificent National Parks." *Maclean's* 23 March 1963: 25 and 41–6
Bourinot, J.G. "Notes of a Ramble through Cape Breton." *New Dominion Monthly* Montreal 1868

Brinley, Gordon. *Away to Cape Breton.* Toronto: McClelland and Stewart 1936

Browne, N. Milton. "A Great Sanctuary in Nova Scotia." *Forest and Outdoors* 1935: 804–8

Carson, Rachel. *Silent Spring.* Boston: Houghton Mifflin Co. 1962

Cautley, R.W. "Report on Examination of Sites for a National Park in the Province of Nova Scotia." In R.H. MacDonald. *Transportation in Northern Cape Breton.* Appendix A. Ottawa: Parks Canada 1979: 49–76

Chrétien, Hon. Jean. "Our Evolving National Parks System." *The Canadian National Parks: Today and Tomorrow.* Vol. 1. Ed. J.G. Nelson and R.C. Scace. Calgary: National and Provincial Parks Association of Canada and the University of Calgary 1969: 7–14

Clarke, C.H.D. "Fluctuations in Populations." *Journal of Mammalogy* 30, no. 1 (February 1949): 21–5

Cooper, Laurie Armstrong, and Douglas Clay. *History of Logging and River Driving in Fundy National Park: Implications for Ecological Integrity of Aquatic Ecosystems.* Alma, NB: Parks Canada, Atlantic Region 1997

Cottam, Clarence, and Elmer Higgins. "DDT: Its Effects on Fish and Wildlife." Washington: US Fish and Wildlife, Department of Interior 1946

Cowan, Ian McTaggert. "A Naturalist-Scientist's Attitudes towards National Parks." *Canadian Audubon* 26 (May 1964): 93–6

Dacre, Douglas. "Expropriation: The Fear and the Facts." *Maclean's* 8 November 1958: 22–3, 81–2, 84–6

Edwards, R.Y., and C. David Fowle. "The Concept of Carrying Capacity." *Transactions of the Twentieth North American Wildlife Conference.* Washington, DC: Wildlife Management Institute 1955: 589–602

"Fundy National Park: Canada's Newest National Playground." *Canada-West Indies Magazine* 91, no. 6 (June 1951): no pagination

Guide Book to Cape Breton, Royal Province of Nova Scotia or New Scotland, Dominion of Canada. London: Letts, Son and Co. 1883

Hall, David G. "Our Food Supply, Wildlife Conservation, and Agricultural Chemicals." *Transactions of the Seventeenth North American Wildlife Conference.* Washington, DC: Wildlife Management Institute 1952: 26–33

Harkin, James B. "Canadian National Parks and Playgrounds." *Annual Report* of the Historic Landmarks Association of Canada 1921: 36–9

– *The History and Meaning of the National Parks in Canada, Extracts from the Papers of the Late Jas. B. Harkin, First Commissioner of the National Parks of Canada.* Compiled by Mabel B. Williams. Saskatoon: H.R. Larson Publishing Co. 1987

– "Our Need for National Parks." *Canadian Alpine Journal* 9 (1918): 98–106

– "Wildlife Sanctuaries." *Rod and Gun in Canada* October 1919

Heppes, J.B. "Terra Nova National Park." *Newfoundland Journal of Commerce* 35, no. 7 (July 1968): 16–22

Hewitt, C. Gordon. *The Conservation of the Wildlife of Canada*. New York: Scribner's 1921

Hiller, J.K., and M.F. Harrington, eds. *The Newfoundland National Convention 1946–1948*. Vol.2. St John's: Memorial University of Newfoundland and McGill-Queen's University Press 1995

Jenkins, F.T. *Report on Forest Survey of the Proposed National Park Area, Bonavista Bay, Newfoundland*. St John's: Government of Newfoundland 1955

Keith, Todd L. *The Cumulative Effects of Development and Land Use at Prince Edward Island National Park*. Parks Canada Technical Reports in Ecosystem Science no. 0002. Halifax: Parks Canada, Atlantic Region 1996

Kennedy, Howard, D. Roy Cameron, and Roland C. Goodyear. *Report of the Royal Commission on Forestry*. St John's: Government of Newfoundland 1955

Krumholz, Louis A. "The Use of Rotenone in Fisheries Research." *Journal of Wildlife Management* 12, no. 3 (July 1948): 305–17

LaForest, G.V., and M.K. Roy. *The Kouchibouguac Affair: The Report of the Special Inquiry on Kouchibouguac National Park*. Fredericton: no publisher given 1981

Lawson, Jessie, and Jean Maccallum Sweet. *This Is New Brunswick*. Toronto: Ryerson Press 1951

Maclean, Norman. *A River Runs through It and Other Stories*. Chicago: University of Chicago Press 1976

MacMillan, A.S. "A Dream Come True: Story of the Development of Tourist Industry in Northern Inverness and Victoria Counties." 1952. Reprinted as "Cabot Trail: A Political Story." *Cape Breton's Magazine* 62 (1993): 2, 66–70

Maxwell, Lilian. *'Round New Brunswick Roads*. Toronto: Ryerson Press 1951

Morley, Margaret Warner. *Down North and Up Along*. New York: Dodd, Mead, and Co. 1900

Mullen, C.E. *Literature Review, Fundy National Park*. Halifax: Parks Canada, 1974

Odum, Eugene. *Ecology: The Link between the Natural and Social Sciences*. 2nd ed. New York: Holt, Rinehart and Winston 1975 [1963]

Pimlott, D.H., C.J. Kerswill, and J.R. Bider. *Scientific Activities in Fisheries and Wildlife Resources*. Special study #15. Ottawa: Science Council of Canada 1971

Rubio, Mary, and Elizabeth Waterston, eds. *The Selected Journals of L.M. Montgomery*. Vol. 1–3. Toronto: Oxford University Press 1985, 1987, 1992

Scobie, Maureen. "Stories from the Clyburn Valley." *Cape Breton's Magazine* 49 (1988): 1–20

Solman, V.E.F., J.P. Cuerrier, and W.C. Cable. "Why Have Fish Hatcheries in Canada's National Parks." *Transactions of the Seventeenth North American Wildlife Conference*. Washington, DC: Wildlife Management Institute 1952: 226–33

Stead, Robert J.C. *Canada's Maritime Playgrounds.* Ottawa: Department of Mines and Resources 1938

Storer, Tracy I. "DDT and Wildlife." *Journal of Wildlife Management* 10, no. 3 (July 1946): 181–3

Transactions of the North American Wildlife Conference. Washington, DC: Wildlife Management Institute 1938–60

Vernon, C.W. *Cape Breton at the Beginning of the ... Twentieth Century.* Toronto: Nation Publishing 1903

Walworth, Arthur. *Cape Breton: Isle of Romance.* Toronto: Longmans, Green and Co. 1948

Warner, Charles Dudley. *Baddeck and That Sort of Thing.* Boston: J.R. Osgood 1874

Webb, Walter Prescott. *The Great Plains.* Boston: Ginn 1931

Williams, Mabel. *Guardians of the Wild.* London, Ont.: Thomas Nelson and Sons 1936

"Wreck Cove Hydro-Electric Investigation." *Cape Breton's Magazine* 9 (October 1974): 4–11

SECONDARY SOURCES

Albanese, Catharine L. *Nature Religion in America: From the Algonkian Indians to the New Age.* Chicago: University of Chicago Press 1990

Alderson, Lucy, and John Marsh. "J.B. Harkin, National Parks and Roads." *Park News* 15, no. 2 (Summer 1979): 9–16

Allardyce, Gilbert. "'The Vexed Question of Sawdust': River Pollution in Nineteenth-Century New Brunswick." *Dalhousie Review* 52 (1972): 177–89. Reprinted in *Consuming Canada: Readings in Environmental History.* Ed. Chad Gaffield and Pam Gaffield. Toronto: Copp Clark 1995: 119–30

Altmeyer, George. "Three Ideas of Nature in Canada, 1893–1914." *Journal of Canadian Studies* 11, no. 3 (August 1976): 21–6

Bankes, N.D. "Constitutional Problems Related to the Creation and Administration of Canada's National Parks." *Managing Natural Resources in a Federal State.* Ed. J. Owen Saunders. Essays from the Second Banff Conference on Natural Resources Law. Toronto: Carswell Press 1986: 212–34

Barclay, James A. *Golf in Canada: A History.* Toronto: McClelland and Stewart 1992

Barrell, John. *The Dark Side of the Landscape.* Cambridge: Cambridge University Press 1980

Barry, Sandra. "The Art of Remembering: The Influence of Great Village, Nova Scotia, on the Life and Works of Elizabeth Bishop." *Nova Scotia Historical Review* 11, no. 1 (June 1991): 2–37

"Becoming Martin Luther King, Jr. – Plagiarism and Originality: A Round Table." *Journal of American History* 78, no. 1 (June 1991): 11–123

Belasco, Warren James. *Americans on the Road: From Autocamp to Motel, 1910–1945*. Cambridge, Mass.: MIT Press 1979

Bella, Leslie. *Parks for Profit*. Montreal: Harvest House 1986

Berger, John. *About Looking*. New York: Pantheon 1979

Bermingham, Ann. *Landscape and Ideology: The English Rustic Tradition, 1740–1860*. Berkeley: University of California Press 1986

Blodgett, Peter. "Striking a Balance: Managing Concessions in the National Parks, 1916–1933." *Journal of Forest and Conservation History* 34, no. 2 (April 1990): 60–8

Bocking, Stephen. "A Vision of Nature and Society: A History of the Ecosystem Concept." *Alternatives* 20, no. 3 (1993): 12–18

Botkin, Daniel. *Discordant Harmonies: A New Ecology for the Twenty-First Century*. New York: Oxford University Press 1990

Bourdieu, Pierre. *Distinction: A Social Critique of the Judgment of Taste*. Trans. Richard Nice. Cambridge, Mass: Harvard University Press 1984

Boyd, Kenneth J. *Expropriation in Canada: A Practitioner's Guide*. Aurora, Ont.: Canada Law Book 1988

Boyer, Paul. *Urban Masses and Moral Order in America, 1820–1920*. Cambridge, Mass.: Harvard University Press 1978

Brown, Dona. *Inventing New England: Regional Tourism in the Nineteenth Century*. Washington: Smithsonian 1995

Brown, Robert Craig. "The Doctrine of Usefulness: Natural Resources and National Park Policy in Canada, 1887–1914." *The Canadian National Parks: Today and Tomorrow*. Vol. 1. Ed. J.G. Nelson and R.C. Scace. Calgary: National and Provincial Parks Association of Canada and the University of Calgary 1969: 94–110

Bryson, Bill. *A Walk in the Woods*. Toronto: Doubleday Canada 1997

Brym, Robert J., and James Sacouman, eds. *Underdevelopment and Social Movements in Atlantic Canada*. Toronto: New Hogtown Press 1979

Buell, Lawrence. *The Environmental Imagination: Thoreau, Nature Writing, and the Formation of American Culture*. Cambridge, Mass.: Belknap Press of Harvard University Press 1995

Burke, Edmund. *A Philosophical Enquiry into the Origin of Our Ideas of the Sublime and Beautiful*. Ed. Adam Phillips. Oxford: Oxford University Press 1990 [1757]

Burrill, Gary, and Ian McKay, eds. *People, Resources, and Power: Critical Perspectives on Underdevelopment and Primary Industries in the Atlantic Region*. Fredericton: Acadiensis Press for the Gorsebrook Research Institute 1987

Burzynski, Michael. *A Guide to Fundy National Park*. Vancouver and Toronto: Douglas and McIntyre, in co-operation with Parks Canada 1985

Bush, Edward T. *A History of Hydro-Electric Development in Canada*. Ottawa: Historic Sites and Monuments Board 1986

Canadian Parks Service. *Keepers of the Flame: Implementing Fire Management in the Canadian Parks Service*. Ottawa: Canadian Parks Service 1989

Cart, Theodore W. "A 'New Deal' for Wildlife: A Perspective on Federal Conservation Policy, 1933–1940." *Pacific Northwest Quarterly* 62 (July 1972): 113–20

Chiasson, Father Anselme. *Cheticamp: History and Acadian Tradition*. Trans. Jean Doris LeBlanc. St John's: Breakwater Press 1986

Colpitts, Nancy. "Alma, New Brunswick and the Twentieth Century Crisis of Readjustment: Sawmilling Community to National Park." Master's thesis, Dalhousie University 1983

– "Sawmills to National Park: Alma, New Brunswick, 1921–1947." *Trouble in the Woods: Forest Policy and Social Conflict in Nova Scotia and New Brunswick*. Ed. L. Anders Sandberg. Fredericton: Acadiensis Press 1992: 90–109

Cook, Ramsay. "Landscape Painting and National Sentiment in Canada." *Historical Reflections* 1, no. 2 (1974): 263–83

Cosgrove, Denis, and Stephen Daniels, eds. *The Iconography of Landscape: Essays on the Symbolic Representation, Design, and Use of Past Environments*. Cambridge: Cambridge University Press 1988

Creighton, Wilfred. *Forestkeeping: A History of the Department of Lands and Forests in Nova Scotia, 1926–1969*. Halifax: Nova Scotia, Department of Lands and Forests 1988

Creighton, Wilfred, with Kenneth Donovan (ed.). "Wilfred Creighton and the Expropriations: Clearing Land for the National Park, 1936." *Cape Breton's Magazine* no. 69 (1995): 1–19

Cronon, William. *Nature's Metropolis*. New York: W.W. Norton and Co. 1991

– "The Trouble with Wilderness; or, Getting Back to the Wrong Nature." *Uncommon Ground: Toward Reinventing Nature*. Ed. William Cronon. New York: W.W. Norton and Co. 1995: 69–90

– "The Uses of Environmental History." *Environmental History Review* 17, no. 3 (Fall 1993): 1–22

Daniels, Stephen. *Fields of Vision: Landscape Imagery and National Identity in England and the United States*. Cambridge: Polity Press 1993

Davis, Donald F. "Dependent Motorization: Canada and the Automobile to the 1930s." *Journal of Canadian Studies* 21, no. 3 (Autumn 1986): 106–32

Dearden, Philip. "Philosophy, Theory, and Method of Landscape Evaluation." *Canadian Geographer* 29, no. 3 (1985): 263–5

Dearden, Philip, and Rick Rollins, eds. *Parks and Protected Areas in Canada*. Toronto: Oxford University Press 1993

Demars, Stanford. "Romanticism and American National Parks." *Journal of Cultural Geography* 11, no. 1 (Fall/Winter 1990): 17–24

Dilsaver, Lary M., ed. *America's National Park System: The Critical Documents*. Maryland: Rowman and Littlefield Publishers 1993

- "Stemming the Flow: The Evolution of Controls on Visitor Numbers and Impact in National Parks." *The American Environment: Interpretations of Past Geographies*. Ed. Lary M. Dilsaver and Craig E. Colten. Maryland: Rowan and Littlefield Publishers 1992: 235–55

Dubasek, Marilyn. *Wilderness Preservation: A Cross-Cultural Comparison of Canada and the United States*. New York: Garland Publishing 1990

Dunlap, Thomas R. *DDT: Scientists, Citizens, and Public Policy.* Princeton: Princeton University Press 1981

- "Ecology, Nature, and Canadian National Park Policy: Wolves, Elk, and Bison as a Case Study." *To See Ourselves/To Save Ourselves: Ecology and Culture in Canada*. Ed. Rowland Lorimer et al. Proceedings of the Annual Conference of the Association for Canadian Studies, 31 May–1 June 1990. Montreal: Association for Canadian Studies 1991: 139–47
- *Saving America's Wildlife*. Princeton: Princeton University Press 1988
- "That Kaibab Myth." *Journal of Forest History* 32 (1988): 60–8
- "Wildlife, Science, and the National Parks, 1920–1940." *Pacific Historical Review* (May 1990): 187–202

Easterbrook, Gregg. *A Moment on the Earth: The Coming Age of Environmental Optimism*. New York: Viking 1995

Evernden, Neil. *The Natural Alien: Humankind and Environment*. Toronto: University of Toronto Press 1985

- *The Social Creation of Nature*. Baltimore: Johns Hopkins University Press 1992

Febvre, Lucien, in collaboration with Lionel Bataillon. *A Geographical Introduction to History.* Trans. E.G. Mountford and J.H. Paxton. New York: Barnes and Noble 1966 [1932]

Flores, Dan. "Place: An Argument for Bioregional History." *Environmental History Review* 18, no. 4 (Winter 1994): 1–18

Forbes, Ernest R. "Cutting the Pie into Smaller Pieces: Matching Grants and Relief in the Maritime Provinces during the 1930s." *Acadiensis* 17, no. 1 (Autumn 1987): 34–55

Foster, Janet. *Working for Wildlife: The Beginning of Preservation in Canada*. 2nd ed. Toronto: University of Toronto Press 1998 [1978]

Freemuth, John C. *Islands under Siege: National Parks and the Politics of External Threats*. Lawrence: University Press of Kansas 1994

Gabrielson, Ira N. *Wildlife Refuges*. New York: Macmillan 1943

Gaffield, Chad, and Pam Gaffield. "Introduction." *Consuming Canada: Readings in Environmental History.* Ed. Chad Gaffield and Pam Gaffield. Toronto: Copp Clark 1995: 1–7

Gillis, R. Peter, and Thomas R. Roach. "The American Influence on Conservation in Canada 1899–1911." *Journal of Forest and Conservation History* 30, no. 4 (October 1986): 160–74

Glacken, Clarence. *Traces on the Rhodian Shore: Nature and Culture in Western Thought From Ancient Times to the End of the Eighteenth Century.* Berkeley: University of California Press 1967

Glassfore, Larry A. *Reaction and Reform: The Politics of the Conservative Party under R.B. Bennett, 1927–1938.* Toronto: University of Toronto Press 1992

Golley, Frank B. *A History of the Ecosystem Concept in Ecology: More Than a Sum of the Parts.* New Haven: Yale University Press 1993

Gosling, F.G. *Before Freud: Neurasthenia and the American Medical Community, 1870–1910.* Urbana: University of Illinois Press 1987

Granatstein, J.L. *The Ottawa Men: The Civil Service Mandarins, 1933–1957.* Toronto: Oxford University Press 1982

Grant, J.L., and G. Wall. "Visitors to Point Pelee National Park: Characteristics, Behaviour, and Perceptions." *Recreational Land Use: Perspectives on Its Evolution in Canada.* Ed. Geoffrey Wall and John Marsh. Ottawa: Carleton University Press 1982: 117–26

Gray, James H. *R.B. Bennett: The Calgary Years.* Toronto: University of Toronto Press 1991

Gwyn, Richard. *Smallwood: The Unlikely Revolutionary.* 2nd ed. Toronto and Montreal: McClelland and Stewart 1972 [1968]

Hall, C. Michael, and John Shultis. "Railways, Tourism and Worthless Lands: The Establishment of National Parks in Australia, Canada, New Zealand and the United States." *Australian-Canadian Studies* 8, no. 2 (1991): 57–74

Hampton, H. Duane. "Opposition to National Parks." *Journal of Forest History* 25, no. 1 (1981): 36–45

Hawkins, John. *The Life and Times of Angus L.* Windsor, NS: Lancelot Press 1969

Hays, Samuel P. *Conservation and the Gospel of Efficiency: The Progressive Conservation Movement 1890–1920.* Cambridge, Mass.: Harvard University Press 1959

– "From Conservation to Environment: Environmental Politics in the US since WWII." *Environmental Review* 6 (Fall 1982): 14–41

Hays, Samuel P., in collaboration with Barbara Hays. *Beauty, Health and Permanence: Environmental Politics in the United States, 1955–1985.* Cambridge: Cambridge University Press 1987

Henderson, Gavin. "James Bernard Harkin: The Father of Canadian National Parks." *Borealis* Fall 1994: 28–33

Hennessey, Catherine, and Edward MacDonald. "Arthur Newbery and the Greening of Queen Square." *Island Magazine* 28 (Fall/Winter 1990): 25–9

Horne, Fred. *Human History: Prince Edward Island National Park.* Charlottetown: Parks Canada 1979

Hummel, Don. *Stealing the National Parks: The Destruction of Concessions and Park Access.* Bellevue, Wash.: Free Enterprise Press 1987

Hunt, William R. *Stef: A Biography of Vilhjalmur Stefansson, Canadian Arctic Explorer.* Vancouver: University of British Columbia Press 1986

Hussey, Christopher. *The Picturesque: Studies in a Point of View.* London: Putnam 1927

Hutcheon, Linda. *As Canadian as ... Possible ... Under the Circumstances!* Toronto: ECW Press and York University 1990

– *Irony's Edge: The Theory and Politics of Irony.* London and New York: Routledge 1995

Huxley, Bob. "Golf Courses in National Parks." *Park News* 17, no. 1 (Spring 1981): 14–16

Jarrell, Richard A. "British Scientific Institutions and Canada: The Rhetoric and the Reality." *Transactions of the Royal Society of Canada* series 4, vol. 20 (1982): 524–47

Jasen, Patricia. *Wild Things: Nature, Culture, and Tourism in Ontario, 1790–1914.* Toronto: University of Toronto Press 1995

Johnson, Ronald Clifford Arthur. "The Effect of Contemporary Thought upon Park Policy and Landscape Change in Canada's National Parks, 1885–1911." Ph.D. thesis, University of Minnesota 1972

Johnstone, Kenneth. *The Aquatic Explorers: A History of the Fisheries Research Board of Canada.* Toronto: University of Toronto Press 1977

Joudrey, George Neil. "The Public Life of A.S. MacMillan." Master's thesis, Dalhousie University 1966

Judd, Richard W. "Reshaping Maine's Landscape: Rural Culture Tourism and Conservation, 1890–1929." *Journal of Forest and Conservation History* 32, no. 4 (October 1988): 180–90

Kant, Immanuel. *Critique of Aesthetic Judgement.* Trans. J.C. Meredith. Oxford: Oxford University Press 1911

Kellert, Stephen R. "Historical Trends in Perceptions and Uses of Animals in Twentieth Century America." *Environmental History Review* 9, no. 1 (Spring 1985): 19–33

Killan, Gerald. *Protected Places: A History of Ontario's Provincial Parks System.* Toronto: Dundurn Press with the Ontario Ministry of Natural Resources 1993

Kline, Marcia B. *Beyond the Land Itself: Views of Nature in Canada and the United States.* Cambridge, Mass.: Harvard University Press 1970

Knighton, Jose. "Eco-porn and the Manipulation of Desire." *Wild Earth* Spring 1993

Langston, Nancy. *Forest Dreams, Forest Nightmares: The Paradox of Old Growth in the Inland West.* Seattle: University of Washington Press 1995

Least Heat-Moon, William. *PrairyErth (a deep map).* Boston: Houghton Mifflin 1991

Leibhardt, Barbara. "Interpretation and Causal Analysis: Theories in Environmental History." *Environmental History Review* 12, no. 3 (Spring 1988): 23–36

Levine, Lawrence. *Highbrow/Lowbrow: The Emergence of Cultural Hierarchy in America.* Cambridge, Mass.: Harvard University Press 1988

Lie, Kari. "The Spruce Budworm Controversy in New Brunswick and Nova Scotia." *Alternatives* 9, no. 2 (Spring 1980): 5–13

Livingston, John A. *Rogue Primate: An Exploration of Human Domestication.* Toronto: Key Porter Books 1994

– *The Fallacy of Wildlife Conservation.* Toronto: McClelland and Stewart 1981

Loewen, Candace. "Terra Nova: New Province, New Park." *Archivist* 16, no. 2 (March-April 1989): 10–11

Lopez, Barry. *Arctic Dreams: Imagination and Desire in a Northern Landscape.* New York: Bantam Books 1986

Lotenburg, Gail. "Wildlife Management Trends in the Canadian and US Federal Governments, 1870–1995." Unpublished report for Parks Canada 1995

Lothian, W.F. *A Brief History of Canada's National Parks.* Ottawa: Parks Canada 1987

– *A History of Canada's National Parks.* 4 vol. Ottawa: Parks Canada 1977–81

MacEachern, Alan. "No Island Is an Island: A History of Tourism on Prince Edward Island, 1870–1939." Master's thesis, Queen's University 1991

– "Rationality and Rationalization in Canadian National Parks Predator Policy." *Consuming Canada: Readings in Environmental History.* Ed. Chad Gaffield and Pam Gaffield. Toronto: Copp Clark 1995: 197–212

– "Why Martin Luther King Didn't Spend His Summer Vacation in Canada." *Globe and Mail* 14 January 1995

MacFarlane, Richard. "Parks Canada: The Failure of National Parks Planning Procedures." *Park News* 14, no. 3 (Fall 1978): 37–42

MacLean, Rick. "Leaving Behind a Bitter Legacy: Expropriated Land for National Parks." *Atlantic Insight* 10, no. 1 (January 1988): 38

Majka, Mary. *Fundy National Park.* Fredericton: Brunswick Press 1977

Major, Kevin. *Terra Nova National Park: Human History Study.* Ottawa: Parks Canada, Environment Canada 1983

March, William. *Red Line: The Chronicle-Herald and the Mail-Star, 1875–1954.* Halifax: Chebucto Agencies 1986

Marsh, George Perkins. *Man and Nature.* Ed. David Lowenthal. Cambridge, Mass.: Belknap Press of Harvard University Press 1965 [1864]

Marsh, John. "Postcard Landscapes: An Exploration in Method." *Canadian Geographer* 39, no. 3 (1985): 265–7

Martin, Calvin Luther. *In the Spirit of the Earth: Rethinking History and Time.* Baltimore: Johns Hopkins University Press 1982

May, Elizabeth. *Budworm Battles.* Halifax: Four East 1982

McCombs, W. Douglas. "Therapeutic Rusticity: Antimodernism, Health and the Wilderness Vacation, 1870–1915." *New York History* October 1995: 409–28

McIntosh, Robert P. *The Background of Ecology: Concepts and Theory.* Cambridge: Cambridge University Press 1985

McKay, Ian. "Tartanism Triumphant: The Construction of Scottishness in Nova Scotia, 1933–1954." *Acadiensis* 21, no. 2 (Spring 1992): 5–47

McNamee, Kevin. "From Wild Places to Endangered Spaces: A History of Canada's National Parks." *Parks and Protected Areas in Canada: Planning and*

Management. Ed. Philip Dearden and Rick Rollins. Toronto: Oxford University Press 1993: 17–44

Meeker, Joseph. "Red, White, and Black in the National Parks." 1973. *On Interpretation: Sociology for Interpreters of Natural and Cultural History.* Ed. Gary E. Machlis and Donald R. Field. Corvallis: Oregon State University Press 1992: 196–205

Meinig, D.W., ed. *The Interpretation of Ordinary Landscapes: Geographical Essays.* New York: Oxford University Press 1979

Merchant, Carolyn. "The Theoretical Structure of Ecological Revolutions" *Environmental Review* 11, no. 4 (Winter 1987): 265–74

Mighetto, Lisa. *Wild Animals and American Environmental Ethics.* Tucson: University of Arizona Press 1991

Mitchell, W.J.T., ed. *Landscape and Power.* Chicago: University of Chicago Press 1994

Nash, Roderick. "The American Invention of National Parks." *American Quarterly* 22 (Fall 1970): 726–35

– *The Rights of Nature: A History of Environmental Ethics.* Madison: University of Wisconsin Press 1989

– "The State of Environmental History." *The State of American History.* Ed. Herbert J. Bass. Chicago: Quadrangle Books 1970: 249–60

– "Wilderness and Man in North America." *The Canadian National Parks: Today and Tomorrow.* Ed. J.G. Nelson and R.C. Scace. Proceedings of a conference organized by the National and Provincial Parks Association of Canada and the University of Calgary, 9–15 October 1968. vol. 1. Calgary: University of Calgary Press 1969: 66–93

– *Wilderness and the American Mind.* 3rd ed. New Haven: Yale University Press 1982 [1967]

"The National Parks: A Forum on the 'Worthless Lands' Thesis, a Roundtable." *Journal of Forest History* 27 (July 1983): 130–45

Neary, Peter. "Party Politics in Newfoundland, 1949–1971: A Survey and Analysis." *Newfoundland in the Nineteenth and Twentieth Centuries: Essays in Interpretation.* Ed. James Hiller and Peter Neary. Toronto: University of Toronto Press 1980: 205–45

Nelson, J.G., and R.C. Scace, eds. *The Canadian National Parks: Today and Tomorrow.* Proceedings of a conference organized by the National and Provincial Parks Association of Canada and the University of Calgary, 9–15 October, 1968. 2 vol. Calgary: University of Calgary Press 1969

Nicol, J.I. "The National Parks Movement in Canada." *The Canadian National Parks: Today and Tomorrow.* Ed. J.G. Nelson and R.C. Scace. Proceedings of a conference organized by the National and Provincial Parks Association of Canada and the University of Calgary. 9–15 October 1968. Vol. 1. Calgary: University of Calgary Press 1969: 35–52

Nicolson, Marjorie Hope. *Mountain Gloom and Mountain Glory: The Development of the Aesthetics of the Infinite*. Ithaca: Cornell University Press 1959

Oates, Stephen B. *Let the Trumpets Sound: The Life of Martin Luther King, Jr*. New York: Harper and Row Publishers 1982

Overton, James. *Making a World of Difference: Essays on Tourism, Culture, and Development in Newfoundland*. St John's: Institute of Social and Economic Research, Memorial University of Newfoundland 1996

Parenteau, Bill, and L. Anders Sandberg. "Conservation and the Gospel of Economic Nationalism: The Canadian Pulpwood Question in Nova Scotia and New Brunswick." *Environmental History Review* 19, no. 2 (Summer 1995): 57–83

Parkin, John Hamilton. *Bell and Baldwin: Their Development of Aerodromes and Hydrodromes at Baddeck, Nova Scotia*. Toronto: University of Toronto Press 1964

Parr, Joy. "Gender History and Historical Practice." *Canadian Historical Review* 76, no. 3 (September 1995): 354–76

Perdue, Charles L., Jr., and Nancy J. Martin-Perdue. "'To Build a Wall around These Mountains': The Displaced People of Shenandoah." *Magazine of Albermarle, Virginia County History* 49 (1991): 48–71

Pick, Daniel. *Faces of Degeneration: A European Disorder, c.1848-c.1918*. Cambridge: Cambridge University Press 1989

Pickersgill, J.W. *My Years with Louis St. Laurent: A Political Memoir*. Toronto: University of Toronto Press 1975

Power Bratton, Susan. "National Park Management and Values." *Environmental Ethics* 7 (Summer 1985): 117–33

Pyne, Stephen J. *Fire in America: A Cultural History of Wildland and Rural Fire*. Princeton: Princeton University Press 1982

Reeder, Carolyn, and Jack Reeder. *Shenandoah Heritage: The Story of the People before the Park*. Washington: Potomac Appalachian Trail Club 1978

Reilly, Robin. "Planning for New National Parks in Atlantic Canada: The Experience with Ship Harbour, Nova Scotia (1965–1973)." *Park News* 17, no. 1 (Spring 1981): 5–10

Restino, Charles. "The Cape Breton Island Spruce Budworm Infestation: A Retrospective Analysis." *Alternatives* 19, no. 4 (1993): 29–36

"A Round Table: Environmental History." *Journal of American History* 76, no. 4 (March 1990): 1087–1147

Rounthwaite, H. Ian. "The National Parks of Canada: An Endangered Species?" *Saskatchewan Law Review* 46, no. 1 (1981–2): 43–71

Rowe, Frederick W. *The Smallwood Era*. Toronto: McGraw-Hill Ryerson 1985

Runte, Alfred. *National Parks: The American Experience*. 2nd ed. Lincoln: University of Nebraska Press 1987 [1979]

– *Yosemite: The Embattled Wilderness*. Lincoln: University of Nebraska Press 1990

Russell, Edmund P., III. "'Speaking of Annihilation': Mobilizing for War against Human and Insect Enemies, 1914–1945." *Journal of American History* March 1996: 1505–29

Sackett, Andrew. "Inhaling the Salubrious Air: Health and Development in St. Andrews, N.B., 1880–1910." *Acadiensis* 25, no. 1 (Autumn 1995): 54–81

Sadler, Barry. "Mountains as Scenery." *Canadian Alpine Journal* 57 (1974): 51–3

Sandberg, L. Anders. "Forest Policy in Nova Scotia: The Big Lease, Cape Breton Island, 1899–1960." *Trouble in the Woods: Forest Policy and Social Conflict in Nova Scotia and New Brunswick*. Ed. L. Anders Sandberg. Fredericton: Acadiensis Press 1992: 66–89

Sax, Joseph. *Mountains without Handrails: Reflections on the National Parks*. Ann Arbor: University of Michigan Press 1980

Schama, Simon. *Landscape and Memory*. Toronto: Random House 1995

Sears, John F. *Sacred Places: American Tourist Attractions in the Nineteenth Century*. Oxford: Oxford University Press 1989

Sellars, Richard West. *Preserving Nature in the National Parks: A History*. New Haven: Yale University Press 1997

Shaw, Marilyn. *Mount Carleton Wilderness*. Fredericton: Fiddlehead Poetry Books and Goose Lane Editions 1987

Simonian, Lane. *Defending the Land of the Jaguar: A History of Conservation in Mexico*. Austin: University of Texas Press 1995

Singer, Peter. *Animal Liberation: A New Ethic for Our Treatment of Animals*. New York: Avon 1977

Smallwood, Hon. Joseph R. "Joey." *I Chose Canada*. Toronto: Macmillan of Canada 1973

Stone, Christopher D. "Should Trees Have Standing? Toward Legal Rights for Natural Objects." *Southern California Law Review* 45 (Spring 1972): 450–501

Sumner, Lowell. "Biological Research and Management in the National Park Service: A History." *George Wright Forum* Autumn 1983: 3–27

Sutherland, Dufferin. "'The Men Went to Work by the Stars and Returned by Them': The Experience of Work in the Newfoundland Woods during the 1930s." *Newfoundland Studies* 7, no. 2 (1991): 143–72

– "'We Are Only Loggers': Loggers and the Struggle for Development in Newfoundland, 1929–1959." Ph.D. thesis, Simon Fraser University 1995

Taylor, C.J. "Legislating Nature: The National Parks Act of 1930." *To See Ourselves/To Save Ourselves: Ecology and Culture in Canada*. Ed. Rowland Lorimer et al. Proceedings of the Annual Conference of the Association for Canadian Studies, 31 May–1 June 1990. Montreal: Association for Canadian Studies 1991: 125–37

– *Negotiating the Past: The Making of Canada's National Historic Parks and Sites*. Montreal: McGill-Queen's University Press 1990

Theberge, John B. "Ecology, Conservation, and Protected Areas in Canada." *Parks and Protected Areas in Canada: Planning and Management*. Ed. Philip Dearden and Rick Rollins. Toronto: Oxford University Press 1993: 137–53

Thomas, Keith. *Man and the Natural World: Changing Attitudes in England 1500–1800*. London: Penguin Books 1983

Thomas, Paul. "The Kouchibouguac National Park Controversy: Over a Decade Strong." *Park News* 17, no. 1 (Spring 1981): 11–13

Tillotson, Shirley. "Gender, Recreation, and the Welfare State in Ontario, 1945–1961," Ph.D. thesis, Queen' University 1991

Todhunter, Rodger. "Banff and the Canadian National Park Idea." *Landscape* 25, no. 2 (1981): 33–9

Todd, Eric C.E. *The Law of Expropriation and Compensation in Canada*. Toronto: Carewell Co. 1976

Turner, Frederick. *Beyond Geography: The Western Spirit against the Wilderness*. New York: Viking Press 1980

Turner, Frederick Jackson. "The Significance of the Frontier in American History." *Annual Report of the American Historical Association for 1893*. Washingon, DC 1894: 197–227

Turner, R.D., and W.E. Rees. "A Comparative Study of Parks Policy in Canada and the United States." *Nature Canada* 2, no. 1 (1973): 31–6

Van Kirk, Sylvia. "The Development of National Park Policy in Canada's Mountain National Parks, 1885–1930." Master's thesis, University of Alberta 1969

Waiser, Bill. *Park Prisoners: The Untold Story of Western Canada's National Parks: 1915–1946*. Saskatoon: Fifth House 1995

– *Saskatchewan's Playground: A History of Prince Albert National Park*. Saskatoon: Fifth House 1989

Walker, James W. St G. *Racial Discrimination in Canada: The Black Experience*. Ottawa: Canadian Historical Association Historical Booklet #41 1985

Wall, Geoffrey, and R. Wallis. "Camping for Fun: A Brief History of Camping in North America." *Recreational Land Use: Perspectives on Its Evolution in Canada*. Ed. Geoffrey Wall and John Marsh. Ottawa: Carleton University Press 1982: 341–53

Wein, R.W. *Historical and Prehistorical Role of Fire in the Maritime National Parks*. Fredericton: University of New Brunswick Press 1978

Weiskel, Thomas. *The Romantic Sublime: Studies in the Structure and Psychology of Transcendence*. Baltimore: Johns Hopkins University Press 1976

White, Richard. "American Environmental History: The Development of a New Historical Field." *Pacific Historical Review* 54 (August 1985): 297–335

Williams, Raymond. "Ideas of Nature." *Problems in Materialism and Culture*. London and New York: Verso 1980: 67–85

Wilson, Alexander. *The Culture of Nature: North American Landscape from Disney to the Exxon Valdez*. Toronto: Between the Lines 1991

Wilson, E.O. *Biophilia*. Cambridge, Mass.: Harvard University Press 1984

– *Naturalist*. Washington, DC: Island Press 1994

Winks, Robin W. *The Blacks in Canada: A History.* Montreal: McGill-Queen's University Press 1981

Worster, Donald. *An Unsettled Country: Changing Landscapes of the American West*. Albuquerque: University of New Mexico Press 1994

– "Appendix: Doing Environmental History." *The Ends of the Earth: Perspectives on Modern Environmental History.* Ed. Donald Worster. New York: Cambridge University Press 1988: 289–307

– *Nature's Economy: A History of Ecological Ideas*. 2nd ed. Cambridge: Cambridge University Press 1985 [1977]

Wright, R. Gerald. *Wildlife Research and Management in the National Parks*. Urbana: University of Illinois Press 1992

Young, R.A. "'and the people will sink into despair': Reconstruction in New Brunswick, 1942–1952." *Canadian Historical Review* 69, no. 2 (1988): 127–66

Zaslow, Morris. *The Northward Expansion of Canada, 1914–1967*. Toronto: McClelland and Stewart 1988

Zukin, Sharon. *Landscapes of Power: From Detroit to Disney World*. Berkeley: University of California Press 1991

Index